Springer Geochemistry/Mineralogy

For further volumes:
http://www.springer.com/series/10171

Mervyn S. Paterson

Materials Science for Structural Geology

 Springer

Mervyn S. Paterson
Research School of Earth Science
Australian National University
Canberra
Australia

ISSN 2194-3176   ISSN 2194-3184   (electronic)
ISBN 978-94-007-5544-4   ISBN 978-94-007-5545-1   (eBook)
DOI 10.1007/978-94-007-5545-1
Springer Dordrecht Heidelberg New York London

Library of Congress Control Number: 2012948187

Printed on acid-free paper

Springer is part of Springer Science+Business Media (www.springer.com)

Dedicated to the memory of Katalin Paterson

Preface

Interest in the plastic deformation of rocks arises mainly from its application in the Earth sciences, especially in structural geology and tectonics. Its experimental study has consequently been pursued mainly in laboratories associated with geology or geophysics departments or institutes. However, the physical mechanisms involved are of considerable interest in materials science and some workers in the field of rock deformation have had a materials science affiliation.

There are a number of aspects to the study of rock deformation, including

- Establishment of the rheological laws for the rocks, that is, the stress-strain-time relationships under particular conditions,
- Mechanisms of the deformation and the relationship between these and the specific rheological laws and microstructural features of the rock,
- Role of environmental variables (pressure, temperature, chemical activities, pore pressure) on the deformation behaviour,
- Development of fabric and multi-grain structural features on scales accessible in the laboratory, e.g. crystallographic preferred orientations, cleavages, micro-folding, metamorphic differentiation, etc.,
- Relation of the deformation aspects to petrological and other aspects of the rock, involving questions of influence of stress or strain on phase transformation and stabilities, rates of metamorphic change, preservation of the setting of radio-active clocks.

The study of the plastic deformation of minerals and rocks is also of considerable interest as a topic in materials science. The knowledge of the deformation behaviour of materials involving ionic and covalent bonding is extended by including minerals, and, in particular, the silicate minerals comprise an especially interesting group of materials in which the silicon–oxygen bond plays a special role, particularly in regard to the influence of water. Our subject will in fact be presented here mainly as an aspect of the materials science of rocks and minerals, a topic both of intrinsic interest and of underlying importance for application in geology and geophysics.

In the application of the results of laboratory studies on rheology of rocks to natural situations in the earth, the extrapolation to the geological timescale presents severe difficulties. Limitations of time and equipment restrict the strain rates that can be reasonably investigated in the laboratory to a range of 10^{-2} or 10^{-3} s^{-1} to about 10^{-6} or 10^{-7} s^{-1} if strains of at least a few percent are required, whereas the strain rates of interest in the earth are generally less than 10^{-10} s^{-1} and, in the majority of applications, probably in the range 10^{-12}–10^{-14} s^{-1}. Thus measurements made over a range of three or four orders of magnitude have to be extrapolated over a further range of six to eight orders of magnitude, a daunting prospect. Several difficulties are involved in such an extrapolation. First, there tends to be considerable scatter of results of tests on rock specimens and so the accuracy of the rheological relations established in the laboratory tends to be inadequate for such a long extrapolation; however, careful selection of material for uniformity, and testing of a sufficient number of specimens can minimise the uncertainties from scatter. More serious is the question of determining whether the nature of the deformation behaviour studied in the laboratory is the same as that involved in the natural situation so that the same rheological laws can be expected to apply; it therefore has to be established that the deformation process is the same in both cases, which requires particular study of the microstructural imprints of these mechanisms. The study of the mechanisms of deformation therefore takes on a particular importance in experimental rock deformation, not only for its intrinsic interest in a materials science sense, but also for establishing the nature of the microstructural evidence that must be correlated with that observed in the naturally deformed rocks in order to justify extrapolation of the laboratory results to the natural situation on the grounds that the same processes are dominant in both situations.

The nature of the extrapolation problem can be illustrated with reference to a so-called deformation mechanism map. Such a diagram illustrates that there are a number of modes or mechanisms of deformation, each with its own flow law; the particular mode that dominates is determined by the combination of stress (or strain rate) and temperature that is imposed on the rock. It follows that extrapolation of a particular flow law is only justified within the stress (or strain-rate) and temperature domain within which that mechanism predominates. It is also to be emphasised that a particular deformation mechanism map only applies for a particular combination of environment and structural state of the specimen; for example, changing the grain-size or the activities of minor chemical components may change both the values of the parameters in a given flow law and the limits of the domain within which it is valid. Extrapolation to geological conditions is therefore a difficult matter and one in which it can only be expected that approximate conclusions or limits can be arrived at after detailed considerations of evidence bearing on mechanisms of deformation.

Since the mid-nineteenth century when Sorby and other pioneers introduced microscopical studies of the materials of technology, any comprehensive approach to the mechanical properties of materials has included some attention to the deformation processes at the microstructural level. At first the focus tended to be on processes at the scale of the crystals in polycrystalline materials, and it is of particular historical interest in the present context that the earliest studies in crystal plasticity were carried out with the minerals, halite and calcite. However, work on metals soon came to dominate the scene for several reasons, including the ductility and technical importance of metals and the development of methods for growing large single crystals and for revealing their crystallographic details by X-ray diffraction. It thus came about that it was in the field of metallurgy that many of the basic concepts of crystal plasticity and of the flow of polycrystals were developed. Consequently, it was from this field that many basic concepts were later carried over for application to other materials such as rocks, although independent lines of approach were also developed in particular areas as interest arose, for example, in soils and polymers.

Such was the situation up to the 1950s. However, during the recent decades the scope of research on mechanical and related properties and its application has broadened out considerably and has come to encompass as well a wide variety of non-metallic materials in a more integrated approach. Consequently the term "materials science" has commonly replaced "metallurgy" when speaking of the study of the fabrication and properties of materials of potential technical application. There is now considerable knowledge of the properties of materials in such distinctive categories as metals, semiconductors, ceramics, oxides in general, glasses, plastics and elastomers, nanomaterials, soils and ice, and there are valuable insights to be gained from comparative studies between the various groups. The category of minerals and rocks, although the subject of some engineering and geological interest since time immemorial, is one of the latest to "come of age" in materials science extending again the range of this subject into yet further regions of distinctive behaviour while still demonstrating much that is common with other materials. The first chapter summarises some of the characteristic aspects of minerals and rocks as materials.

Much of the present text was written in the 1980s. However, the basic material science concepts of relevance to the study of rock deformation were already established by this time and so it is felt that this presentation is still of relevance. A limited amount of updating has been done and some later references added.

I am very indebted to Ian Jackson for careful reading of the text and to John Fitz Gerald for assistance in the preparation of the micrographs.

Contents

Chapter 1
The Nature of Minerals and Rocks as Materials

1.1 General

The classification, genesis, and geological role of the various minerals and rocks are well covered in standard textbooks of mineralogy and petrology (Deer et al. 1992; Mueller and Saxena 1977; Best 2003; Nesse 2000; Perkins 2002; Philpotts and Ague 2009; Vernon and Clarke 2008; Wenk and Bulakh 2004; Yardley 1989). Here we review briefly some of the characteristics of minerals and rocks that are relevant from a materials science point of view. Minerals show a wide range in physical and chemical properties, and rocks have the additional complexity of the textural and structural variety of polycrystalline materials. However, there are some generalizations that can be made. Thus, most minerals are electrically insulating, optically transparent in thin sections, and brittle under ordinary atmospheric conditions, and many are silicates in chemical constitution.

In the next two sections, we look at aspects of minerals and rocks that are of particular relevance to their mechanical properties. In the case of minerals, this involves the nature of the chemical bonding and of the defects in their crystal structures. In the case of rocks, particular importance attaches to the nature of the grain boundaries in intact rock and of pore structure in porous rock.

1.2 The Constitution of Minerals

1.2.1 Atomic Bonding

In their physical and chemical nature minerals cover a wide range. However, comparing them with materials that are well studied because of technological importance, they tend to be nearest to the oxides. The majority of rock-forming minerals are, in fact, either oxides or oxy-salts, with the silicates forming the

M. S. Paterson, *Materials Science for Structural Geology*,
Springer Geochemistry/Mineralogy, DOI: 10.1007/978-94-007-5545-1_1,
© Springer Science+Business Media Dordrecht 2013

largest and most distinctive class, while the sulfides, of special importance in ore deposits, form a rather separate class in respect of their properties. The mechanical properties of minerals are mainly determined by their crystal structure, the nature of their bonding, and the properties of their crystal defects. The crystal structures are well known from X-ray diffraction studies and need not be elaborated here (Bragg and Claringbull 1965; Deer et al. 1992; Wenk and Bulakh 2004). In this section, we consider crystal bonding and defects (Barrett et al. 1973, Chap. 2; Frank-Kamenetskaya et al. 2004; Hammond 2001; Kelly et al. 2000; Kittel 1976, Chap. 3; Phillips 1975; Putnis 1992; Rohrer 2001; Sands 1969; Schmalzried 1995).

Bonding is atomic interaction that arises mainly through the valence electrons and is related to the electron density distribution. The overall strength of bonding is represented in the cohesive energy, the difference in total energy of the atoms when separated and when assembled in the crystal structure. The mechanical properties are sensitive to the nature of the bonding. In minerals, this nature generally falls somewhere in the continuous range between the limiting cases of ionic or heteropolar bonding and covalent or homopolar bonding. In ideal *ionic bonding,* valence electrons are completely transferred from the more electropositive to the more electronegative atoms and the interaction between them can be fairly satisfactorily viewed as a classical electrostatic one between spherical ions. In ideal *covalent bonding*, in contrast, valence electrons are equally shared between the atoms and tend to be distributed spatially in such a way as to define directional bonds; the bonding results from the electrons occupying lower energy states than when localized on either atom and require a quantum–mechanical explanation.

The theoretical analysis of bonding may be approached either in terms of individual atom–atom bonds, as in molecular orbital and valence bond theories, or in terms of the total assemblage of atoms of the crystal, as in energy band theory (Adler 1975; Coulson 1961; Harrison 1980; Kittel 2005; Madelung 1978, Chap. 8; Marfunin 1979, Chaps. 3 and 4; Martin 2004; Weaire 1975). Energy band theory explains why most minerals are insulators. This property arises from the existence of a band gap separating the valence band of energy states, fully occupied by the valence electrons, from the next band of possible but unoccupied energy states known as the conduction band. For strongly ionic crystals the band gap tends to be wide (order of 10 eV, $\gg$ kT) but it can be relatively narrow for crystals with a considerable degree of covalency in the bonding (for example, 3.6 eV for ZnS and 0.3 eV for PbS). Another important property of the electron energy distribution is the Fermi energy or chemical potential of the electrons, that is, the energy level of an extra electron in equilibrium with the existing electron population of the crystal; for the pure insulator, the Fermi energy lies in the middle of the band gap but its level can be strongly influenced by impurities.

Although the cases of pure covalent bonding (C, Ge, Si) and of highly ionic bonding (alkali halides) are well characterized, it has proved to be difficult to specify quantitatively the degree of *ionicity* in intermediate cases. The initial approach of Pauling was through electronegativity, that is, the relative ability of an atom to attract electrons, and various measures of this property have been proposed, most corresponding more or less to the sum of the ionization energy and the electron affinity (Flynn 1972; Kittel 2005; Samsonov 1968). Clearly, the degree of ionic character (ionicity) will tend to increase with increasing difference in electronegativities of anion and cation but there is no general agreement on how to express the ionicity quantitatively in terms of electronegativities or of other parameters. The usual approach has been to try to identify measures of the homopolar and ionic components of the bonding based on observed properties that reflect the distribution of electron density; thus the definitions of Pauling (1960, 1971), Phillips (1968, 1969, 1970, 1973), van Vechten (1969a, b), Kowalczyk et al. (1974), Hübner and Leonhardt (1975), Stewart et al. (1980) and Guo et al. (1999) are variously derived from thermochemically determined electronegativities, dielectric constant, spectroscopic parameters and "orbital electronegativities" calculated from ionization potentials and electron affinities.

A generally useful scale for comparing ionicities appears to be that of Phillips and Van Vechten, initially only applied to the binary compounds $A^N B^{8-N}$ (N = group in the periodic table) but extended to other compounds by Levine (1973a, b); as well as the references above, see Ramakrishnan (1974) for a simple exposition and Madelung (1978, pp. 331–352) for further discussion of this and other measures; also Catlow and Stoneham (1983). The Phillips-Van Vechten ionicities are obtained as follows:

1. Resolving the crystal potential into a covalent component, corresponding to an average of the properties of the two atoms, and an ionic component, corresponding to their difference, an average band gap E_g is defined as $E_g = (E_c^2 + E_i^2)^{\frac{1}{2}}$, where E_c, E_i are components of the band gap expressing the covalent and ionic character, respectively.
2. E_g is obtained from the observed static dielectric constant.
3. E_c is determined from an empirical formula relating it to the interatomic spacing, using the purely covalent C and Si to calibrate this relationship; thence $E_i^2 = E_g^2 - E_c^2$ is obtained.
4. The ionicity parameter f_i is defined as $f_i = E_i^2/E_g^2$, $0 \leq f_i \leq 1$.

Selected values of f_i are given in Table 1 from Levine (1973a, b).

The general trends and relative positions are similar to those given by Pauling (1960), Kowalczyk et al. (1974), and Hubner and Leonhardt (1975), although there are a few discrepancies.

An alternative approach to determining the degree of ionicity in bonding is to start with the electron density distribution from an accurate X-ray structure determination and resolve it into spherically averaged components ("pseudo atoms") centered on each site and a non-spherical component representing the covalent bonding; this enables one to determine a net or residual charge for each atom, from which an ionicity may be derived by comparing this charge with the formal charge corresponding to full ionization according to the chemical valency (for example, Stewart 1976). In this way the net charge on Si in α-quartz has been determined to be about 1 electron unit, compared with a formal charge of 4 for

Table 1 Phillips and van Vechten ionicities (after Levine 1973a, b)

CaF_2	0.97	FeO	0.87	SiO in Mg_2SiO_4	$(0.62)^b$
CsCl	0.96	MgO	0.84	SiO_2	0.57
AgCl	0.96^a	Al_2O_3	0.80	GeO_2 (quartz)	0.51
KCl	0.95	GeO_2 (rutile)	0.73	ClO in $NaClO_3$	0.49
NaCl	0.94	TiO_2	0.69	PO in $AlPO_4$	0.48
NaI	0.93	Fe_2O_3	0.68	InSb	0.33
CaO	0.92	AlO in $AlPO_4$	0.65	SiC	0.20
LiF	0.91	PbS	0.63	Si	0
		ZnS	0.62		
		BeO	0.62		

[a] Phillips' (1970) value is 0.86 for AgCl by the same method, the difference being that Levine (1973a) has assumed a greater contribution of d-electrons to the valence band; for the other $A^N B^{8-N}$ compounds listed above, Phillips figures are very close to Levine's.
[b] Estimated by analogy with ionicities of 0.63 and 0.58 determined for the SiO bond in magnesian olivine and quartz, respectively, by Tossell (1977) using the ionicity scale of Kowalczyk et al. (1974) and observations on X-ray spectra.

fully ionic bonding (Stewart et al. 1980). Therefore, Stewart et al., defining an ionicity parameter equal to (net charge)/(formal charge), have concluded that the Si–O bond in quartz has only 25 % ionic character. However, Pauling (1980) has argued, on the basis of his view of bonding, that a net charge of 1 on the Si corresponds to 50 % ionic character for the Si–O bond, which would also correspond more nearly with the 57 % given on the dielectric scale (Table 1), although still not with the nearly 100 % ionicity deduced by Rosenberg et al. (1978) from observations on Compton scattering. The Si–O bond may have slightly greater ionicity in the stishovite structure (Kirfel et al. 2001).

In spite of the difficulty and subjectivity involved in defining ionicity, some conclusions of a relative nature can be reached from Table 1 that are generally consistent with the various views and measures of ionicity:

1. The substances listed in the first column are highly ionic and can usually be treated as purely so in a first approximation.
2. Covalent character is likely to have some influence for substances listed in the second column, perhaps only minor in the first few cases but more significant in the last few, although probably not overshadowing the ionic component.
3. The substances in the third column can be expected to be strongly influenced in their properties by the covalent aspect of the bonding, this being predominant in the semiconductors.

1.2.2 Crystal Defects

There are many types of defects in mineral crystals, all of which represent some sort of departure from the perfect periodicity of the ideal crystal. For brief surveys, see Bollmann (1970), Barret et al. (1973, Chap. 3), Fine (1975), Schmalzried

(1981), Tilley (1986, 1998, 2008), and Kelly et al. (2000). Crystal defects can be conveniently classified dimensionally as point, line, interfacial, and bulk defects.

Point defects are generally regarded as being of particular importance for transport properties such as diffusion and electrical conductivity. Basically, they arise either at a normal atom site through the absence of the atom (*vacancy*) or the substitution of another type of atom (*substitutional*; usually a foreign atom), or at a site not normally occupied through the insertion of an extra atom (*interstitial*; may be foreign or not). The defects involving foreign atoms are called *extrinsic* defects, as opposed to *intrinsic* or native defects. In compounds, combinations of these single defects occur, especially if stoichiometry or charge balance is to be preserved; thus, in binary compounds a *Schottky pair* is a pair of vacancies, one of each type, and a *Frenkel pair* is a combination of a native interstitial and its corresponding vacancy (sometimes restricted to the cation).

The principal line defect is the dislocation, of central importance in crystal plasticity and discussed later (Chap. 6). Stacking faults, twin boundaries, antiphase boundaries, and crystallographic shear surfaces are the simplest of the interfacial or extended defects. Crystallographic shear surfaces (Wadsley defects) can, however, occur in such concentrations as to be regarded as an aspect of the structure itself rather than a perturbation, especially if periodically spaced as in Wadsley shear phases, common in certain non-stoichiometric oxides, such as those of W, Ti, and Nb (Fine 1975; O'Keeffe and Hyde 1985, 1996). High concentrations of point defects can also be clustered and ordered, as in the case of the vacancies in wüstite, $Fe_{1-x}O$. These assemblages can be regarded as volume defects. Other volume defects are cavities, precipitates, and precipitation segregations, for example, Guinier–Preston zones (Guinier 1938; Preston 1938).

The presence of defects can change the electronic structure and properties of a crystal significantly. A small perturbation of the crystal potential may be insufficient to split off any electron energy levels to positions outside the bands of the perfect crystal but, when the perturbation exceeds a certain amount, localized energy states can appear in the band gaps of insulators, corresponding to attenuated rather than propagating wave functions (Adler 1975; Flynn 1972, p. 190; Madelung 1978, p. 378). Such localized states can be treated as approximating a hydrogen-like situation if the perturbing potential is not too great, in which case the localized levels lie near to the band from which they are split off and are said to be "shallow" levels within the band gap. As the defect states fall deeper in the gap, electrons occupying them become more strongly localized and treatment by a hydrogen-like approximation is no longer valid. Defects such as substitutional impurities of higher nuclear charge tend to give rise to occupied localized levels falling below the conduction band of the pure crystal and are known as donors; the converse acceptor defects, deficient in nuclear charge, tend to give rise to unoccupied localized levels above the valence band. When a donor level is unoccupied or acceptor level occupied, the defect is said to be ionized or charged. In strongly ionic materials, impurity states tend to be highly localized because of the relatively low dielectric constant and the impurity defects are often ionized in the ground state (for further discussion on the charge state of impurities in ionic crystals, see Flynn 1972, p. 579).

In view of the roles of the various types of crystal defects in mechanical properties, their concentrations can be of particular interest. The dislocation content of a crystal depends on its history and degree of annealing, that is, on the kinetics of rearrangement and annihilation of the dislocations and not on equilibrium thermodynamic considerations; in fact, except possibly at temperatures in the neighborhood of the melting point, a crystal in thermodynamic equilibrium is predicted to be free of dislocations (Friedel 1964; Nabarro 1967, p. 688). A similar situation probably applies for many volume defects. However, point defects can exist in finite concentrations in crystals in thermodynamic equilibrium, and the equilibrium can often be attained in laboratory times at temperatures well below the melting point. These concentrations are governed by chemical equilibrium considerations (*"defect chemistry"*) which can be lengthy and complicated but which involve basically the following steps (Flynn 1972, pp. 207–216; Kröger 1974, Chaps. 9–14; Libowitz 1975; Madelung 1978, pp. 397–406; Mrowec 1980, Chap. 1; Smyth 2000; van Gool 1966):

1. Identify all relevant defect species, atomic and electronic, and the reactions between them that are thought to be of interest, paying particular attention to the states of charge and taking into account ionization of defects, formation of associates, changes in occupancy of valence and conduction bands, and interactions with the environment. Physical insights play an important part at this stage.
2. Apply the law of mass action to each reaction; that is, for the reaction

$$aA + bB + \ldots = cC + dD + \ldots$$

Obtain

$$\left(a_A^a a_B^b \ldots / a_C^c a_D^d \ldots\right) = K = \exp\left(-\Delta G^0 / RT\right)$$

or

$$a \ln a_A + b \ln a_B + \ldots - c \ln a_C - d \ln a_D \ldots = \ln K = -\Delta G^0 / RT$$

where a_A, a_B, ... are the activities of the defect species A, B, ..., K is the thermodynamic equilibrium constant, ΔG^0 is the standard Gibbs function for the reaction, R is the gas constant, and T is the absolute temperature (see, for example, Atkins 1986, pp. 213–218), noting that:

a. $a_A = \gamma_A x_A \ldots$ where $\gamma_A \ldots$ are the activity coefficients and $x_A \ldots$ the mole fractions of $A \ldots$
b. if molar concentrations $c_A \ldots$ are used in place of mole fractions $x_A \ldots$, the values of $\gamma_A \ldots$, K and ΔG^0 must be adjusted accordingly, now expressing ΔG^0, for example, in kJ m^{-3} instead of kJ mol^{-1} and taking care about definition of standard states (Atkins 1986, p. 186).

c. the notation $[A]$... is variously used for both mole fractions and molar concentrations (cf. Kröger 1974, p 34).

It is to be further noted that in any ionization process in which an electron is interchanged between a localized state for the defect and the collective population of delocalized electrons, the chemical potential for the latter is given by the Fermi level, to be used in determining ΔG^0

3. Write down the condition for electrical neutrality of the crystal, taking into account the effective charge of each defect and including electrons added to the conduction band and electron holes added to the valance band by the reactions (the effective charge of a defect is the difference in charge of a crystal with the defect and of the perfect crystal; the cryptic symbols •, prime, and x are commonly used for positive, negative, and zero effective charges to avoid confusion with the usual absolute or formal charge states designated +, −, and 0).
4. Solve the simultaneous equations of (2) and (3) in order to obtain whatever is being sought in the way of relationships between defect concentrations or expressions for unknown concentrations in terms of experimentally controlled quantities such as temperature and the activities of components in the environment. In carrying through these calculations, it is often convenient to introduce approximations for particular ranges of the variables, first by taking into account the relative magnitudes of the equilibrium constants K, some of which may be assumed to be negligibly small compared with others even if absolute values are not known, and second by only considering the most abundant defect of each charge sign when writing the condition of electroneutrality (Brouwer's approximation). The latter approximation often enables particular ranges to be defined in which the logarithm of a defect concentration can be expressed as a simple power of the activity or partial pressure of a component in the environment (for example, the oxygen fugacity).

For an example of such considerations in the case of olivine, see Stocker (1978a, b).

1.2.3 Order–Disorder

In some crystal structures, it can happen that certain types of sites in the unit cell can be occupied nearly equally favorably by two different species of atom. It is then possible that, from unit cell to unit cell or among sites within a unit cell, these sites can be occupied by the two species non-randomly, in which case the crystal structure is said to be *ordered*. For reasons of entropy, there tends to be a fairly sharp transition from the ordered to the disordered state as the temperature is increased, the sharpness of the transition being a result of its "cooperative" nature (Christian 1975, p. 206 et seq; Massalski 1983). If the ordering consists simply of a statistical tendency to order in the neighborhood of a given site, it is known as *short-range ordering* but of it persists over large domains it is known as *long-range ordering*. Mistakes or faults in the long-range ordering give rise to *antiphase* boundaries separating *antiphase* domains. Long-range ordering introduces a longer periodicity into the structure, to which the term *superlattice* is attached. For an example of long-range ordering in a mineral, see Redfern et al. (2000)

When long-range ordering is not complete, the degree of ordering can be described by the Bragg–Williams long-range order parameter S, defined by

$$S = \frac{p - r}{1 - r} \tag{1.1}$$

where p is the probability that a given site is occupied by a given species of atom and r is the fraction of this species that could potentially occupy this site. The parameter S can thus vary from zero for complete disorder to unity for complete order. The degree of short-range ordering, in contrast, is defined by the Bethe short-range order parameter s, defined by

$$s = \frac{q - q_r}{q_m - q_r} \tag{1.2}$$

where q is the actual fraction of unlike atoms on neighboring sites and q_r, q_m are the corresponding fractions at maximum randomness and maximum order, respectively.

1.3 The Constitution of Rocks

1.3.1 Rocks as Aggregates of Grains

Rocks, as aggregations of mineral grains, have mechanical properties mainly determined by the following factors:

1. The properties of the mineral grains themselves, as surveyed in the previous section, assuming that the deformation is penetrative to the grain scale.
2. The mutual interactions of the mineral grains, especially as it affects the intragranular deformation.
3. The manner in which the grains are stuck together, including the nature of the grain boundaries and of any binding cements.
4. The ease of access of fluids, involving both porosity and permeability.
5. The existence of a hierarchy of structures on different scales, as, for example, in conglomerates, oolites, layered complexes, or chemically segregated rocks.

Strain compatibility between grains undergoing plastic deformation is a complicated issue in rocks because of the tendency to low symmetry in the mineral grains, the existence of contrasting properties between different types of grains in polyphase rocks, and the variety of possible accommodating mechanisms that may arise. This topic will be discussed later (Chaps. 5–7).

In respect of intergranular attachment, rocks present a wide spectrum as they range from zero-porosity monomineralic polycrystalline bodies to granular masses that are little more than consolidated soils. The *intergranular region* in rocks thus shows great variety and is often of the utmost importance for mechanical properties. At the one extreme, it may consist of clean, intact grain boundary between identical or differing phases, with structural properties akin to those of grain boundaries in metals. At the other extreme, the grains are separated by a more or less continuous layer of a cementing material, itself often polycrystalline but much finer grained. Between these two extremes are boundaries containing precipitates, films and voids, or cracks in various extents at the contact between the two grains.

1.3.2 Simple Grain Boundaries

While a clean, intact grain boundary between identical phases may be uncommon in practice, the concept of it provides a point of reference for many discussions and forms a basis for explaining properties that are insensitive to the presence of impurities. The study of such grain boundaries has advanced considerably since around 1970, the general trends being covered in the following reviews and collections of papers: Bollmann (1970, 1982), Hirth and Balluffi (1973), Smith and Pond (1976), Ashby et al. (1978), Chadwick and Smith (1976), Johnson and Blakely (1979), Priester (1980), Balluffi et al. (1981), Hahn and Gleiter (1981), Clarke (1987), Doherty et al. (1997), Farkas (2000), Flewitt and Wild (2001). In the present context, the succinct review of Balluffi et al. (1981) is especially relevant. We shall now attempt to summarize briefly the main concepts that have evolved. It should be recalled first that the complete geometric specification of a planar segment of grain boundary involves not only the relative orientation and translation of the two joining crystal lattices but also the orientation and position of the grain boundary itself, requiring, in general, nine parameters. There have been two main conceptual approaches to the analysis of the grain boundary structure, the earlier one a formal geometrical approach with emphasis on periodicities and dislocations, the other, more recent, an analytical or atomic approach derived from computer modeling of atomic interactions, with an emphasis on vacancies.

The geometrical approach. Since the grain boundary is a region where the influence of the two periodic structures of the grains overlaps or interacts, this region, or "core" of the grain boundary, can itself be expected to have a periodic structure in some degree. This structure is described with reference to the following periodic lattices:

1. The individual crystal lattices of the grains.
2. The CSL, coincidence site lattice (Friedel 1926; Kronberg and Wilson 1949):
   Given the crystal lattices defined in each grain but extended through all space,

the CSL consists of those lattice points of the two coexisting crystal lattices that coincide in space. The fraction of the lattice points that coincide is designated by $1/\Sigma$, where Σ is an integer. A low value of Σ and therefore a relatively small CSL unit cell represents a rather special orientational and translational relationship between two grains. For more general relationships a CSL may or may not exist.

3. The O-lattice (Bollmann 1970): Given the crystal lattices defined in each grain as before, an O-lattice is formed by any set of equivalent points within the unit cells of the two coexisting crystal lattices that coincide in space. These points are then called O-points (analogous to the pattern of light areas in a transmission Moiré pattern in two dimensions). An O-lattice exists for all relationships between two grains. If a CSL exists, it represents a special case of an O-lattice in which the O-points are crystal lattice points. Other special cases arise where the relationship between the two crystal lattices is a rotation about a common low-index axis or a matching of planes across the boundary. Thus, the O-lattice is fundamental to all considerations of grain boundary core structure. However, it is to be understood that it is only the O-points that define the O-lattice and that we have to consider additionally the *pattern* of crystal lattice points that fall within the O-lattice unit cells. This pattern may vary from one cell to the next and may or may not be periodic. Thus, there are special relationships between the two crystal lattices for which there is a finite number of different patterns within the unit cells (called "pattern elements" by Bollmann), or even a single one, and the total pattern is then periodic. In the latter case, the number N' of crystal unit cells per period of the pattern is a measure of the overall degree of coincidence of all points within the crystal lattice unit cells, N' being smaller the better this fit, in a manner somewhat analogs to that in which Σ is a measure of the degree of crystal lattice site coincidence but with more profound implications for the energy of a grain boundary configuration.

4. The DSC lattice ("displacement shift complete") (Sutton and Balluffi 1995): In defining the DSC lattice, it is assumed that the two crystals are in a special or optimal relationship (having either a CSL of low Σ or, at least, an O-lattice of low N') so that there is a periodicity in the pattern of crystal lattice points, as just discussed. The DSC lattice is then derived from all displacements connecting pairs of crystal lattice points in this pattern which preserves the overall pattern; it is the coarsest lattice that can accommodate all the crystal lattice points of the two crystals (note: not all the DSC lattice sites or equivalent points are occupied by crystal lattice points, the pattern itself having the periodicity not of the DSC lattice but of a multiple or sublattice of the CSL or the O-lattice). The DSC lattice has a sort of reciprocal relationship to the corresponding CSL or O-lattice in the sense that the higher is Σ or N' the smaller is the spacing of the DSC lattice.

So far we have considered the relationships between the two crystals that are to share a boundary. Now we consider the location and structure of the boundary itself. A grain boundary can be viewed as a particular plane located within the

O-lattice. For best fit, and therefore presumably minimum energy, this plane will pass as much as possible through O-lattice points. Since the O-lattice points represent sites of a maximum degree of coincidence of the two crystal structures, and since the regions halfway between these points [that is, the intersections with the cell walls in the sense of Wigner-Seitz cells (Wigner and Seitz 1933)] represent regions of maximum misfit, the grain boundary can be viewed as made up of regions of good fit separated by dislocation lines having a Burgers vector equal to a crystal lattice vector and a spacing corresponding to the periodicity of the O-lattice (the elements of dislocation theory are summarized in Chap. 6). These dislocations are known as *primary grain boundary dislocations*. Such a description of a grain boundary corresponds well with observation in the case of small-angle boundaries but it is more a geometrical formalism for high-angle grain boundaries where the dislocation cores will overlap, even allowing for some relaxation along the grain boundary to maximize the areas of good fit. However, it is generally held that, to minimize energy, the structure of the core of high-angle grain boundaries will adjust itself to conform as far as possible to the nearest low energy configuration, such as that of a CSL of low Σ, and that the remaining misfit will be concentrated in a further network of dislocations known as *secondary grain boundary dislocations*. Burgers vectors of the latter dislocations are vectors of the DSC lattice, thus revealing the significance of its introduction (note that in imagining the formation of such a dislocation, the Volterra cut must be made in the grain boundary itself). Such dislocations have also been imaged by TEM although their smaller Burgers vectors make the imaging more difficult. Both types of grain boundary dislocation are described as *intrinsic* dislocations since they represent aspects of the intrinsic structure of the ideal or equilibrium grain boundary core formed between two perfect crystals. It may also be noted that steps or ledges in the boundary can be intrinsically associated with grain boundary dislocations.

Just as a perfect dislocation can have defects introduced into it, so defects can be added to the ideal grain boundary, in particular, vacancies and other dislocations known as *extrinsic* dislocations. Extrinsic dislocations can be imagined as arising when a boundary is formed between two imperfect crystals or when a crystal dislocation is moved from the grain interior into the boundary. In the latter case the dislocation may dissociate into other grain boundary dislocations having Burgers vectors of both the crystal lattice and the DSC lattice if it is energetically favorable to do so. If a sufficient number of dislocations is introduced, the structure of the grain boundary core may be regarded as being transformed by incorporating a regular array of the new dislocations as intrinsic grain boundary dislocations. One of the principal differences between intrinsic and extrinsic dislocations lies in the absence of long range stress fields around the former on account of their regular spacing, while extrinsic dislocations can show a stress field that appreciably penetrates the adjacent grains.

The analytical or atomistic approach. The geometrical approach to the grain boundary core structure just outlined takes little account of the nature of the crystal structure and bonding type and has naturally found most application in metals where the structure is simple and the bonding non-directional and non-polar. In the atomistic approach the interatomic forces between atom pairs form the starting point for a computer simulation of the grain boundary core structure. The energy of a representative group of atoms constituting the core structure is calculated assuming a pairwise interaction potential and the atomic configuration found that minimizes this energy with respect to variations in atomic positions and to slight relative translations of the two crystals (see, for example, Balluffi et al. 1981; Farkas 2000; Gleiter 1979; Guyot and Simon 1976; Hahn and Gleiter 1981; Harrison et al. 1976; Karki and Kumar 2007; Mishin and Farkas 1998; Priester 1980). Although such calculations have many limitations, when applied to metals they have generally indicated that the grain boundary core is very narrow, of the order of two atomic distances, that there exist energy minima for certain relative orientations of the grains, and that there are often tendencies for delocalization of vacancies and dislocation cores with accompanying reduction in the free volume associated with these defects. Also it has been found that the deduced structure of the grain boundary core can generally be described in terms of the linking of polyhedral groups of atoms (Ashby et al. 1978; Pond et al. 1979). Less progress has been made with similar calculations for the more complicated situations involving non-metallic compounds of relevance to mineral systems but it may be expected that the above general findings for metals will probably again apply and, in particular, that polyhedral grouping will be an important concept in describing the structure of the grain boundary core. However, as with metals (Gleiter 1979), it may be that for more refined treatment the electronic energy structure of the grain boundary will also have to be taken into account.

Some idea of the likely properties of grain boundaries specific to mineral systems can be gained from reviews on grain boundaries in ceramics by Kingery (1974) and Balluffi et al. (1981). On the whole, similar geometrical properties can be expected to those for metals, with perhaps a greater tendency to faceting. Also the width of the grain boundary core can again be expected to be very narrow; an estimate by Ricoult and Kohlstedt (1983), based on extrapolation from electron diffraction studies on low-angle boundaries, indicate that for high-angle grain boundaries in olivine the width will be less than about 1 nm, that is, similar to the unit cell dimensions. Specific studies on minerals include Wirth (1986), Johnson et al. (2004), Hiraga et al. (2004), Kuntcheva et al. (2006), and Drury and Pennock (2007). However, mineral grain boundaries are likely to have electrical charges associated with them and a greater variety in core structure can be expected than in metals because of the greater range in types of bonding and interatomic potentials. Finally, impurity segregation and precipitation at grain boundaries will be of particular importance in mineral systems, both because of their relative impurity and because of strong electrical interactions between grain boundaries and impurities.

Similar notions can probably be extended to the *interphase boundaries* between different minerals, which exist in all but the simplest rocks. One may speculate that the polyhedral grouping of atoms in a narrow core region will again be important, perhaps with affinities to intermediate structures, but the geometrical properties of the interface region will be complicated by the lack of commensurateness between the two crystal lattices (Warrington 1980). For references on interphase boundaries, see the three-yearly proceedings of the International Conferences on Intergranular and Interphase Boundaries. Farver and Yund (1995) and Keller, Hauzenberger and Abart (2007) deal with diffusion in interphase boundaries.

1.3.3 Solute Segregation, Extra Phases, and Void Space at Grain Boundaries

In considering the *segregation* of impurities or heterogeneity in chemical composition in or near grain boundaries, a distinction must first be made between equilibrium and non-equilibrium situations. At equilibrium, segregation appears to be confined to the core regions of grain boundaries, of the order of 1 nm in thickness or less. Observations have been mainly carried out on metals, involving measurements on grain boundary energies and direct analyses on grain boundary fracture surfaces using Auger electron spectroscopy (Hondros 1976; Seah and Hondros 1973). However, a few observations on ceramics and minerals, such as aluminium oxide (for example, Johnson 1977) suggest that similar effects may be expected in rocks although the quantitative aspects may be rather different because of differences in defect formation energies and, in ionic materials, the concentration of electrical charge in grain boundaries tends to be an important additional factor (Hall 1982; Kingery 1974). Similar considerations can also be expected to apply to individual dislocations and to low angle or subgrain boundaries.

The reason for grain boundary segregation is envisaged to lie in the existence of a variety of sites different from those in the interior of the grains, a situation analogs to that at the surface of a crystal although less pronounced. Thus, thermodynamical considerations similar to those for surfaces have been applied (Hondros and Seah 1977; McLean 1957), involving the equivalent of a Langmuir adsorption isotherm expressing the concentration of a species in the boundary in terms of the concentration of available sites and an energy of segregation or binding energy; typically 0–20 kJ mol^{-1} for solute impurities in metals (Seah and Hondros 1973), comparable to that for physisorption (Atkins 1986, p. 772). The segregation commonly amounts to less than a monolayer, especially at high temperatures. However, multilayers of two or more atoms thickness have also been observed, analogous to multilayer adsorption of gases on surfaces (Seah and Hondros 1973), For non-metallic materials space charge effects may lead to some broadening of the profile of segregation (Hall 1982; Yan et al. 1983) but it is not clear that this effect is generally very important (Johnson 1977); for example, Tiku and Kröger (1980) find the effect on conductivity in Al_2O_3 to be small. There have been suggestions that in some cases the segregation may lead to new structures in the grain boundary core (Guttmann 1977) and atomistic aspects have been considered (Vitek and Wang 1982). The extent of segregation can be expressed by a grain boundary enrichment ratio relative to the grain interiors. Enrichment ratios from unity to 10^5 have been shown to be inversely related to the solid solubility in the grains (Bender et al. 1980; Dobson et al. 2007; Johnson 1977; Kingery et al. 1979; Seah and Hondros 1973; Unertl et al. 1977). The enrichment ratio may also vary with the nature of the boundary (Balluffi 1979).

In contrast to the highly localized nature of equilibrium segregation in grain boundaries, systems that are not in equilibrium may show gradients in chemical composition that extend much further into the grains, even tens or hundreds of microns (Westbrook 1969). Such gradients can generally be viewed as frozen-in transients in systems that are in course of adjustment to changing conditions, most commonly involving the precipitation of excess solute as the solid solubility decreases during cooling (Kingery 1974); presumably the precipitation occurs in the grain boundaries because of easier nucleation there (Clemm and Fisher 1955). The process has been modeled by Cai (1991) and Xu (1987).

Minor phases present in the boundary regions between the major phases can be of special importance in the experimental deformation of rocks. These accessory phases may be of igneous origin, be formed during metamorphism or weathering, or represent cements in sedimentary rocks. They commonly give rise to a fluid phase at elevated temperatures, either because of dehydration of hydrous phases or because of partial melting related to a relatively low solidus in the system of phases present. Interest attaches to the role of such fluid phases in grain boundaries both from the point of view of interpreting high temperature experimental observations and for possible application to natural systems such as migmatites or low velocity regions in the lithosphere.

The influence of a small amount of *melt* on the flow properties of a rock depends on the distribution of the melt, which in turn is related to its amount and to the relative solid–solid and solid–liquid interfacial energies, γ_{SS} and γ_{SL}. If a simple grain boundary intersects a pocket of melt and thermodynamic equilibrium is established, the grain-melt interfaces will take on spherical curvatures and the dihedral angle θ where these interfaces meet the grain boundary (Fig. 1.1) is governed, in the absence of anisotropy, by the following considerations (Bulau et al. 1979; Raj 1981; Raj and Lange 1981; Smith 1948, 1952; Wray 1976):

1. When $\gamma_{SL} \leq \gamma_{SS}/2$, the dihedral angle $\theta = 0$. The melt will then tend to penetrate the total grain boundary area and so have a maximum direct mechanical effect. However, taking into account also the maximization of the interfacial curvature, complete wetting of the grain boundaries at equilibrium for a fixed grain size will not occur unless the relative volume of melt exceeds about 21 %, depending on the exact grain shapes (Wray 1976); below this fraction, the melt will tend to concentrate in the 3-grain and 4-grain junctions and to be not interconnected. The case $\theta = 0$ is only to be expected where there is very close chemical affinity between solid and melt.

2. When $\gamma_{SS}/2 < \gamma_{SL} < \gamma_{SS}$, then $\cos\theta = \gamma_{SS}/2\gamma_{SL}$ and $0 < \theta < 60°$. The melt will tend to take up a prismatic configuration in the 3-grain junctions (grain edges) and the volume fraction at which equilibrium interconnection occurs will decrease from about 21 % to about 0.6 % as θ increases from 0 to 60° (Wray 1976). In contrast, the volume fraction at which complete wetting or separation

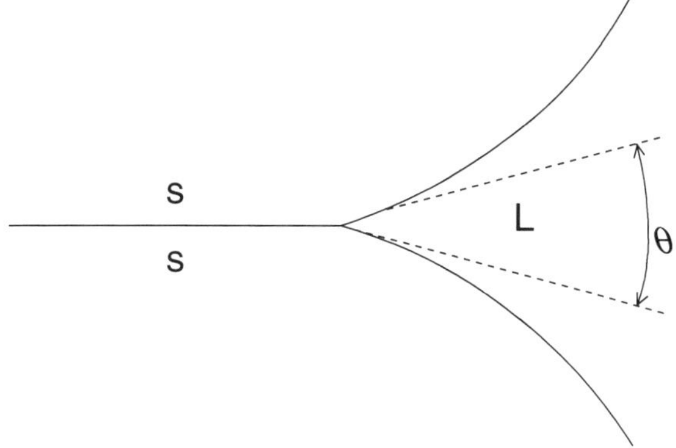

Fig. 1.1 Defining the dihedral angle θ between the solid–liquid (S–L) interfaces and their junction with a solid–solid interface (S–S)

of the grains occurs will become larger than 21 % when θ exceeds zero. The case $0 < \theta < 60°$ might be expected to be common where the melt is moderately similar chemically to the solid and γ_{SL} is somewhat less than γ_{SS} because of greater ease of atomic configurational adjustment and the liquid–solid boundary being the predominating factor. The observations of Waff and Bulau (1979, 1992) on annealed olivine-basalt mixtures fall in this category, the values of γ_{SS} and γ_{SL} being about 0.9 and 0.5 J m^{-2}, respectively (Cooper and Kohlstedt 1982), and the grain boundaries themselves being shown to be free of melt within a resolution of 2 μm (Vaughan and Kohlstedt 1982).

3. When $\gamma_{SL} > \gamma_{SS}$, then $\theta > 60°$ and the melt will tend to be segregated in tetrahedral pockets at 4-grain junctions (grain corners). The volume fraction for interconnection of the melt at equilibrium increases as θ increases, up to about 30 % for $\theta = 180°$, and substantially larger amounts of melt are needed to surround the grains completely for all values of θ in this range. The case $\theta > 60°$ is likely to arise when melt and solid are strongly dissimilar chemically.

In practice, long annealing times may be necessary to achieve the equilibrium configurations (~ 200 h at 1,240 °C for grain size ~ 50 μm in Waff and Bulau's study). Many specimens in laboratory studies may therefore have non-equilibrium distributions of melt in the form of widespread grain boundary films and irregularly shaped pockets of melt occurring wherever the melt was formed initially.

Similar considerations apply to interfaces between solid phases, although torque terms may sometimes have to be included in the dependence of interfacial energy on interface orientation (Cahn 1982). Here, however, equilibrium may be approached even more slowly than for the liquid–solid case and there may be more situations where anisotropy cannot be neglected.

The nature of void or *pore structure* associated with grain boundaries is of paramount importance in determining the permeability of a rock but it can also influence the mechanical properties; for example, at low temperatures, voids may act as nuclei for microcracks and at high temperatures they may contain fluids that promote diffusional creep. Void space may be of many origins and have many configurations, the configurations tending more to reflect the history of formation than the influence of the thermodynamic factors discussed above and so making generalization difficult. Commonly voids will represent space occupied during the genesis of the rock by aqueous fluids since lost or space formed by microcracking under external or internal stresses, and they may be re-filled in various degrees by material deposited later. Microcracks may also be transgranular and other pores such as fluid inclusions may be found within grains. For further information, see David et al. (1999).

References

Adler D (1975) The imperfect solid: transport properties. In: Hannay NB (ed) Treatise on solid state chemistry. Plenum Press, New York, pp 237–332

Ashby MF, Spalpen F, Williams S (1978) The structure of grain boundaries described as a packing of polyhedra. Acta Metall 26:1647–1663

Atkins PW (1986) Physical chemistry, 3rd edn. Oxford University Press, Oxford, 857 pp

Balluffi RW (1979) Grain boundary structure and segregation. In: Johnson WC, Blakely SM (eds) Interfacial segregation. American Society for Metals, Metals Park, Ohio, pp 193–236

Balluffi RW, Bristowe PD, Sun CP (1981) Structure of high-angle grain boundaries in metals and ceramic oxides. J Am Ceram Soc 64:23–34

Barrett CR, Nix WD, Tetelman AS (1973) The principles of engineering materials. Prentice-Hall, Englewood Cliffs

Bender B, Williams DB, Notis M (1980) Investigation of grain-boundary segregation in ceramic oxides by analytical electron microscopy. J Am Ceram Soc 63:542–546

Best MG (2003) Igneous and metamorphic petrology, 2nd edn. Oxford Publishers, Malden, 729 pp

Bollmann W (1970) Crystal defects and crystalline interfaces. Springer, Berlin 254 pp

Bollmann W (1982) Crystal lattices, interfaces, matrices: an extension of crystallography. W Bollmann, Geneva 360 pp

Bragg WL, Claringbull GF (1965) Crystal structures of minerals. George Bell and Sons Ltd, London 409 pp

Bulau JR, Waff HS, Tyburczy JA (1979) Mechanical and thermodynamic constraints on fluid distribution in partial melts. J Geophys Res 84:6102–6108

Cahn JW (1982) Transitions and phase equilibria among grain boundary structures. J de Phys, 43, Colloq. C6, C6-199 to C196-212

Cai W-p (1991) A new kinetic model for non-equilibrium grain boundary segregation. J Phys: Condens Matter 3:609–612

Catlow CRA, Stoneham AM (1983) Ionicity in solids. J Phys C: Solid State Phys 16:4321–4338

Chadwick GA, Smith DA (eds) (1976) Grain boundary structure and properties. Academic Press, London 388 pp

Christian JW (1975) Transformations in metals and alloys. Part I equilibrium and general kinetic theory, 2nd edn. Pergamon Press, Oxford 586 pp

Clarke DR (1987) Grain boundaries in polycrystalline ceramics. Ann Rev Mater Sci 17:57–74

Clemm PJ, Fisher JC (1955) The influence of grain boundaries on the nucleation of secondary phases. Acta Metall 3:70–73

Cooper RF, Kohlstedt DL (1982) Interfacial energies in the olivine-basalt system. In: Akimoto S, Manghnani MH (eds) High-pressure research in geophysics. D Riedel Publishing Company, Dordrecht, pp 217–228

Coulson CA (1961) Valence, 2nd edn. Oxford University Press, London, 404 pp

Deer WA, Howie RA, Zussman J (1992) An introduction to the rock-forming minerals. Longman, London 696 pp

Dobson DP, Alfredson M, Holzapfel C, Brodholt JP (2007) Grain-boundary enrichment of iron on magnesium silicate perovskite. Eur J Min 19:617–622

Doherty RD, Hughes DA, Humphries FJ, Jonas JJ, Juul Jensen D, Kassner ME, King WE, McNelley TR, McQueen HJ, Rollett AD (1997) Current issues in recrystallization: a review. Mater Sci Eng A238:215–274

Drury MR, Pennock GM (2007) Subgrain rotation recrystallization in minerals. Mater Sci Forum 550:95–104

Farkas D (2000) Atonistic theory and computer simulation of grain boundary structure and diffusion. J Phys: Condens Matter 12:R497–R516

Farver JR, Yund RA (1995) Interphase boundary diffusion of oxygen and potassium in K-feldspar/quartz aggregates. Geochim Cosmochim Acta 59:3697–3705

Fine ME (1975) Introduction to chemical and structural defects in crystalline solids. In: Hannay NB (ed) Treatise on solid state chemistry, vol 1. The chemical structure of solids. Plenum Press, New York, pp 283–333

Flewitt PEJ, Wild RK (eds) (2001) Grain boundaries: their microstructure and chemistry. John Wiley Ltd, Chichester 338 pp

Flynn CP (1972) Point defects and diffusion. Clarendon Press, Oxford 826 pp

Frank-Kamenetskaya OV, Rozhdestvenskaya IV (2004) Atomic defects and crystal structure of minerals (trans: Furstenko RV). Yanis Publishers, St Petersburg, 187 pp

Friedel G (1926) Lecons de Cristallographie, 2nd edn. Blanchard, Paris

Friedel J (1964) Dislocations. Pergamon Press, Oxford 491 pp

Gleiter H (1979) Recent developments in the understanding of the structure and properties of grain boundaries in metals. Kristall und Technik 14:269–284

Guinier A (1938) Structure of age-hardened aluminium-copper alloys. Nature 142:569–570

Guo YY, Kuo CK, Nicholson PS (1999) The ionicity of binary oxides and silicates. Solid State Ionics 123:225–231

Guttmann M (1977) Grain boundary segregation, two dimensional compound formation, and precipitation. Metall Trans 8A:1383–1401

Guyot P, Simon JP (1976) Symmetrical high-angle tilt boundary energy calculation in aluminium and lithium. Phys Stat Sol (a) 38, 207–216

Hahn W, Gleiter H (1981) On the structure of vacancies in grain boundaries. Acta Metall 29: 601–606

Hall EL (1982) Application of the analytical electron microscope to the study of grain boundary chemistry. J de Phys, 43, Colloq C6, C6-239 to C236-250

Hammond C (2001) The basics of crystallography and diffraction, 2nd edn. Oxford University Press, Oxford, 331 pp

Harrison WA (1980) Electronic structure and the properties of solids. W H Freeman, San Francisco 582 pp

Harrison RJ, Bruggeman GA, Bishop GH (1976) Grain boundary structure and properties. In: Chadwick GA, Smith DA (eds) Academic Press, London, pp 45–91

Hiraga T, Anderson IM, Kohlstedt DL (2004) Grain boundaries as reservoirs of incompatible elements in the Earth's mantle. Nature 427:699–703

Hirth JP, Balluffi RW (1973) On grain boundary dislocations and ledges. Acta Metall 21:929–942

Hondros ED (1976) Grain boundary segregation: assessment of investigative techniques. In: McLean D (ed) Grain boundary structure and properties. Academic Press, London pp 265–299

Hondros ED, Seah MP (1977) The theory of grain boundary segregation in terms of surface adsorption analogues. Met Trans A 8A:1363–1371

Hübner K, Leonhardt G (1975) Ionicity and electrical conductivity in transition-metal oxides. Phys Stat Sol (b) 68, K175–K179

Johnson WC (1977) Grain boundary segregation in ceramics. Metall Trans 8A:1413–1422

Johnson WC, Blakely JM (eds) (1979) Interfacial segregation. American Society for Metals, Metals Park, Ohio, 440 pp

Johnson CL, Hütch MJ, Buseck PR (2004) Nanoscale waviness of low-angle grain boundaries. Proc Nat Acad Sci USA 101:17936–17939

Karki BB, Kumar R (2007) Computer simulation of grain boundary structures in minerals. ICCES 3:35–41

Keller LM, Hauzenberger CA, Abart R (2007) Diffusion along interphase boundaries and its effect on retrograde zoning patterns of metamorphic minerals. Contr Mineral Petrol 154:205–216

Kelly A, Groves GW, Kidd P (2000) Crystallography and crystal defects. Wiley, Chichester, Revised, 470 pp

Kingery WD (1974) Plausible concepts necessary and sufficient for interpretation of ceramic grain-boundary phenomena: I. Grain-boundary characteristics, structure and electrostatic potential. II. Solute segregation, grain-boundary diffusion and general discussion. J Am Ceram Soc 57:74–83

Kingery WD, Mitamura T, van der Sande JB, Hall EL (1979) Boundary segregation of Ca, Fe, La and Si in magnesium oxide. J Mater Sci 14 Letters, 1766–1767

Kirfel A, Krane H-G, Blaha P, Schwartz K, Lippmann T (2001) Electron-density distribution in Stishovite, SiO_2: a new high-energy synchrotron-radiation study. Acta Cryst A57:663–677

Kittel C (1976) Introduction to solid state physics, 5th edn. Wiley, New York 608 pp

Kittel C (2005) Introduction to solid state physics, 8th edn. Wiley, Hoboken, 704 pp

Kowalczyk SP, Ley L, McFeely FR, Shirley DA (1974) An ionicity scale based on X-ray photoemission valence-band spectra of $A^N B^{10-N}$ type crystals. J Chem Phys 61:2850–2856

Kröger FA (1974) The chemistry of imperfect crystals, vol 2, 2nd edn. Imperfection chemistry of crystalline solids. North-Holland, Amsterdam, 988 pp

Kronberg ML, Wilson FH (1949) Secondary recrystallization in copper. Trans Met Soc AIME 185:501–514

Kuntcheva BT, Kruhl JH, Kunze K (2006) Crystallographic orientations of high-angle boundaries in dynamically recrystallized quartz: first results. Tectonophysics 421:331–346

Levine BF (1973a) d-electron effects on bond susceptibilities and ionicities. Phys Rev B7:2591–2600

Levine BF (1973b) Bond susceptibilites and ionicities in complex crystal structure. J Chem Phys 59:1463–1486

Libowitz GG (1975) Defect equilibria in solids. In: Hannay B (ed) Treatise in solid state chemistry, vol 1. The chemical structure of solids. Plenum Press, New York, pp 335–385

Madelung O (1978) Introduction to solid-state theory (trans: Taylor BC). Springer, Berlin, 486 pp

Marfunin AS (1979) Physics of minerals and inorganic materials (trans: Egorova NG, Mischenko AG). Springer, Berlin, 340 pp

Martin RM (2004) Electronic structure: basic theory and practical methods. Cambridge University Press, Cambridge 624 pp

Massalski TB (1983) Structure of solid solutions. In: Cahn RW (ed) Physical metallurgy. North-Holland Physic Pub., Amsterdam, pp 159–184

McLean D (1957) Grain boundaries in metals. Oxford University Press, London 346 pp

Mishin Y, Farkas D (1998) Atomistic simulation of [001] symmetrical tilt grain boundaries in NiAl. Phil Mag A 78:29–56

Mrowec S (1980) Defects and diffusion in solids. An introduction (trans: Marcinkiewicz S), Elsevier Amsterdam, (also PWN: Warsaw), 466 pp

Mueller RF, Saxena SK (1977) Chemical petrology. Springer, New York 394 pp

Nabarro FRN (1967) Theory of crystal dislocations. Clarendon Press (also Dover, New York, 1987), Oxford, 821 pp

Nesse WD (2000) Introduction to mineralogy. Oxford University Press, Oxford 442 pp

O'Keeffe M, Hyde BG (1985) An alternative approach to non-molecular crystal structures, with emphasis on the arrangement of cations. Struct Bond 61:77–144

O'Keeffe M, Hyde BG (1996) Crystal structures. Mineralogical Society of America, Washington

Olgaard DL, Rodriguez Rey A, and David C (eds) (1999) Imaging, analysing and modelling pore structure in geomaterials. Phys Chem Earth, Part A: Solid Earth and Geodesy 24, 551–644

Pauling L (1960) The nature of the chemical bond, 3rd edn. Cornell UP, Ithaca 644 pp

Pauling L (1971) Discussion on the chemical bond. Phys Today 24(February):9–13

Pauling L (1980) The nature of silicon-oxygen bonds. Am Min 65:321–323

Perkins D (2002) Mineralogy, 2nd edn. Prentice-Hall Inc., New Jersey, 483 pp

Phillips JC (1968) Dielectric definition of electronegativity. Phys Rev Lett 20:550–553

Phillips JC (1969) Covalent bonding in crystals, molecules, and polymers. University of Chicago Press, Chicago 267

Phillips JC (1970) Ionicity of the chemical bond in crystals. Rev Mod Phys 42:317–356

Phillips JC (1973) Bonds and bands in semiconductors. Academic Press, New York

Phillips JC (1975) Chemical bonds in solids. In: Hannay NB (ed) Treatise in solid state chemistry, vol 1: the chemical structure of solids. Plenum Press, New York, pp 1–41

Philpotts AR, Ague JJ (2009) Principles of igneous and metamorphic petrology, 2nd edn. Cambridge University Press, Cambridge, 686 pp

Pond RC, Smith DA, Vitek V (1979) Computer simulation of <110> tilt boundaries: structure and symmetry. Acta Metall 27:235–241

Preston GD (1938) Structure of age-hardened aluminium-coper alloys. Nature 142:570

Priester L (1980) Approche géométrique des joints de grains. Intérêt et limite, Revue Phys Appl 15:789–830

Putnis A (1992) Introduction to mineral sciences. Cambridge University Press, Cambridge 457 pp

Raj R (1981) Morphology and stability of the glass phase in glass-ceramic systems. J Am Ceram Soc 64:245–248

Raj R, Lange FF (1981) Crystallization of small quantities of glass segregated in grain boundaries in ceramics. Acta Metall 29:1993–2000

Ramakrishnan TV (1974) Ionicity and covalence in crystals. In: Rao, CNR (ed) Solid state chemistry. Marcel Decker, New York, pp 187–214

Redfern SAT, Artioli G, Rinaldi R, Henderson CMB, Knight KS, Wood BJ (2000) Octahedral cation ordering in olivine at high temperature. II: an in situ neutron powder diffraction study on synthetic Mg FeSiO$_4$(Fa50). Phys Chem Min 27:630–637

Ricoult DL, Kohlstedt DL (1983) Structural width of low-angle grain boundaries in olivine. Phys Chem Miner 9:133–138

Rohrer GS (2001) Structure and bonding in crystalline materials. Cambridge University Press, Cambridge 548 pp

Rosenberg M, Martino F, Reed WA, Eisenberger P (1978) Compton-profile studies of amorphous and single crystal SiO$_2$. Phys Rev B18:844–850

Samsonov GV (ed) (1968) Handbook of the physicochemical properties of the elements (trans: from Russian). Plenum Press, New York, pp 941

Sands DE (1969) Introduction to crystallography. Dover Publications, 165 pp

Schmalzried H (1981) Solid state reactions, 2nd edn. VCH, Weinheim, 254 pp

Schmalzried H (1995) Chemical kinetics of solids. VCH, Weinheim, 422 pp

Seah MP, Hondros ED (1973) Grain boundary segregation. Proc Roy Soc (London) A335: 191–212

Smith CS (1948) Grains, phases and interfaces: an interpretation of microstucture. Trans AIME 175:15–51

Smith CS (1952) Grain shapes and other metallurgical applications of topology. In: Metal interfaces. 1961 ASM seminar. American Society for Metals, Metals Park, Ohio, pp 65–108

Smith DA, Pond RC (1976) Bollman's O-lattice theory: a geometrical approach to interface structure. Int Met Rev 21:61–74

Smyth DM (2000) The defect chemistry of metal oxides. Oxford University Press, Oxford 304 pp

Stewart RF (1976) Electron population analysis with rigid pseudoatoms. Acta cryst A32:565–574

Stewart RF, Whitehead MA, Donnay G (1980) The ionicity of the Si-O bond in low-quartz. Am Min 65:324–326

Stocker RL (1978a) Influence of oxygen pressure on defect concentrations in olivine with fixed cationic ratio. Phys Earth Planet Int 17:118–129

Stocker RL (1978b) Point defect formation parameters in olivine. Phys Earth Planet Int 17:108–117

Sutton AP, Balluffi RW (1995) Interfaces in crystalline materials. Oxford University Press, New York 819 pp

Tiku SK, Kröger FA (1980) Effects of space charge, grain boundary segregation, and mobility differences between grain boundary and bulk on the conductivity of polycrystaliine Al_2O_3. J Am Ceram Soc 63:183–189

Tilley RJD (1986) Defect crystal chemistry and its applications. Kluwer Academic Publishers, Dordrecht, 256 pp

Tilley RJD (1998) Principles and applications of chemical defects. Taylor and Francis, London, 320 pp

Tilley RJD (2008) Defects in solids. Wiley-Interscience, New York, 529 pp

Tossell JA (1977) A comparison of silicon-oxygen bonding in quartz and magnesian olivine from X-ray spectra and molecular orbital calculations. Am Min 62:136–141

Unertl WN, de Jonghe LC, Tu YY (1977) Auger spectroscopy of grain boundaries in calcium-doped sodium beta-alumina. J Mater Sci 12:739–742

van Gool W (1966) Principles of defect chemistry of crystalline solids. Academic Press, New York 148 pp

van Vechten JA (1969a) Quantum dielectric theory of electronegativity in covalent systems. I. Electronic dielectric constant. Phys Rev 182:891–905

van Vechten JA (1969b) Quantum dielectric theory of electronegativity in covalent systems. II. Ionization potentials and interband transition energies. Phys Rev 187:1007–1020

Vaughan PJ, Kohlstedt DL (1982) Distribution of the glass phase in hot-pressed, olivine-basalt aggregates: an electron microscope study. Contr Mineral Petrol 81:253–261

Vernon RH, Clarke GL (2008) Principles of metamorphic petrology. Cambridge University Press, Cambridge 460 pp

Vitek V, Wang GJ (1982) Atomic structure of grain boundaries and intergranular segregation. J de Phys, 43, Colloq C6, C6-147 to C146-160

Waff HS, Bulau JS (1979) Equilibrium fluid distribution in an ultramafic partial melt under hydrostatic stress conditions. J Geophys Res 84:6109–6114

Waff HS, Faul UH (1992) Effects of crystalline anisotropy on fluid distribution in ultramafic partial melts. J Geophys Res 97:9003–9014

Warrington DH (1980) Formal geometrical aspects of grain boundary structure. In: Grain-boundary structure and kinetics. 1979 ASM materials science seminar. American Society for Metals, Metals Park, Milwaukee, Ohio pp 1–12

Weaire D (ed) (1975) Energy bands. In: Treatise on solid state chemistry, vol 1: the chemical structure of solids. Plenum Press, New York, pp 43–114

Wenk H-R, Bulakh A (2004) Minerals. Their constitution and origin. Cambridge University Press, Cambridge 646 pp

Westbrook JH (1969) Solute segregation at interfaces. In: Gifkins RC (ed) Interfaces: proceedings of the international conference, Melbourne, Butterworths, London, pp 283–305

Wigner E, Seitz F (1933) On the constitution of metallic sodium. Phys Rev 43:804–810

Wirth R (1986) High-angle grain boundaries in sheet silicates (biotite/chlorite): a TEM study. J Mater Sci Lett 5:105–106

Wray PJ (1976) The geometry of two-phase aggregates in which the shape of the second phase is determined by its dihedral angle. Acta Metall 24:125–135

Xu T (1987) Non-equilibrium grain-boundary segregation kinetics. J Mater Sci 22:337–345

Yan MF, Cannon RM, Bowen HK (1983) Space charge, elastic field, and dipole contributions to equilibrium solute segregation at interfaces. J Appl Phys 54:764–778

Yardley BWD (1989) An Introduction to metamorphic petrology. Longman Scientific and Technical, Harlow, Essex

Chapter 2
Thermodynamics

2.1 General

Thermodynamics is the theory of the interaction of heat and work and of their relationship to the physical properties and processes in material systems, dealt with at the macroscopic scale. One of its principal uses is therefore to provide constraints on the constitutive equations that describe the state of systems or the processes occurring in them. The foundation of thermodynamics consists of a minimal number of postulates or empirical laws drawn from experience. However, it has also been proposed that it can be regarded as being rooted in some universal and fundamental concepts of symmetry or invariance under transformation that apply to physical laws (Callen 1974). The scope of thermodynamics has traditionally been limited mainly to systems in equilibrium but has more recently been extended to deal also with non-equilibrium situations. We shall give here a brief summary of the principal results of these two branches of the theory.

2.2 Equilibrium Thermodynamics

We first recall the elements of classical equilibrium thermodynamics (or thermostatics). We take the concepts of heat Q and work W to be understood from classical physics or physical experience. However, because of path dependences, these quantities are inadequate for the proper description of the state of a physical system at any instant, for which further concepts are required (for general texts, see Callen 1960; Denbigh 1971; Guggenheim 1985; Callen 1985; for general texts, see Pippard 1957).

Thermodynamics therefore begins with the introduction of the macroscopic concept of the energy of a system, which is defined in terms of its conservation, through the First Law, as function of state called the *internal energy U*.

M. S. Paterson, *Materials Science for Structural Geology*,
Springer Geochemistry/Mineralogy, DOI: 10.1007/978-94-007-5545-1_2,
© Springer Science+Business Media Dordrecht 2013

Only changes in the energy of a system can be measured and these changes derive from work done on the system or from heat or substance added to the system. The internal energy is thus an extensive variable of fundamental importance in describing the state of the system. Other extensive or additive variables of similar importance are the amount of substance in the system and the dimensions of the system, if mechanical work is involved, or analogous parameters associated with other forms of work (electrical, etc.). For simplicity in exposition we shall here only consider systems having as the other extensive variables the volume V and the amounts of substance (moles) n_i of each of i components ($n_i = N_i/L$ where N_i is the number of molecules or entities of substance i and L is the Avogadro number). The state of such a system is then completely specified by U, V and n_i if it is at equilibrium.

However, we also wish to consider systems in which changes of state (transitions or processes) are occurring and to establish criteria of equilibria. Further, it is well known that real physical processes are irreversible or dissipative in some fundamental sense and we need a criterion for determining the direction of change. In order to deal with these aspects, another extensive variable and function of state called the *entropy S* is introduced through the Second Law, according to which, in an isolated system, entropy is unchanged ($\Delta S = 0$) in a reversible process and increases ($\Delta S > 0$) in an irreversible process. This law can be restated to give a criterion of equilibrium, namely, that in an isolated system the entropy is a maximum at equilibrium.

Some insight into the nature of entropy can be obtained from the molecular point of view of statistical mechanics, in which entropy is given by $S = k \ln g$ where k is the Boltzmann constant and g is the number of quantum states accessible to the system and assumed to be equally probable (Kittel and Kroener 1980, Chap. 2), that is, the number of different microscopic possibilities or configurations under which the given thermodynamic state can be realized. Macroscopically, it can only be stated that the entropy is a function of the other extensive variables,

$$S = S(U, \ V, \ n_i) \tag{2.1}$$

This relation serves as a *fundamental relation* from which all other properties of the thermodynamic system in equilibrium can be derived, since the specification of the extensive variables fully characterizes the state of the system. From consideration of the differential of S in the case of the reversible addition of an amount of heat ΔQ to the system at constant V and n_i it follows that $\Delta S = \Delta Q/T$, or $S = \int \Delta Q/T$ where the integration path is a reversible path; in the irreversible case, $\Delta S > \Delta Q/T$ but no other statement can in general be made, that is, we cannot in general define an entropy exactly in a system out of equilibrium unless some restrictive statements are made about the nature of the system.

The fundamental relation in the "entropic form" (2.1) can be rewritten in the "energetic form"

$$U = U(S, \ V, \ n_i) \tag{2.2}$$

From the differential of U the following (energetic) intensive variables are defined:

$$\text{absolute temperature} \qquad T \equiv \left(\frac{\partial U}{\partial S}\right)_{V,n_i}$$

$$\text{pressure} \qquad p \equiv -\left(\frac{\partial U}{\partial V}\right)_{S,n_i}$$

$$\text{(electro) chemical potential}$$

$$\text{of the ith component} \qquad \mu_i \equiv \left(\frac{\partial U}{\partial n_i}\right)_{S,V,n_{j\neq i}}$$

The intensive variables enable one to deal with the coupling of the system to its environment. Using them the relation (2.2) can be rewritten in the Gibbs form

$$dU = TdS - pdV + \sum_i \mu_i dn_i \tag{2.3a}$$

or in the Euler form

$$U = TS - pV + \sum_i \mu_i n_i \tag{2.3b}$$

If follows from (2.3) that the above criterion of equilibrium can be restated in terms of minimizing the internal energy at constant S, V, n_i.

In practice it is often convenient to use one or both the intensive variables T, p as independent variables instead of the respective conjugate extensive variables S, V used in writing (2.2). Such transformation leads to the definition of the additional functions,

$$\text{Helmholtz (free) energy} \quad A \equiv A(T,\ V,\ n_i) = U - TS$$

$$\text{enthalpy} \qquad H \equiv H(S,\ p,\ n_i) = U + pV$$

$$\text{Gibbs (free) energy} \qquad G \equiv G(T,\ p,\ n_i) = U + pV - TS,$$

to equivalent forms of the fundamental relations (2.3),

$$dA = -SdT - pdV + \sum \mu_i dn_i \tag{2.4}$$

$$dH = TdS + Vdp + \sum \mu_i dn_i \tag{2.5}$$

$$dG = -SdT + Vdp + \sum \mu_i dn_i \tag{2.6}$$

and to corresponding extremum conditions for equilibrium under constraint of the specified independent variables (for example, the Gibbs energy G is a minimum at equilibrium when T, p and n_i are the independent variables). There are many mathematical relations between the various parameters or functions of state so far introduced and others derivable from them, which constitute much of the useful content of thermodynamic theory and are summarized in textbooks (for example Callen 1960, Chaps. 3 and 7; Denbigh 1971, pp. 89–98). One such relation of particular importance is the Gibbs–Duhem relation expressing an interdependence among the intensive variables,

$$S dT - V dp + \sum n_i d\mu_i = 0 \qquad (2.7)$$

In the application of the thermodynamic principles to a real physical situation, it is necessary to know the explicit form of the fundamental relation, in any of its equivalent versions (2.1)–(2.6), in order to fully express the physics of the situation, although the search for this explicit form is not strictly part of thermodynamics itself. The physics of the situation can alternatively be introduced in the form of equations of state, which are explicit relations between the independent extensive variables and the intensive variables, such as the ideal gas laws $V = nRT/p$ and $U = 3nRT/2$ ($R = Lk$).

Since systems at equilibrium are homogeneous within regions free from internal walls it is sufficient to discuss the total amounts S, U, V, n_i, $G, \ldots \ldots$ of the extensive properties within the homogeneous regions. However, in chemical thermodynamics, it is often useful to normalize the extensive properties to the total amount of substance n ($= \Sigma n_i$) in the system to give, respectively, the molar quantities S_m, U_m, V_m, x_i, $G_m, \ldots \ldots$ (where $S_m = S/n \ldots$ and $x_i =$ mole fraction of the i'th component); further, the molar quantities may be partitioned among the components of substance as the partial molar quantities, noting that the partial molar Gibbs energy is identical to the chemical potential.

2.3 Non-Equilibrium Thermodynamics

We now turn to systems out of equilibrium. Here, the thermodynamic treatment rests on less well-defined foundations than does classical equilibrium thermodynamics. There exist, in fact, a number of distinct approaches of diverse aims, but two major branches of theory can be distinguished, characterized by Germain (1974) as the "ambitious attitude" and the "cautious attitude". The former, variously labeled as rational thermodynamics or continuum thermodynamics, claims to attempt the broadest possible analysis. It aims not to depend on generalization from classical equilibrium thermodynamics but to deal *ab initio* with processes (described by constitutive relations) rather than with states. The physical specification of the system at any instant involves not only the values of the measurable parameters at that instant but also their histories at all previous instants, expressed as functionals such as hereditary integrals (Malvern 1969, pp. 256, 319). The concepts of temperature and entropy are introduced as primitive quantities. No attempt will be made to expound this approach here (see Coleman 1964; Day 1972; Noll 1974; Truesdell 1984).

The alternative and more conservative approach is to retain the concept of local state in giving the physical specification of a system at any instant. The postulate that a local state exists is often taken as being equivalent to assuming some form of local equilibrium (Glansdorff and Prigogine 1971, p. 14) but it can have a wider meaning (Lavenda 1978, p. 77). Attempts to justify this postulate usually point to the relaxation time for fluctuations at the atomic scale being short compared with the timescale of the macroscopic processes to which the theory is applied. Where the local state cannot be fully specified in terms of measurable macroscopic variables it is assumed that there exist additional internal or hidden variables (for example, dislocation density) which complete the description of the local state. The importance of the postulate of a local state is that it enables many concepts to be carried over from classical equilibrium thermodynamics, such as the concepts of entropy and energy as scalar potentials and the Gibbs–Duhem and the Gibbs relations (2.3a) and (2.7). The production of entropy envisaged by the Second Law for irreversible processes is then discussed with a view to placing constraints on the laws governing these processes, especially in relation to their stability. The applications have commonly been confined to processes in systems not very far from equilibrium, the theory for which is termed the *linear* thermodynamics of irreversible processes. The theoretical situation for more general applications is less well developed, and still to a considerable extent the subject of research. Thus there have been attempts to develop the theory of the nonlinear thermodynamics of irreversible processes applicable to systems far from equilibrium; see for example, Glansdorff and Prigogine (1971), who introduce the concept of dissipative structures; also Lavenda (1978). Other applications have been to processes such as friction or ideal plasticity where dissipation is equally important no matter how slowly the process proceeds, for example, Kestin (1966) and Nemat-Nasser (1974). The remainder of these notes will concern linear thermodynamic theory under the assumption of the existence of a local state, as expounded by Denbigh (1951), Meixner and Reik (1959), de Groot and Mazur (1962), Katchalsky and Curran (1965), Prigogine (1967), Fisher and Lasaga (1981), Kuiken (1994), Martyushev and Seleznev (2006), Holyst (2009), Kleidon (2009), and others.

The starting point for the *linear thermodynamics of irreversible processes* is the Second Law and the concept of entropy production in an irreversible process. In any irreversible change in a system, the rate of change in entropy is made up of a part due to entropy flow from the surroundings and a part due to changes within the system. The latter part is known as the rate of entropy production, or simply the entropy production, and designated σ per unit volume; according to the Second Law it must be positive. In the energy representation of the evolution of the system, the corresponding quantity is $T\sigma$, which is sometimes called a dissipation function or potential since it represents the rate at which irrecoverable energy or work must be supplied or done to maintain the process. The dissipation function can be written in the form

$$T\sigma \equiv \sum_{\alpha} J_{\alpha} X_{\alpha} > 0 \quad \alpha = 1, 2, 3 \ldots \tag{2.7}$$

where J_α are the rates of change of local extensive parameters and X_α are conjugate intensive parameters, the summation being over all the component processes contributing to the entropy production. There can be some ambiguity in the factoring of the terms of (2.7) into the J_α and X_α but no serious confusion arises if consistent rules are followed (Miller 1974).

Where heat flow, diffusion, and chemical reactions are to be taken into account, the dissipation function can be expressed in the following terms (Katchalsky and Curran 1965, Chap. 7)

$$T\sigma = J_Q \cdot T\,\mathrm{grad}\frac{1}{T} \ + \ J_i \cdot T\,\mathrm{grad}\left(-\frac{\mu_i}{T}\right) \ + \frac{\mathrm{d}\xi}{\mathrm{d}t}A \qquad (2.8a)$$

or

$$T\sigma = J_S \cdot \mathrm{grad}(-T) \ + \ J_i \cdot \mathrm{grad}(-\mu_i) \ + \frac{\mathrm{d}\xi}{\mathrm{d}t}A \qquad (2.8b)$$

where J_Q, J_S and J_i are vectors representing the currents or rates of flow of heat, entropy, and substance i, respectively, $\mathrm{d}\xi/\mathrm{d}t$ is a scalar representing the rate of advancement of the chemical reaction (ξ is the extent of chemical reaction), and $A = -\sum v_i\mu_i$ is the chemical affinity driving the reaction (v_i are the stoichiometric coefficients for the reaction). In cases involving viscous flow, electric and magnetic effects, or other dissipative processes, further terms can be added. The factors J_Q, J_S, J_i and $\mathrm{d}\xi/\mathrm{d}t$ in (2.8a) can be clearly identified as the extensive parameters J_α in (2.7) and their multiplying factors are then the respective conjugate intensive parameters X_α. The seeming ambiguity regarding the intensive parameter conjugate to J_i arises from the different ways in which the total dissipation is partitioned between the different terms in (2.8a) and (2.8b). The form (2.8a) is appropriate where flow of heat and diffusion of substance are being individually and simultaneously measured in the presence of a temperature gradient, in which case it is evident that the intensive parameter driving the diffusion is to be taken to be $T\,\mathrm{grad}\,(\mu_i/T)$. The form (2.8b) is appropriate when the entropy changes associated with the movement in the temperature gradient of the measurable heat and of the substance (through its heat content) are brought together in the factor J_S; although the latter quantity is not directly measurable, the form (2.8b) is useful in indicating that, in the absence of a temperature gradient, the intensive parameter driving the isothermal diffusion can be taken to be $\mathrm{grad}(-\mu_i)$.

In the entropy representation, the interest centers on σ, the rate of entropy production itself, instead of on $T\sigma$. Again σ is written as a sum of terms that are the products of extensive and intensive parameters and, somewhat confusingly, the same symbols are often used as were used in (2.7) for $T\sigma$; further, σ or $\sigma/2$ is often taken as a dissipation function or potential. In both representations, the J_α factors are commonly termed "fluxes" or "flows", although these terms are scarcely appropriate for quantities such as $\mathrm{d}\xi/\mathrm{d}t$, and the X_α are variously termed the "thermodynamic forces" or "affinities", regardless of whether being in energy or entropy representation and in spite of the $1/T$ factor subsumed in the X_α in the latter case. The applications of the linear thermodynamics of irreversible processes now follow from consideration of the quantities in the terms of $T\sigma$ and of the inequality expressed in (2.7).

The physics of the processes is expressed in the relationships between the forces and fluxes, X_α, J_α, known as the phenomenological, constitutive, or kinetic relations or as the thermodynamic equations of motion; their role is in many respects analogous to that of the equations of state relating extensive and intensive quantities in equilibrium thermodynamics. One of the main activities of thermodynamic theory has been to place constraints on the relations between the thermodynamic variables and to discuss criteria for stability and stationary states. However, whereas the constraints governing the quantities entering the equilibrium equations of state can generally be stated independently of the particular nature of these equations, it is very difficult to establish laws of general validity governing the quantities entering the phenomenological equations for an arbitrary non-equilibrium situation. Substantial progress has only been made for certain classes of situations, chiefly for those close to equilibrium. We therefore restrict considerations to the latter and in particular to the situations in which the relationships between forces and fluxes can be written in the linear form

$$J_\alpha = \sum_\beta L_{\alpha\beta} X_\beta \qquad \beta = 1, 2, 3\ldots \tag{2.9}$$

where $L_{\alpha\beta}$ are constants, often called the phenomenological or kinetic coefficients. In writing the phenomenological relations in the form (2.9), the quantities X_β, J_α are treated as scalars but in practice they can represent scalar quantities or the Cartesian components of vector or tensor quantities. The relations (2.9) take into account the possibility of coupling effects between non-conjugate forces and fluxes, for example, coupling between heat flow and diffusion or between the diffusion of different species. It is now possible, using (2.9), to write (2.7) in the form

$$T\sigma = \sum_{\alpha,\beta} L_{\alpha\beta} J_\alpha J_\beta \geq 0 \tag{2.10}$$

which has important consequences.

The principal initial success of the linear thermodynamics of irreversible processes lies in the enunciation of the Onsager (1931a) reciprocal relations

$$L_{\alpha\beta} = L_{\beta\alpha} \tag{2.11}$$

which express a symmetry between coupling effects. Note that some elaboration of (2.11) is needed when magnetic fields or rotational effects are present or the forces or fluxes are mixed in respect of being odd or even in tensorial character; see Casimir (1945) and Meixner and Reik (1959). Much has been written concerning the statistical mechanical derivation of (2.11) using a fundamental principle of microscopic reversibility; for critical discussion and references, see Lavenda (1978, Chap. 2): also Callen (1960). However, (2.11) can also be treated as an empirical axiom of macroscopic theory adequately supported by experimental observation (Miller 1974). Curie's principle concerning the symmetry relationships between cause and effect (Curie 1894; Paterson and Weiss 1961) is also often invoked to constrain further the possible values of $L_{\alpha\beta}$, this principle being paraphrased to state that no direct coupling occurs between processes described respectively by quantities of odd and even tensorial character, for example, between chemical reaction (scalar) and diffusion (vector). Then, we can put $L_{\alpha\beta} = 0$ where α, β $(\alpha \neq \beta)$ refer to two such processes, although this does not prevent indirect interference occurring (Prigogine 1967, p. 89). Other possible constraints on the $L_{\alpha\beta}$ are discussed by Fisher and Lasaga 1981.

One can expect non-equilibrium thermodynamics to be concerned also with the likely paths to be followed by processes, with the nature of stationary states, and with questions of stability, in analogy with topics in equilibrium thermodynamics such as the criteria of equilibrium and conditions governing phase transitions. Actually, when not ignored, these non-equilibrium topics appear to be the subject of considerable debate and research, and only a few general remarks are appropriate here (for a summary on stability considerations, see Prigogine 1980). It seems that a general principle of fundamental and far-reaching importance is the principle of least dissipation of energy (Lavenda 1978, Chap. 6; Onsager 1931a, b). In the particular case of linear phenomenological laws, it follows that a system will evolve in the direction of diminishing rate of entropy production, towards a state characterized by a minimum rate of entropy production (Prigogine 1967). However, in a completely unconstrained situation this minimum rate will be zero, reached when all irreversible processes have stopped and equilibrium is attained. In order to maintain a stationary or steady non-equilibrium state, that is, one that no longer evolves with time, it is therefore necessary to constrain at least one term in the dissipation function $T\sigma$ to be nonzero, that is, to hold at least one of the thermodynamic forces X_α at a constant, nonzero value. It follows that in the stationary state all unconstrained parameters X_α, J_α will become zero. Thus the stationary state in any system in which linear processes are occurring is that in which the rate of entropy production is a minimum under certain auxiliary conditions such as specified nonzero values for at least one of the X_α or J_α, and this state will be stable.

For corresponding considerations in nonlinear and far from-equilibrium situations, including the occurrence of stable "dissipative" structures, see (Glandsdorff and Prigogine 1971; Lavenda 1978; Prigogine 1980; Fisher and Lasaga 1981 and Ross 2008).

References

Callen HB (1960) Thermodynamics: an introduction to the physical theories of equilibrium thermostatics and irreversible thermodynamics. Wiley, New York 376 pp

Callen HB (1974) A symmetry interpretation of thermodynamics. In: ed, Foundations of continuum mechanics, London, Macmillan, pp 51-78; also in: foundation of physics 54 pp 423–443 (1974)

Callen HB (1985) Thermodynamics and an introduction to thermostatics, 2nd edn. Wiley, New York 493 pp

Casimir HBG (1945) On Onsager's principle of microscopic reversibility. Rev Mod Phys 17:343–350

Coleman BD (1964) Thermodynamics of materials with memory. Arch Rat Mech Anal 17:1–46

Curie P (1894) Sur la symétrie dans les phénomènes physiques, symétrie d'un champ electrique et d'un champ magnetique. J de Phys 3:393–415

Day WA (1972) The thermodynamics of simple materials with fading memory. Springer, Berlin 134 pp

De Groot SR, Mazur P (1962) Non-equilibrium thermodynamics. North Holland, Amsterdam 510 pp

Denbigh KG (1951) The thermodynamics of the steady state. Methuen, London 103 pp

Denbigh KG (1971) The principles of chemical equilibrium, 3rd edn. Cambridge University Press, Cambridge 494 pp

Fisher GW, Lasaga AC (1981) Irreversible thermodynamics in petrology. In: Geochemical processes. Reviews in mineralogy vol 8, Mineral Soc Amer pp 171–209

Germain P (1974) The role of thermodynamics in continuum mechanics. In: Foundations of continuum thermodynamics, London, Macmillan, pp 317–333

Glansdorff P, Prigogine I (1971) Thermodynamic theory of structure, stability and fluctuations. Wiley, London 306 pp

Guggenheim EA (1985) Thermodynamics. An advanced treatment for chemists and physicists, 7th edn. North-Holland, Amsterdam, 390 pp

Holyst R (2009) Challenges in thermodynamics: irreversible processes, nonextensive entropies, and systems without equilibrium states. Pure Appl Chem 81:1719–1726

Katchalsky A, Curran PF (1965) Nonequilibrium thermodynamics in biophysics. Harvard University Press, Cambridge Mass 248 pp

Kestin J (1966) A course in thermodynamics 2 vols (revised 1979), Washington, Hemisphere Publ Co, 725 + 617 pp

Kittel C, Kroener H (1980) Thermal physics, 2nd edn. Freeman, San Francisco, 473 pp

Kleidon A (2009) Nonequilibrium thermodynamics and maximum entropy production in the Earth system. Naturwissenschaften 96:653–677

Kuiken GDC (1994) Thermodynamics of irreversible processes applications to diffusion and rheology. Wiley, Chichester 458 pp

Lavenda BH (1978) Thermodynamics of irreversible processes. MacMillan, London 182 pp

Malvern LE (1969) Introduction to the mechanics of a continuous medium. Prentice-Hall, Englewood Cliffs 713 pp

Martyushev LM, Seleznev VD (2006) Maximum entropy production principle in physics, chemistry and biology. Phys Rep—Rev Sect Phys Lett 426:1–45

Meixner J, Reik HG (1959) Thermodynamik der irreversiblen Prozess. In: Handbuch der Physik. Vol III/2 Prinzipien der Thermodynamik und Statistik, Springer, Berlin, pp 413–523

Miller DG (1974) The Onsager relations: experimental evidence. In: Foundations of continuum mechanics, London, MacMillan, pp 185–214

Nemat-Nasser S (1974) On nonequilbrium thermodynamics of viscoelasticity and viscoplasticity. In: International symposium on the foundations of continuum thermodynamics, Bussaco, Portugal, 1973, London, MacMillan, pp 259–281

Noll W (1974) The foundations of mechanics and thermodynamics: selected papers by W Noll. Springer, Berlin 324 pp

Onsager L (1931a) Reciprocal relations in irreversible processes—1. Phys Rev 37:405–426

Onsager L (1931b) Reciprocal relations in irreversible processes—2. Phys Rev 38:2265–2279

Paterson MS, Weiss LE (1961) Symmetry concepts in the structural analysis of deformed rocks. Geol Soc Am Bull 72:841–882

Pippard AB (1957) Elements of classical thermodynamics. Cambridge University Press, Cambridge 165 pp

Prigogine I (1967) Introduction to thermodynamics of irreversible processes, 3rd edn. Interscience Publishers, New York, 147 pp
Prigogine I (1980) From being to becoming: time and complexity in the physical sciences. W H Freeman, San Francisco 272 pp
Ross J (2008) Thermodynamics and fluctuations far from equilibrium. Springer, New York 210 pp
Truesdell C (1984) Rational thermodynamics, 2nd edn. Springer, New York 578 pp

Chapter 3
Rate Processes

3.1 Introduction

3.1.1 General Considerations

We now proceed to some general considerations of processes in material systems. A rate process in any system may be defined as a course of change in the system as a function of time. Very broadly, three types of rate processes may usefully be distinguished; reactions, transport processes, and deformations.

Reactions are changes in the nature of the components of substances in a system. They may be simple transformations of the type $A \to B$, such as polymorphic phase transitions and solid-state recrystallization, involving one "reactant" A and one "product" B. Or they may be chemical reactions involving two or more reactants and/or products. In all cases, the course of change with time can be represented by a scalar quantity.

Transport processes are those that bring about a spatial redistribution of any of the extensive quantities that characterize the system (see Chap. 2 for the thermodynamic definition of extensive and intensive variables and of other concepts introduced here). Examples are heat flow and diffusion. Such processes are normally considered as occurring in a continuous system, or continuum, in contrast to the homogeneous finite system of classical thermodynamics. The course of change with time is now, in general, represented by a vector quantity.

Deformation processes in which energy dissipation occurs are also rate processes insofar as the rate of deformation is a significant variable. In this case, the course of change with time is, in general, a second rank tensor and the system may be viewed as finite and homogeneous or as a continuum depending on whether homogeneous or inhomogeneous deformation is concerned. We shall not consider deformation processes further in this chapter but shall return to them in the later chapters.

M. S. Paterson, *Materials Science for Structural Geology*,
Springer Geochemistry/Mineralogy, DOI: 10.1007/978-94-007-5545-1_3,
© Springer Science+Business Media Dordrecht 2013

In more complex cases, two or more of the above types of rate processes may be identifiable as occurring simultaneously. For example, in heterogeneous chemical reactions, both reaction and diffusion are involved. However, in such cases, one of the constituent processes is usually rate controlling; normally, the slowest one if the processes are sequentially dependent on each other.

3.1.2 Theory of Rate Processes

The theoretical treatment of rate processes can be approached from two points of view, the thermodynamic and the kinetic. However, while at first sight the two approaches may appear distinct, they are necessarily eventually related or equivalent.

In the *thermodynamic approach*, the change in the system is viewed, somewhat formally, as a response to the system not being at equilibrium. The general considerations of non-equilibrium thermodynamics, as sketched out in Chap. 2, are applied. A specific process is discussed in terms of a measure of the departure from equilibrium, a measure of the response of the system to this departure, and a quantitative relationship between these two measures, involving some material parameters. The linear thermodynamics of irreversible processes provides constraints on this relationship in cases where the system is not very far from equilibrium. Nonlinearity appears when systems are far from equilibrium and new phenomena may then be involved, as exemplified by the transition from laminar to turbulent flow in fluid dynamics or by oscillations and dissipative structures in chemical reactions (Glansdorff and Prigogine 1971). No attempt will be made here to deal with nonlinear systems, although they may be important in connection with some actual geological structures (Fisher and Lasaga 1981).

The treatment of processes taking into account the atomic structure of matter and the existence of fluctuations in the fine-scale distribution of energy leads one into statistical mechanics. This approach, often termed *kinetic theory*, especially in connection with gases, views the change in the system directly in terms of instantaneous or spontaneous elementary events, such as reactions between colliding molecules or motions of diffusing atoms, and of the macroscopic summation of these events in time. In this approach, the primary quantities in the macroscopic description are the concentrations of the entities involved in the elementary events, whereas in the thermodynamic approach the primary quantities are the activities which add a "thermodynamic weighting" to the concentrations through the inclusion of the activity coefficients in order to make practical application of ideal thermodynamic laws.

Associated with the kinetic approach and in parallel with the thermodynamic approach, there has also been the establishment of a phenomenological or empirical framework in what may be termed *empirical kinetics*. This development has often occurred earlier and independently of the thermodynamic approach and is illustrated in empirical relations such as the Arrhenius law and Fick's laws of diffusion. Compared with the thermodynamic/formal kinetic approaches, the formalisms of empirical kinetics tend to be more specifically adapted to individual processes and perhaps less concerned with the proper identification of driving forces, that is, with the true "dynamics". However, we shall attempt to bring out connections between the two approaches.

3.1.3 Temperature and Pressure Dependence

The rate of change in a system is observed to be not solely dependent on the degree to which the system departs from equilibrium, as defined in a classical or homogeneous system by a given variable not being an extremum or in a continuous system by a variable being nonuniform. The rate may also depend on the values of other variables even if they are not changing with time themselves. Most notable of these variables is the temperature. Where the influence of temperature is significant, the process is said to be *thermally activated*. Other processes may appear to be athermal, at least to a first approximation, as in the case of martensitic transformations or crystal plasticity at relatively low temperatures, but such cases become less common at higher temperatures. Pressure may also enter as a significant variable influencing the rate of a process, especially if substantial volume changes are involved.

The dependence of the rate of thermally activated processes on the temperature can normally be expressed in the exponential form

$$rate = Ae^{-\frac{Q}{RT}} \tag{3.1}$$

where Q is a parameter with the dimensions of energy per unit amount of substance, called the *activation energy*, A is a constant, T is the thermodynamic or absolute temperature, and R is the gas constant, 8.314 J K^{-1} mol^{-1} ($R = Lk$, where L is the Avogadro number and k the Boltzmann constant). The symbol E or E_a is often used instead of Q for the activation energy and is in many respects preferable, but Q has been widely used in the materials science literature, especially for the empirically determined quantity, and is retained here for this reason. As an empirical expression, the factoring of the multiplier of $1/T$ in (3.1) into Q/R is, of course, arbitrary, but it serves to facilitate comparison with theoretically derived formulae in which kT appears as the fundamental temperature from statistical thermodynamics.

In a wide range of rate processes, the relation (3.1), known as the law of Arrhenius (1889), is found empirically to apply, with constant Q, over a substantial interval of temperature. This law was originally established empirically, but it can also be shown to be plausible from a statistical mechanical point of view. However, the proper theoretical justification of the form (3.1) requires a specific treatment for each process (cf. Flynn 1972, Chap. 7 for diffusion).
Some properties of the exponential temperature dependence can be seen in the plot of $\exp(-Q/RT)$ versus T for different values of Q shown in Fig. 3.1. Particularly

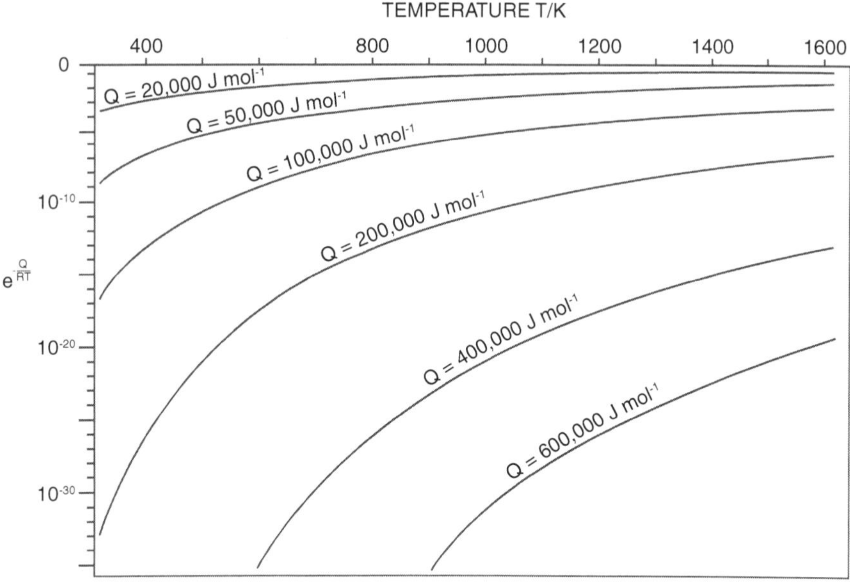

Fig. 3.1 Trends in activation rates with temperature T and activation energy Q

striking features are the rapid decrease in order of magnitude of the absolute value and the increase in temperature sensitivity (slope) at given temperature as Q is increased. The increase in temperature sensitivity as the temperature decreases at fixed Q is also evident. These properties are less obvious in the usual experimental plots of $\ln(rate)$ versus $1/T$, which give straight lines for fixed values of Q.

The pressure may also influence the rate of a process, although its effect tends to be less marked than that of the temperature. To take the influence of pressure into account, (3.1) can be rewritten in the form

$$rate = Ae^{-\frac{Q_0 + pV^*}{RT}} \tag{3.2}$$

where Q_0 is the activation energy at zero or other reference pressure, p is the pressure relative to the reference pressure, and V^* is a parameter having the dimensions of volume per unit amount of substance and known as the activation volume; V^* is commonly found to be constant for measurements over an appreciable range of pressure. Again, the form (3.2) is supported empirically but it can be rationalized in a similar way to that given for (3.1) by postulating that the threshold level of energy for activating the elementary microscopic events must include an amount sufficient to provide the work required for the momentary increases in volume occurring locally during the events.

It is to be noted that, as introduced here, the activation energy Q or $Q_0 + pV$ in (3.1) and (3.2) is a kinetically rather than a thermodynamically defined quantity and so, strictly, is not to be identified immediately with one of the thermodynamic molar energy functions such as U_m, A_m, H_m, and G_m (Chap. 2), although it is sometimes called the activation enthalpy because of a formal similarity to an enthalpy. Only in connection with a particular model in which the process is specified in thermodynamic terms, as in the transition state theory, can Q be properly identified with a thermodynamic energy function or potential. If temperature and pressure are the independent variables, the appropriate quantity to use is the molar Gibbs energy, in which case the specification of Q as an activation enthalpy presumes that the entropy term in the Gibbs energy has been subsumed in the pre-exponential constant. In any case, in the determination of Q from an "Arrhenius plot" of $\ln(rate)$ versus $1/T$, the slope gives only the enthalpy part. When the concept of activation energy is used in describing experimental situations where no single activated process has been identified or where it is possible that several are involved, it is appropriate that the term be qualified as the *empirical, experimental, and/or apparent activation energy*. In this case, it is simply a measure of the sensitivity of the measured macroscopic rate to change in temperature.

3.2 Reaction Kinetics

3.2.1 Thermodynamic Approach

A reaction can be represented in general in the form

$$v_A A + v_B B + \cdots \rightarrow v_P P + v_Q Q + \cdots \tag{3.3a}$$

where $A, B, \ldots$ are the reactants and $P, Q, \ldots$ are the products. The symbols $A, B, \ldots$ and $P, Q, \ldots$ serve first to identify the components involved in the reaction, and second to represent a unit amount of the component that is consumed or produced when the reaction occurs. The dimensionless quantities v_i are termed the stoichiometric coefficients, which normally will be integers without a common divisor. It is conventional to take the v_i to be negative for reactants and positive for products, in which case we have the following algebraic balance:

$$0 = v_A A + v_B B + \cdots + v_P P + v_Q Q + \cdots \tag{3.3b}$$

A simple phase transformation or similar change can also be represented by the form (3.3), with only one "reactant" and one "product", both v_i being of unit magnitude.

In a thermodynamic approach, one can consider a time interval dt during which $-dn_i = -v_i d\xi$ of each reactant is consumed and $dn_i = v_i d\xi$ of each product produced, where ξ is a quantity termed the advancement or extent of reaction; thus $dn_i/dt = v_i d\xi/dt$ is the rate of increase in the amount of each component i during the reaction and is a measure of the rate of reaction (in mol s^{-1} in SI.

units). The quantity $d\xi/dt$ is called the rate of reaction (Atkins 1986, p. 651). When the system is taken to be a unit volume, dn_i/dt becomes the rate of change in the amount-of-substance concentration of the component i, designated dc_i/dt or $d[i]/dt$, a measure that can be extended to continuous systems (SI. units mol m^{-3} s^{-1}). When dn_i/dt or dc_i/dt is used to specify the rate of reaction care must be taken to specify the component i to which it refers, an elaboration not required when the "true rate" $d\xi/dt$ or $(dc_i/dt)/v_i$ is used.

In a closed system, an equilibrium state will exist at certain proportions of reactants and products, as determined thermodynamically by the equilibrium constant $K = \prod_i (a_i)^{v_i}$, where $a_i = \gamma_i c_i$ is the activity of the ith component and γ_i its activity coefficient. At equilibrium the rate of reaction is zero. When the departure from equilibrium is small, it may be expected that the rate of reaction will be proportional to the degree of departure from equilibrium, giving the linear thermodynamic relation

$$\dot{\xi} = L_k A \tag{3.4}$$

where $\dot{\xi} = d\xi/dt$ is taken as the rate of reaction, L_k is the phenomenological/ kinetic/thermodynamic coefficient, and A is a thermodynamic force (affinity) driving the reaction, defined in chemical reactions at constant temperature and pressure as $A = -(\partial G/\partial \xi) = -\sum_i v_i \mu_i$, where the μ_i are the chemical potentials of the components. In transformations at constant temperature and pressure, A is simply the difference in chemical potentials of the untransformed and transformed phases. In SI units, A is in J mol^{-1} and so L_k is in mol J^{-1} s^{-1}; if the rate of reaction is alternatively specified in terms of a rate of change of molar concentration, then the phenomenological coefficient is in mol^2 m^{-3} J^{-1} s^{-1}. "Near to equilibrium" can be usefully defined by the condition $A \ll RT$, a condition that can be rationalized on the basis of the theory of fluctuations or statistical thermodynamics using the approximation that $\exp(A/RT)$ is linear in A if $A \ll RT$.

3.2.2 Kinetic Approach

In empirical kinetics, relations are sought between the rate of reaction and the concentrations of the components rather than between the rate of reaction and a quantity (affinity) based on the chemical potentials of the components. The empirical kinetic relations are therefore of the form

$$rate = kf(c_i) \tag{3.5}$$

where the rate of reaction is specified by $\dot{\xi}$ or, more commonly, by the rate of change in concentration of one of the reactants, k is a parameter called the empirical or kinetic rate coefficient or rate constant (an optional suffix can be added to distinguish it from the Boltzmann constant, if necessary), and $f(c_i)$ is a function of the concentrations c_i of the components involved. The form of the function $f(c_i)$ is determined empirically. Usually, it is found possible to write it as a product of the concentrations of the reactants only, each raised to a suitable power (0, 1, 2...). In this case, the kinetics of the reactions are said to be of first order if the corresponding power is unity, and so forth, the sum of the powers of the terms involved being called the overall order. The concept of order is particularly used in connection with chemical reactions (Atkins 1986, Chap. 28), but it can also be usefully applied to processes such as recovery and phase transformation.

If the reaction rate is found to depend on the concentration of only one reactant and to be first order in it, that is, $f(c_i) = c$, where c is the concentration of that component, and if the rate of reaction is specified by dc/dt, then we have the particularly simple form

$$\frac{1}{c}\frac{dc}{dt} = k \tag{3.6}$$

where the rate coefficient k is identical to the specific rate of the process and has the dimensions s^{-1}. In the more general form (3.5), the dimensions of k depend on the form of $f(c_i)$. The temperature and pressure dependence shown explicitly in (3.1) and (3.2) are normally incorporated in the rate coefficient k as

$$k = k_0 e^{-\frac{Q}{RT}} \tag{3.7}$$

where the pre-exponential factor k_0 has the same dimensions as k, being again a frequency in the simple case (3.6) or whenever $f(c_i)$ has the same dimensions as ξ.

3.2.3 Statistical Approach to Thermal Activation

From a statistical thermodynamics point of view, the nature of a thermally activated reaction is often depicted as in Fig. 3.2, showing a change from an assemblage of reactants 1 to an assemblage of products 2 via what is assumed in transition state theory to be a thermodynamically definable intermediate state corresponding to the energy peak. It is assumed that temperature and pressure are the independent variables, so that Gibbs energies are used to define the states. The forward change involves an activation barrier G_m and it is often useful to picture it, in a simple-minded way, as being governed by a relation

$$\dot{\xi}_+ = f_+(c_i)v_+ e^{-\frac{G_m}{RT}} \tag{3.8}$$

there being at the same time a corresponding tendency for the reverse change to occur at the rate

$$\dot{\xi}_- = f_-(c_i)v_- e^{-\frac{G_m + \Delta G_m}{RT}} \tag{3.9}$$

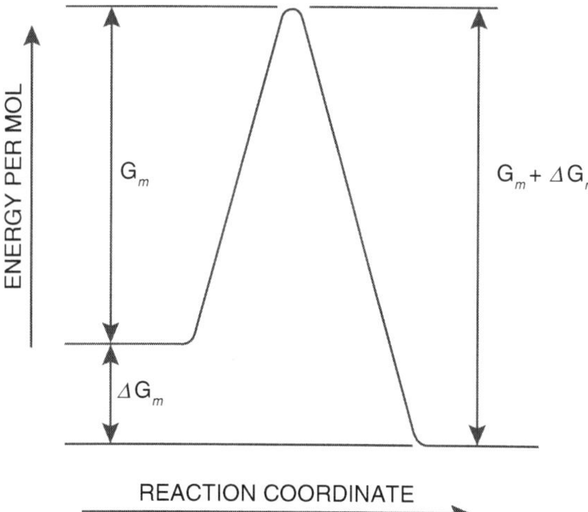

Fig. 3.2 Gibbs energy change during a reaction

where the $f(c_i)$ are now taken to represent an amount or concentration of potential reaction species subject to fluctuations of energy, v is the frequency of the fluctuations (the suffixes $+,-$ distinguish the forward and reverse cases, respectively), and $\exp(-G_m/RT)$ is the probability that a given fluctuation will reach G_m. Under the assumptions that the "attempt" rates $f_+(c_i)v_+$ and $f_-(c_i)v_-$ in the two directions are the same (they must be so at equilibrium, when $\Delta G_m = 0$,) the net rate of reaction $\dot{\xi} = \dot{\xi}_+ - \dot{\xi}_-$ will be given by

$$\dot{\xi} = \dot{\xi}_+\left(1 - e^{-\frac{\Delta G_m}{RT}}\right) \tag{3.10a}$$

(See Lasaga 1981 for a more satisfying derivation of 3.10 using transition state theory). In the case $\Delta G \ll RT$, we then have

$$\dot{\xi} \approx \dot{\xi}_+ \frac{\Delta G_m}{RT} = f(c_i)\frac{\Delta G_m}{RT}ve^{-\frac{G_m}{RT}} \tag{3.10b}$$

or substituting (3.8) and dropping the suffixes, the rate coefficient, now also the specific rate, is given by

$$k = \frac{\dot{\xi}}{f(c_i)} \approx \frac{\Delta G_m}{RT}ve^{-\frac{G_m}{RT}} \tag{3.11}$$

This case thus corresponds to the "near to equilibrium" linear case discussed earlier in connection with (3.4).

The above argument is put on a somewhat sounder basis in transition state theory, although still under the assumption that there is a thermodynamically definable intermediate state, known as the activated complex, which is in equilibrium with the reactants in a steady-state reaction. On statistical mechanical grounds (Atkins 1978, Chap. 27; Christian 1975, Chap. 3; Flynn 1972, Chap. 7;

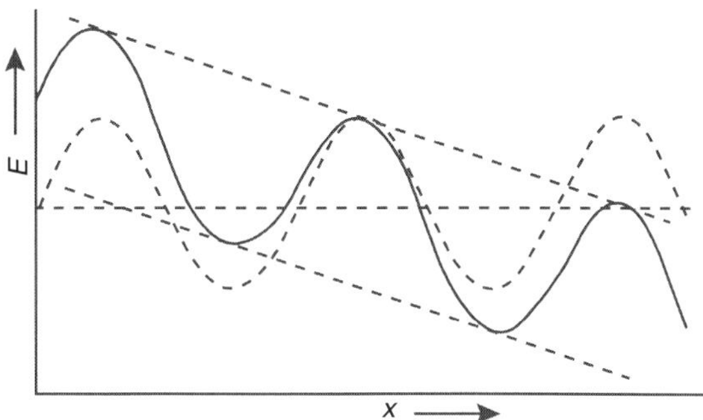

Fig. 3.3 Energy E as a function of distance x in the presence of a stress

Glasstone et al. 1941; Lasaga 1981), it is then calculated that the specific rate or rate coefficient for the forward reaction will be proportional to $k_B T/hK^{\neq}$, where k_B is the Boltzmann constant, h the Planck constant, and $K^{\neq}$ the equilibrium constant for the equilibrium between the reactants and the activated complex, which can be expressed in terms of the partition functions of the participating species including the activated complex; the proportionality constant contains the activity coefficients of the participating species. The factor $k_B T/h$ $(= 2.08 \times 10^{10} T \text{ s}^{-1})$ is a "fundamental frequency" that determines the dynamics of the process. Thus, apart from numerical factors generally of the order of unity, we can identify v in (3.8) and (3.12a) with $k_B T/h$ and $K^{\neq}$ with $\exp(-G_m/RT)$. However, transition state theory has many limitations and other approaches to a theory of reaction kinetics have also been attempted (Christian 1975, Chap. 3; Flynn 1972, Chap. 7).

3.2.4 Stress-Assisted Thermal Activation

We now consider the rate of thermally activated events that are assisted by the action of an applied stress on the entity involved. The existence of a mechanical effect implies that energy can be transferred to the entity by a force acting on it during a displacement in space. Therefore the appropriate reaction coordinate is a distance in space. Figure 3.3 depicts the energy E of the entity versus the distance coordinate x.

We are concerned with an activation event that results in the entity being displaced from the position 1 to position 3 over the barrier 2. The frequency of forward jumps will be

$$v^+ = v_0 \exp\left(-\frac{\Delta E^* - F\,\mathrm{d}x^*}{kT}\right)$$

and reverse jumps

$$v^- = v_0 \exp\left(-\frac{\Delta E^* + F(dx - dx^*)}{kT}\right)$$

where v_0 is the attempt frequency, F the applied force ($F = -dE/dx$ in the absence of the barrier), ΔE^* the height of the barrier, dx^*, dx the displacements from equilibrium position 1 to peak 2 and to the next equilibrium position 3, respectively, k the Boltzmann constant, and T the absolute temperature. The net forward rate of jumping is then

$$v = v^+ - v^- = v_0 \exp\left(-\frac{\Delta E^*}{kT}\right)\left\{\exp\left(\frac{F dx^*}{kT}\right) - \exp\left(-\frac{F(dx - dx^*)}{kT}\right)\right\}$$

$$(3.12a)$$

The expression (3.12a) leads in particular cases to three well-known forms, as follows:

(1) When $dx^* = \frac{1}{2}dx$, that is, the equilibrium positions 1 and 3 are symmetrically positioned about the barrier 2 (Fig. 3.3) we always have

$$v = 2v_0 \sinh\frac{F dx}{2kT}\exp\left(-\frac{\Delta E^*}{kT}\right) \tag{3.12b}$$

(2) When $\frac{1}{2}F dx^* \leq \frac{1}{2}F dx \ll kT$, that is, at relatively low stress and high temperature, we have the approximation

$$v \approx \frac{v_0 F dx}{kT}\exp\left(-\frac{\Delta E^*}{kT}\right) \tag{3.12c}$$

(3) When $\frac{1}{2}F dx \gg kT$, that is, at relatively high stress and low temperature, we have the approximation

$$v \approx v_0 \exp\left(-\frac{\Delta E^* - F dx^*}{kT}\right) \tag{3.12d}$$

The main application of this formalism is in treating the thermally activated motion of dislocations (Sect. 6.4).

3.3 Changes in Crystalline Organization

3.3.1 General

We now consider some particular rate processes that may be viewed as simple reactions, or transformations, involving only one reactant and one product. These are the organizational processes affecting the atomic structure of the crystalline grains and the granular structure of the polycrystalline body. Here it is useful to distinguish, in general, between homogeneous and heterogeneous processes. A homogeneous process is one that proceeds simultaneously in all parts of the body concerned. A heterogeneous process is one that is, at any instant, localized at certain discrete sites or along a discrete front, the location of the activity migrating through the body with time in order to affect all parts. However, the distinction may be to some extent a matter of scale; a process that is heterogeneous when viewed at one scale may appear statistically homogeneous when viewed on another scale.

In rock deformation studies, interest in organizational processes arises for several reasons. First, there is a tendency for the structural disruption caused by the deformation processes themselves to be restored, as in recovery and recrystallization processes. Second, there may be a tendency for the re-equilibration of a body to changed environmental conditions involved with the deformation, leading to phase transitions, precipitation, or other solid-state reactions that are coupled with the deformation of the body. Third, the primary growth or dissolution of crystals may be concerned, especially if a fluid phase is present in pores.

3.3.2 Recovery

The term recovery, in its original metallurgical usage, refers to the restoration or change, without recrystallization, of physical properties such as indentation, hardness, or electrical resistivity during the annealing of a body that has been previously plastically deformed. It is now used more generally to refer to any more or less homogeneous intracrystalline response to the presence of structural defects in excess of an equilibrium concentration that have been introduced by plastic deformation or other cause such as irradiation. Thus, the term is now used in a mechanistic sense for a reorganization within the crystal with respect to structural defects based on the original crystal structure, which may proceed either concurrently with or subsequent to the deformation or other disrupting process.

The structural defects involved in recovery may be point defects or extended defects. In the first case, recovery to an equilibrium state may be achieved in available time if the temperature is sufficiently high. However, in the second case, involving dislocations or stacking faults that would need to be completely eliminated to achieve thermodynamic equilibrium, recovery is usually only partial.

In the case of an excess of point defects, recovery involves their diffusion to the surface or to internal sinks, or there may be mutual annihilation of complementary defects. If only vacancy diffusion is required or the sinks are closely spaced, recovery with respect to point defects can proceed relatively rapidly. Christian (1975, p.137 et seq.) gives some discussion of the kinetic considerations in the case of vacancies in metals.

The most important recovery in deformed crystalline materials involves dislocations, the density of which has increased during deformation. The thermally activated migration of dislocations to a free surface or grain boundary sink is normally too slow to be effective in eliminating dislocations but mutual annihilation may be important in reducing the dislocation density. However, even if the latter process proceeds as far as possible, there will usually be some excess dislocations of one sign remaining. These dislocations, as well as possibly many that could potentially have been annihilated, tend to arrange themselves by glide and climb at elevated temperatures into relatively stable configurations, often represented microstructurally by "polygonization", which minimize the strain energy associated with the long-range stress fields around the dislocations. Recovery involving dislocation climb is very important in high-temperature dislocation creep and is discussed again in Sect. 6.5.3 where the kinetic aspects are dealt with.

3.3.3 Recrystallization

Recrystallization is a solid-state process whereby a new crystalline structure of the same phase replaces that of the original crystal, the degree of reorganization being such that the new crystals can no longer be regarded as structures based on the original orientation of crystallographic axes. If the recrystallization occurs during heat treatment subsequent to a relatively low temperature plastic deformation in which substantial strain hardening has occurred, it is termed static recrystallization (Poirier 1985, Sect. 2.4.7). Alternatively, if the recrystallization occurs concurrently with deformation at elevated temperatures, it is termed dynamic recrystallization. Dynamic recrystallization is of particular interest industrially in "hot working" processes. It is also the recrystallization process that is of most interest in geological studies.

Broadly, there are two ways in which the original crystals can be reorganized in recrystallization, one by grain boundary migration or bulging (migration recrystallization) and the other by local rotation of the crystal structure (rotation recrystallization):

- *Migration recrystallization* has been widely studied in static experiments. The term "nucleation and growth recrystallization" was formerly used for this process but is inappropriate since classical nucleation theory, based on thermal fluctuations, does not apply to recrystallization (Cahn 1983; Christian 1975). To obtain static recrystallization, certain minimum values of strain and temperature are required and the process is driven primarily by the "stored" energy associated with the structural defects resulting from the prior deformation, especially the dislocations. The recrystallization process is thought to involve the bulging out of high-angle grain boundaries to generate strain-free crystal (Bailey and Hirsch 1962). It is also an important process in dynamic recrystallization (Poirier 1985, Sect. 6.3.1).

- *Rotation recrystallization* results from the progressive relative rotation of subgrains until they can be viewed as distinct grains (Poirier 1985, Sect. 6.3.1). It is only observed to occur during deformation as a dynamic recrystallization, occurring concurrently with deformation at elevated temperatures. It seems to have been first reported in quartz (Hobbs 1968) but has now also been recognized in the hot working of metals (Sakai 1989). A model for the process has been proposed by Shimizu (1998, 1999).

The overall extent of recrystallization, measured by the volume fraction of the specimen recrystallized at constant temperature, can often be described by the Avrami relation (Christian 1975, p. 19)

$$\xi = 1 - \exp(-kt^n) \tag{3.13}$$

where t is elapsed time, k, n are parameters, and n commonly being ~ 3–4.

The new grain size upon completion of recrystallization can be greater or smaller than the original grain size. Subsequent to recrystallization, further growth of some grains may occur at the expense of others, resulting in a more of less uniform coarsening of the grain structure. The driving force for this coarsening, termed *grain growth*, is the reduction in grain boundary energy, generally a much smaller quantity than the stored energy from plastic deformation. Grain growth is usually much slower than recrystallization owing to this smaller driving force. Sometimes, however, continued heating after recrystallization of previously deformed materials gives rise to high rates of selective growth of certain grains, to which the terms "exaggerated grain growth" or "secondary recrystallization" are applied. Such an effect seems to occur in marble (Schmid et al. 1980), although in the case of metals it is said to be only observed in sheet material (Christian 1965, p. 737). The kinetics of grain growth can also be importantly influenced by the behavior of impurities segregated in the grain boundaries since they affect the grain boundary energy. Thus, there may be different rates of grain boundary migration depending on whether the impurities migrate with the grain boundary or are left behind (Guillopé and Poirier 1979; Lücke and Stüwe 1971).

3.3.4 Solid-State Transformations and Reactions

Polymorphic phase transitions are relatively simple in that no long-range material transport is necessary. In general, such transformations are heterogeneous and can be classified into two types, distinguished mainly by whether the atom movements during growth of the new phase are individually uncoordinated (as in diffusion) or are cooperative (as in mechanical twinning). In metallurgy, the terms "nucleation and growth" and "martensitic" are commonly used for the two types, respectively, while in mineralogy they are often referred to as "reconstructive" and "displacive". The second type is commonly characterized by significant shape change, leading to the term "shear transformation", and by the tendency for non-hydrostatic stress to play an important role. The label "diffusionless" is also used for the second type (Zhang and Kelly 2009).

The formal kinetic theory of solid-state reactions is usually based on either a nucleation and growth model, where a nucleation stage is identifiable, or on a simple growth model where a nucleation stage does not enter, as in spinodal decomposition. Nucleation is commonly heterogeneous, occurring at grain boundaries, dislocations, etc.; see Flynn (1972), Christian (1975), and Kirkpatrick (1981) for general theory on nucleation. Once an interface is established, the local rate of growth is generally controlled either by processes at the interface itself (interface control) or by the rate at which material is transported to the reaction site (diffusion control). In the case of interface control, after a local steady state is established, the growth tends to be linear with time. In spinodal decomposition cases where a laminar structure grows by edgewise propagation, a linear growth law also tends to apply. However, in common cases of diffusion control, as where the reacting material reaches the interface through a thickening layer of product material or the depletion of the reactants has to be taken into account, the growth tends to be proportional to the square root of time (parabolic growth). Instability in growth rate may also arise at an interface, with important results for the morphology of product phases (c.f. dendritic structures). In a global view, covering both nucleation and whatever type of growth control that pertains, it is often possible, as in the case of recrystallization, to describe the overall kinetics by an Avrami relation of the type (3.13) with a suitable choice of n, the value of which may vary from less than 1 to more than 4 (Christian 1975, p. 542).

The mechanisms of the reactions or diffusion processes are important in determining the actual kinetic coefficients. Crystal defects are usually involved, point defects, dislocations, or planar defects ("shear planes") having roles in particular cases. The defects may play an important part in the formation of an activated complex where such can be recognized or usefully postulated.

In solid-state reactions, two other factors also influence the kinetics to an important degree. The first is local volume or shape change. This gives rise to internal stresses, and the associated strain energy has to be taken into account as work to be provided during the transformation. The magnitude of the internal stresses will be influenced by whether the reaction products are crystallographically coherent with the matrix or not. The stresses may also directly interact with the reaction mechanism if shearing is involved. The second factor is the impingement of reaction zones associated with separate nuclei or sites of reaction. The impingement may involve either a meeting of the actual reaction interfaces or an overlap of the zones of depletion from which diffusion is recurring.

For a selection of papers concerning kinetics of solid-state processes in geological systems, see Hofmann et al. (1974) and Lasaga and Kirkpatrick (1981).

3.3.5 Dissolution and Crystallization

Many processes in rocks proceed more rapidly when a fluid phase is present because of the faster transport of material either by diffusion or by bulk transport in the fluid, and some are only defined in the presence of a fluid. An example of fluid involvement is recrystallizing of dissolved material as new crystals or as additional growth on existing crystals.

The kinetics of dissolution are simplified by the absence of a nucleation barrier but, as with growth, the kinetics can still be either interface controlled or transport controlled (Berner 1981). The kinetics of the interface processes may depend sensitively on the presence of adsorbed impurities, occupying sites that might otherwise be occupied by reactants, or by crystal imperfections that intersect the interface to provide sites where bonding energies will be different; the latter effect is revealed in etch pits located where dislocations or planar defects intersect the interface.

Crystallization is, in many aspects, the converse of dissolution although its overall kinetics can be quite different because of the requirement of nucleation. Again, interfacial imperfections can play an important part, as seen in the role of screw dislocations in providing a persistent step at which growth occurs, revealed in spiral growth patterns, and which obviates some of the nucleation difficulties (Burton et al. 1951). A review for igneous systems has been given by Cashman (1990)

The driving force for dissolution or crystallization derives from the difference between the saturation concentration and the actual concentration of the crystal species in the fluid, although the resulting rates sometimes vary as a high power of this difference rather than linearly (Berner 1981). However, the saturation concentration, or solubility, depends on the curvature of the solid–fluid interface. It is determined thermodynamically through an equilibrium constant or solubility product, K, based on the activities of the relevant components in fluid and solid where $K = \exp(-\Delta G / RT)$ depends on the change in Gibbs energy ΔG per mole crystallizing or dissolving under standard conditions of temperature and pressure. There is also a contribution to G from the interfacial energy if the curvature of the interface is changed. Thus, for a given distribution of solid particle sizes, the fluid may be supersaturated with respect to some of the particles and undersaturated with respect to others; the result is a trend for the smaller particles to dissolve and the larger ones to grow, leading to a coarsening of particle or grain size, an example of a so-called Ostwald ripening process (for example, Chai 1974).

The solubility can also be affected by the state of stress or the defect content in the solid. It is especially affected by a stress component normal to the interface, giving rise to the "pressure solution" effect (Lehner 1990; Paterson 1973; and Thompson 1962). A number of natural structural features have been attributed to pressure solution, although in some cases the influence of heterogeneity on the transport of material may have been more directly responsible, as in the case of stylolites (Renard et al. 2004). Solution transfer processes in which pressure solution may have played a part are potentially of importance in the deformation of rocks containing pore fluids (Rutter 1983).

3.4 Transport Processes in General

Thermodynamically, the relationship between the variables describing any transport process can, in general, be expressed in the form

$$j = \lambda X \tag{3.14}$$

where j is the flux density; X the thermodynamic driving force, and λ a constant. The flux density j is the amount of the extensive property or quantity, Z which passes through a reference unit cross-sectional area in unit time, that is, $j = (\mathrm{d}/\mathrm{d}A)(\mathrm{d}Z/\mathrm{d}t)$ where A is cross-sectional area, and t time. Some writers call j simply the flux but since this term is used by others for the total flow, possible ambiguity is avoided by calling j the flux density. The driving force X can be expressed as the negative gradient of an intensive property or potential ϕ, that is, $X = -\mathrm{d}\phi/\mathrm{d}x$ in a one-dimensional situation where x is the space coordinate (ϕ can, in turn, generally be expressed as the derivative of an extensive quantity with respect to another extensive quantity see Chap. 2). As discussed in Chap. 2, the driving force X conjugate to the flux j is so defined that jX correctly expresses the rate of dissipation per unit cross-sectional area. In effect, this requirement determines the definition of X once j is defined as needed to describe the transport process.

In the empirical approach, the same type of relationship (3.14) is used. However, the quantity X to which j is related is now chosen, not for reasons of thermodynamic necessity, but for empirical reasons of convenience, as some quantity relatively easy to measure, which can be usefully correlated with j. For example, whereas thermodynamically the driving force X for diffusion is the negative gradient in the chemical potential, the concentration is more conveniently measured and its negative gradient is used as X in empirical treatments of diffusion. Apart from this point, however, the formulation of theory for the two approaches is similar.

The parameter λ represents a property of the medium that may be termed a "generalized conductivity". For simplicity, (3.14) is given in 1D form with λ as a scalar constant but, in general, j and X are vectors and λ is a second rank tensor in which is expressed the anisotropy of the medium. Constancy of λ corresponds to linear behavior thermodynamically, a property that is common in transport processes, and λ is then identical with the kinetic coefficient L in the thermodynamic formulation (Chap. 2). The linearity can be rationalized as arising from the circumstance that over atomic distances, the scale on which the elementary transport events occur, the change in driving potential, expressed as a molar energy, is small compared to RT. The value of λ may, however, be found to depend on the actual value of the potential ϕ if measurements are made over a wide range of ϕ.

Where the transport of discrete entities is envisaged, the flux density can be expressed as $j = c_N v$ where v is the velocity of the entities and c_N their concentration (number per unit volume) at the reference cross-section. We can now define the mobility M of the entities as their velocity when unit force is acting on them, so that

$$v = MX \tag{3.15}$$

Then we have $\lambda = c_N M$ and (3.14) becomes

$$j = c_N M X \tag{3.16}$$

In the case of a diffusing substance, the thermodynamically defined mobility is the velocity of transport in unit gradient of chemical potential; this mobility can be related to the diffusion coefficient D in the empirical treatment of diffusion by the Einstein formula

$$M = \frac{D}{RT} \tag{3.17}$$

in the case of ideal solutions, as will be shown in Sect. 3.5, where non-ideality is also discussed. Eq. (3.17) thus relates a thermodynamically defined quantity M to an empirically defined quantity D. Formally, one could also take dc_N/dx as a "force" and obtain an empirically defined "mobility" that would be equal to D/c_N; however, in practice the concept of mobility is normally only used in relation to thermodynamically defined forces.

In the practical analysis of transport processes, a distinction has to be made between transient or evolving situations and steady states. The relationship (3.14), which in terms of the potential ϕ can be written as

$$j = -\lambda \frac{d\phi}{dx}, \tag{3.18}$$

serves to fully describe the steady state, in which the flux density and potential gradient at any given point are unchanged with time. However, steady-state measurements do not reveal all the properties relevant to transport in more general situations; in particular, they give no information about the "capacity" of the medium for the transported property, as defined by $dZ/d\phi$. Thus, in a transient situation the amount per unit volume, ρz, of the property Z at any given point is a function of time involving this generalized capacity (z is the amount of Z per unit mass and ρ the density). Consideration of the fluxes in and out of an elementary volume leads to the continuity or conservation equation

$$\frac{\partial(\rho z)}{\partial t} = -\frac{\partial j}{\partial x} \tag{3.19}$$

However, any change in ρz is related to a change in ϕ through $\Delta(\rho z) = \rho c \Delta \phi$, where c is a generalized specific capacity (capacity per unit mass) of the medium and ρc is the generalized volumetric capacity or amount of Z per unit volume that can be accommodated in the medium per unit increase in ϕ; thus we can write

$$\frac{\partial(\rho z)}{\partial t} = \rho c \frac{\partial \phi}{\partial t} \tag{3.20}$$

Combining (3.18), (3.19), and (3.20) leads to the transient-state equation

$$\frac{\partial \phi}{\partial t} = \frac{1}{\rho c} \frac{\partial}{\partial x}\left(\lambda \frac{\partial \phi}{\partial x}\right) \tag{3.21}$$

If λ can be taken as constant over the range of ϕ concerned, (3.20) becomes

$$\frac{\partial \phi}{\partial t} = a \frac{\partial^2 \phi}{\partial x^2} \quad \text{where } a = \frac{\lambda}{\rho c} \tag{3.22}$$

The parameter a can be called a "generalized diffusivity" in analogy to the thermal diffusivity in transient heat flow, for which (3.22) will be recognized as the governing equation when we put $\phi = T$. Since (3.21) is homogeneous in ϕ the generalized diffusivity always has the same dimensions (SI units $\text{m}^2\ \text{s}^{-1}$) regardless of the particular transport process concerned.

   Alternatively to the above thermodynamic treatment of transport processes, there is the empirical or experimental approach, as mentioned in Sect. 3.2.2. In this approach, the flux is related to a conveniently measurable quantity as a proxy for the gradient of the thermodynamic potential. Fick's law for the treatment of diffusion (Sect. 3.5.2) is an example of the empirical approach; the diffusive flux density j is related to the gradient in concentration c_N of the diffusing species, thus:

$$j = -D \frac{\mathrm{d}c_N}{\mathrm{d}x}$$

where D is the diffusion coefficient.

3.5 Atomic Diffusion

3.5.1 General

A transport process of particular importance at relatively high temperatures is the diffusion of atoms, or small groups of strongly bonded atoms, in a matrix of the same or different material. The topic will be treated briefly both at the phenomological level and in terms of atomic theory. For general references, see Shewman (1989,

1963), Adda and Philibert (1966), Manning (1968, 1974), Flynn (1972), Christian (1975), Crank (1975), Le Claire (1976), Anderson (1981), Kirkaldy and Young (1987), Philibert (1991), Allnatt and Lidiard (1993), Wilkinson (2000), Mehrer (2007), and Cussler (2009).

The region of space through which the diffusion occurs is normally 3-dimensional but in cases of strong anisotropy or of interfacial or pipe diffusion it can be essentially 2D or even 1D. In the following sections, the diffusion equations will be written for 1D diffusion but they are readily generalized to 3Ds, in which case the diffusion coefficient becomes a second rank tensor with symmetry properties appropriate to the material.

3.5.2 One Mobile Component

We deal first with the elementary case of the diffusion of one mobile component in a matrix. Since the change in energy in an elementary volume as a result of adding substances to it while other variables are held constant is determined by the chemical potential of the substance, it can be expected that under isothermal and isobaric conditions, the diffusive flux of the component will be determined by the gradient of its chemical potential and that the relationship will be a simple proportionality if the gradient is not too steep (Sect. 3.4 and Chap. 2). Experience supports this view and we therefore write, for isothermal and isobaric conditions,

$$j = L_d \frac{\mathrm{d}\mu}{\mathrm{d}x} = -cM \frac{\mathrm{d}\mu}{\mathrm{d}x} \tag{3.23}$$

where j is the flux density, that is, the amount of substance passing per unit time t through unit cross-sectional area in a defined frame of reference, μ the chemical potential, x the distance in the same frame of reference, and L_d a phenomenological, kinetic or transport coefficient, which can also be written as the product of the amount-of-substance concentration c and the mobility M (in this section, for convenience, we use c instead of c_N for the concentration). So long as volume changes can be neglected, the immobile matrix serves as a frame of reference. The rate of diffusion of the substance in the matrix can then be described by the parameter M.

In practice, however, information about the rate of diffusion is derived from measurements of concentration profiles and is expressed in terms of a diffusion coefficient D, defined as the ratio of the flux density of substance j to the negative of the gradient of the concentration (the SI units of D are $\mathrm{m}^2\,\mathrm{s}^{-1}$). This definition stems from Fick's first law (Fick 1855)

$$j = -D \frac{\mathrm{d}c}{\mathrm{d}x} \tag{3.24}$$

Note that D may depend on the concentration c.

For nonsteady-state conditions, combining (3.24) with an equation of continuity (Sect. 3.4) leads to Fick's second law,

$$\frac{\partial c}{\partial t} = \frac{\partial}{\partial x}\left(D\frac{\partial c}{\partial x}\right) \quad \text{or} \quad \frac{\partial c}{\partial t} = D\frac{\partial^2 c}{\partial x^2} \tag{3.25}$$

where the second form applies only for D independent of concentration. In the latter case, it is identical with the heat flow equation, for which many solutions are known (Carslaw and Jaeger 1959). It is to be noted, however, that, in contrast to the case of heat flow, the conductivity parameter in (3.24) is identical to the diffusivity parameter in (3.25), that is, in atomic diffusion, the diffusion coefficient and the diffusivity are identical.

Equations (3.25) are commonly used in experimental studies. Starting from a known distribution, the concentration profile is measured after a certain time and compared with the appropriate solution of (3.25) in order to evaluate D (Crank 1975). In the case of a semi-infinite solid with zero initial concentration, placed in contact with a reservoir that maintains a constant concentration c_0 of the diffusing species at the surface, the concentration profile at time t is given by

$$c(x, t) = c_0\left\{1 - erf\left(x/2\sqrt{Dt}\right)\right\}$$

where $erf\ z$ is the error function. In this case, the value of c falls to about $0.5\,c_0$ at $x \approx \sqrt{Dt}$ and to $0.1\,c_0$ at $x = 2\sqrt{Dt}$. It is, therefore, convenient in approximate calculations to use $\sqrt{Dt}$ or $2\sqrt{Dt}$ as a "diffusion distance".

The relationship between the mobility M and the diffusion coefficient D can be obtained if we relate the chemical potential to the concentration through the equation of state

$$\mu = \mu^\circ + RT \ln \gamma c \tag{3.26}$$

where μ° is the chemical potential in a reference state, γ the activity coefficient, R the gas constant, and T the absolute temperature. Substituting this relation into (3.23) and comparing with (3.24) leads to

$$D = RTM\left(1 + \frac{d \ln \gamma}{d \ln c}\right) \tag{3.27}$$

and hence to

$$j = -\frac{cD}{RT\left(1 + \frac{d \ln \gamma}{d \ln c}\right)}\frac{d\mu}{dx} \tag{3.28}$$

The so-called thermodynamic factor $(1 + d \ln \gamma/d \ln c)$ is unity for ideal mixing, as in the case of low concentrations or self-diffusion, in which case, (3.27) becomes the well-known Einstein relation $D = RTM$.

When the diffusing species is chemically distinct from the matrix and its diffusion leads to significant change in chemical composition, D is called an impurity or chemical diffusion coefficient. On the other hand, when an isotopic tracer is used to study self-diffusion or diffusion in a case where the concentration of the tracer is so low that it can be regarded as diffusing in a matrix of uniform chemical composition, D is called a tracer diffusion coefficient and distinguished as D^*. The distinction between the two coefficients is seen to lie in the thermodynamic factor which is unity in the tracer case, so that

$$D = D^* \left(1 + \frac{d \ln \gamma}{d \ln c} \right) \tag{3.29}$$

The relationship (3.29) only applies exactly for the case of one mobile component. However, this case is somewhat artificial since there are commonly other mobile species present, including vacancies, the movement of which also influences D to some degree. The temperature and pressure dependence can be expressed in Arrhenius form as for other rate processes (Sect. 3.1.2),

$$D = D_0 e^{-\frac{Q}{RT}} = D_0 e^{-\frac{Q_0 + pV^\bullet}{RT}} \tag{3.30}$$

where D_0 and Q, or D_0, Q_0, and $V^\bullet$ are empirical constants.

There are additional terms in the diffusion Eqs. (3.24) and (3.25) when μ in (3.23) is generalized to include potential gradients additional to that associated with the concentration, as through (3.26). Additional potential gradients may include those in elastic or electric fields. See Shewman(1963, p. 25) for solution of the diffusion equations in such cases.

For experimental methods of measuring diffusion coefficients, see the references in Sect. 3.5.1. Computer simulation methods have also been developed in recent times (Dohmen and Chakraborty 2007; Mantina et al. 2008, 2009; Miyamoto and Takeda 1983; Watson and Baxter 2007). Some values for diffusion coefficients in silicate minerals can be found.

3.5.3 Multicomponent Diffusion

When the diffusion of more than one component is considered, the phenomenological Eq. (2.9) of Chap. 2 apply as a generalization of (3.23). However, while these equations give a fundamental description of the diffusion, suitable for relating to other thermodynamically based considerations, the gradients in chemical potential cannot be directly measured. Practical diffusion studies are based on measurements of the gradients in chemical or isotopic composition, which can be described in terms of the generalization of Fick's first law proposed by Onsager (1945). For a system of n components under isothermal and isobaric conditions we write

$$j_1 = -D_{11}\frac{dc_1}{dx} - D_{12}\frac{dc_2}{dx} - \cdots\cdots - D_{1n}\frac{dc_n}{dx}$$

$$j_2 = -D_{21}\frac{dc_1}{dx} - D_{22}\frac{dc_2}{dx} - \cdots\cdots - D_{2n}\frac{dc_n}{dx}$$

$$\vdots$$

$$j_n = -D_{n1}\frac{dc_1}{dx} - D_{n2}\frac{dc_2}{dx} - \cdots\cdots - D_{nn}\frac{dc_n}{dx} \qquad (3.31)$$

where, for each component i ($i = 1, 2, \ldots n$), j_i is the flux density in terms of the chosen concentration units with respect to a suitable frame of reference and dc_i/dx is the gradient in concentration with respect to the same frame; the n^2 coefficients D_{ij} then specify the diffusion properties of the system and are sometimes called the practical diffusion coefficients. Quantities relating to the nth component can be omitted in (3.31) if the frame of reference is chosen, so that there is no net flux with respect to it and provided the system is closed, thus making the nth quantities dependent.

There is, in general, no a priori reason to assume that the matrix $[D]$ of the D_{ij} is symmetric in a given frame of reference or that any of its components will be zero, although the off-diagonal terms ($i \neq j$) are commonly smaller than the diagonal terms ($i = j$). However, it is in general possible to transform the frame of reference so as to diagonalize $[D]$, making all D_{ij} ($i \neq j$) equal to zero and so relating the diffusion coefficient for each component uniquely to its concentration gradient in this frame; this property follows from the Onsager reciprocal relations (Cooper 1974). It follows immediately that, in a binary system ($n = 2$,) only one diffusion coefficient is needed to describe the interdiffusion of the two components in a frame of reference for which there is no net flux. However, in a ternary system there are not only two independent coefficients (D_{11}, D_{22}) relating the diffusion of a given substance to its own concentration gradient, but there are also coupling coefficients (D_{12}, D_{21}) relating this diffusion to the concentration gradient of the other independent component. It should also be borne in mind that over a substantial range of compositions of a given chemical system the values of D_{ij} can, in general, be expected to be concentration dependent.

We cannot pursue here the manifold ramifications of multicomponent diffusion; see, for example, Brady (1975a, 1975b), Anderson and Graf (1976), Lasaga (1979), Anderson (1981). However, attention may be drawn to the following points of principle:

(1) A choice of the chemical components considered to constitute the system has to be made in the light of the physics of the situation and the applications in mind. From a thermodynamic point of view, the components will in general be molecular species for which chemical potentials can be defined. In practice, concentration gradients for ionic or isotopic species may be measured and analyzed, but constraints such as electrical neutrality and stoichiometric balance tend eventually to reduce the essence of the situation to the diffusion of

molecular components (Brady 1975b; Lasaga 1979; Lasaga et al. 1977). Crystal defects such as vacancies must also be represented among the components when it is implicit that the crystal lattice is conserved in the diffusion.

(2) A choice must also be made of a reference frame for the fluxes and the concentration gradients, bearing in mind that, with different components diffusing at different rates, volumes may be changing or centers of mass or volume moving relative to external axes and that the values of the diffusion coefficients will depend on the choice of the reference frame (Anderson 1981; Brady 1975a; Crank 1975; De Groot and Mazur 1962). The following are some of the possibilities:

(a) Laboratory frame, such as a notional grid fixed to one end of the specimen. While commonly the initial choice for representing measured concentration profiles, it is often not the most appropriate for subsequent analysis where the movement of components relative to each other is of interest.

(b) Velocity-fixed frames which move with the mean displacement of volume, mass, substance, or a particular component of substance relative to the laboratory frame—called, respectively, a volume-fixed, mass-fixed, molar-fixed, or nth component-fixed frame. The volume-fixed frame is particularly appropriate when the volume of the system remains constant while the center of volume moves relative to the laboratory frame. Diffusion coefficients measured relative to a volume-fixed frame are sometimes called "standard diffusion coefficients" (Hooyman et al. 1953). The nth component-fixed frame is often also called the solvent-fixed frame. If the terms involving the nth component are omitted in (3.31), it is implicit that an appropriate velocity-fixed frame is used.

(c) Inert marker frame based on inert markers embedded in the system or, in the case of crystals, notionally attached to the lattice or unit cell (lattice-fixed frame). This frame is used when it is desired to eliminate from consideration any bulk flow of substance, that is, common displacement of all components. Diffusion coefficients referred to an inert marker frame are sometimes called "intrinsic diffusion coefficients" (Hartley and Crank 1949) since they relate to the intrinsic mobility of the individual component relative to a section through which no bulk flow occurs. For a binary system, the inter-diffusion coefficient D^V referred to a volume-fixed frame is related to the "intrinsic" diffusion coefficients D_1^M, D_2^M of the two components, 1, 2 by $D^V = x_1 D_1^M + x_2 D_2^M$, where x_1, x_2 are the respective mole fractions.

(3) Since the practical measurement of diffusion coefficients is usually done under nonsteady-state conditions, a generalization of Fick's second law is also needed, obtained by combining (3.31) with an equation of continuity. In cases involving chemical diffusion coefficients, where concentration dependence is likely to arise, methods of solution are more complex than for concentration-independent cases such as tracer diffusion, as is illustrated in the Matano treatment of inter-diffusion in a binary system; see Crank (1975) for analysis in binary systems and

Anderson (1981) for extension to ternary systems. The practical difficulties increase greatly with more components and attention has therefore been given to methods of estimating diffusion coefficients using ionic conductances or trace diffusion coefficients (Anderson 1981; Lasaga 1979).

3.5.4 Atomic Theory of Diffusion

We turn now from the macroscopic or phenomenological approach to consider the atom movements involved in diffusion, which are the basis of the atomic or kinetic theory of diffusion. In any material, diffusion arises essentially from the tendency for an atom to move randomly relative to its neighbors. In crystals, this movement involves displacements at a certain jump frequency between structural or interstitial sites. Many mechanisms have been proposed, taking into account any necessary rearrangements to accommodate the atom in its new position. Basically, the various mechanisms involve (1) more or less direct exchange of structural sites, (2) exchange with a defect (especially a vacancy), or (3) movement via interstitial sites (Howard and Lidiard 1964; Manning 1968, 1974).

If the probability that a given atom will jump to a neighboring site is independent of the direction of that site and of the previous jump history, then it can be shown (Flynn 1972, Chap. 6; Manning 1974) that the resultant random walk motion can be expressed through a diffusion coefficient that is the sum of terms of the form $\alpha d^2 \Gamma$, where d is the jump distance, Γ the jump frequency, and α a numerical factor of order unity ($\alpha = 1/6$ for a simple cubic crystal), the sum being taken over all combinations of the different types of sites between which jumps can occur, weighted according to probability of occupation. However, a given jump may be influenced by the previous jump; for example, in the case of a vacancy mechanism there is a bias in favor of a reverse exchange with a vacancy with which an exchange has just occurred. This effect can be allowed for by multiplying the previous terms by a correlation factor f. The factor f will in general be a tensor, depending on diffusion mechanism and on temperature, but for simple crystal structures with one type of jump it is a scalar constant, the calculation of which is given in most treatments of the kinetic theory of diffusion; for example, $f = 0.78$ for diffusion by a simple vacancy mechanism in f.c.c. crystals and $f = 1$ for interstitial mechanisms (Manning 1968). Thus, under the assumption that other factors (drift forces) affecting the relative probabilities of forward and backward jumps are absent, we obtain the theoretical result that the tracer diffusion coefficient D^* for a single mobile component having a single type of site should be given by

$$D^* = \alpha d^2 f \Gamma \tag{3.32}$$

In other cases, the diffusing atom may be subject to a drift force F affecting the relative probabilities of forward and backward jumps and giving rise to a drift velocity v_F proportional to F. The flux can then be written as

$$j = -D^* \frac{\mathrm{d}c}{\mathrm{d}x} + v_F c$$

Thus, defining the single component chemical diffusion coefficient D through $j = -D(\mathrm{d}c/\mathrm{d}x)$, we have

$$D = D^* - v_F c \left(\frac{\mathrm{d}c}{\mathrm{d}x} \right)^{-1} \tag{3.33}$$

This quantity can be identified with the intrinsic diffusion coefficient defined earlier since the position of a site can be taken as fixed relative to an inert marker or lattice frame.

A drift force arises from a gradient in the nonideal part of the chemical potential (note that in the case of ideal mixing there is no drift force, even where there is a concentration gradient, and the tracer and chemical diffusion coefficients are then identical, the diffusion being driven entirely by the gradient in entropy of mixing). A drift force may also arise from an electric field when the diffusing species is charged, from a gravitational potential, or from the presence of a gradient in temperature (thermal diffusion), pressure, stress, or concentration of another component. Taking $\mathrm{d}c/\mathrm{d}x$ to be negative, (3.33) indicates that with a sufficiently large drift force in the backward direction (negative v_F), D can become negative, giving "uphill" diffusion.

Kinetic theory is thus concerned primarily with calculating the jump frequencies Γ and the responses v_F to particular drift forces F. Although quantum-mechanical tunnelling may exist for light atoms at very low temperatures (Flynn 1972, Chap. 7), the jump frequency for a particular type of jump can generally be obtained from transition state theory as

$$\Gamma = v e^{-\frac{E}{kT}} \tag{3.34}$$

where v is the "attempt" frequency of the order of atomic vibrational frequencies ($\sim 10^{-14}$ s^{-1}) and $\exp(-E/kT)$ is a Boltzmann ("success") factor, E being the energy barrier to be surmounted, k the Boltzmann constant, and T the absolute temperature (Lasaga 1981). When the mechanism requires the formation of a particular type of defect, E is the sum of the energies of formation and of migration of this defect. To give a jump frequency of 1 s^{-1}, a temperature of about 100 °C is required if E is 100 kJ mol^{-1} (~ 1 eV atom^{-1}), and temperatures of about 500, 1,200, and 2,000 °C, respectively, are required for values of E of 200, 400, and 600 kJ mol^{-1}.

The effect of the drift force F is to lower the barrier E for forward jumps and to raise it for backward jumps (Fig. 3.3). Since the change in the barrier is generally relatively small ($Fd \ll kT$) the effect can be expressed approximately through

$$D = D^* \left\{ 1 - \frac{Fc}{kT} \left(\frac{\mathrm{d}c}{\mathrm{d}x} \right)^{-1} \right\} \tag{3.35}$$

(Manning 1968, 1974). In the case of one component chemical diffusion, F is equal to $-kT(\mathrm{d}\ln\gamma/\mathrm{d}c)$, thus leading back to (3.28). In a multicomponent system, there will be additional terms relating to the concentration gradients of the other components; in general, these terms will appear as cross-terms $D_{ij}(i \neq j)$ in (3.30) but where only one additional species is involved the coupling effect may conveniently be expressed as an additional factor in the primary diffusion coefficient, as is done for the "vacancy wind" effect in simple cases of diffusion by a vacancy mechanism (Flynn 1972, Chap. 8; Manning 1968, 1974).

From (3.35), it is seen that the main part of the temperature dependence of D, as expressed in an experimental activation energy Q, (3.29), lies in D^*. Through (3.32) and (3.34), Q will therefore contain mainly contributions from E but it may also include a contribution from the temperature dependence of f in more complex crystals. Often it is possible to distinguish a higher temperature, "intrinsic" regime, where diffusion involves thermally generated defects and Q reflects contributions from both defect formation and migration terms in E, from a lower temperature, "extrinsic" regime where defects already present are involved and Q is lower due to there being no defect formation term in E (this concept of an intrinsic regime should not be confused with the intrinsic diffusion coefficient mentioned earlier).

The pre-exponential term D_0 in (3.30) is correspondingly seen to be of the order of $d^2 v \exp(\Delta S/k)$ if α and f are taken to be of the order of unity and $T\Delta S$ is the entropic part of E. If ΔS is small and $d \sim 0.1$ nm, D_0 will be roughly of the order of 10^{-6} m^2 s^{-1}; however, it is difficult in general to estimate ΔS and reported values of D_0 range many orders of magnitude either way from 10^{-6} m^2 s^{-1}.

When the diffusing species is ionized and an electric field E is present, the drift force F is equal to zeF where z is the effective charge number and e the elementary charge so that, in (3.33), $v_F = uzeF$ where u is the electric mobility of the charged species. Through the Einstein relation $D_\sigma = ukT/ez$ (Atkins 1986, p. 675), a theoretical diffusion coefficient D_σ can be obtained from electrical conductivity measurements that give u. The ratio D/D_σ is known as the Haven ratio (Le Claire 1976), the determination of which is helpful in identifying the diffusion mechanism (for example, the Haven ratio is equal to the correlation factor f in the case of a simple vacancy mechanism). Another aspect of ionic diffusion is that the displacement of entities of one sign only sets up an internal electric field, known as a diffusion or Nernst potential (Manning 1968, Chap. 7), which acts on the oppositely charged species. This potential tends to rise to a level sufficient to maintain equal fluxes in opposite directions for identically charged species or equal fluxes in the same direction for oppositely charged species, thus preserving stoichiometry and neutrality. For this reason, it is often more appropriate in a macroscopic treatment to view the neutral components as the diffusing species. However, the problem then arises of relating the diffusion coefficient of the molecular species to the diffusion coefficients of the constituent ionic or atomic species, which may be more easily measured, for example, by isotopic tracer methods.

We approach this problem by considering the diffusion of a simple ionic substance of formula $A_\alpha B_\beta$ in a matrix. We assume that the substance is fully ionized into charged species $A^{\beta+}$ and $B^{\alpha-}$. If one species tends to diffuse faster than the other, a Nernst field will build up, as just mentioned, which will slow down the rate of diffusion of this species. The effect of the Nernst field can be taken into account by using the full electrochemical potential $\tilde{\mu}$ in (3.23), where $\tilde{\mu} = \mu + zF\phi$, μ being the chemical potential defined without taking into account the interaction between the charge and an electric potential, z the charge number of the species, F the Faraday constant, and ϕ the electric potential. If, following Howard and Lidiard (1964), we ignore coupling terms, we can use (3.23) to write the fluxes of A and B as

$$j_A = -L_A \frac{d\tilde{\mu}_A}{dx} = -L_A\left(\frac{d\mu_A}{dx} - \beta FE\right) \tag{3.36}$$

$$j_B = -L_B \frac{d\tilde{\mu}_B}{dx} = -L_B\left(\frac{d\mu_B}{dx} + \alpha FE\right) \tag{3.37}$$

where $E = -d\phi/dx$ is the Nernst electric field and L_A, L_B are the phenomenological coefficients. In order to maintain stoichiometry and electrical neutrality, we also have

$$\beta j_A = \alpha j_B \tag{3.38}$$

The three Eqs. (3.36–3.38) enable us to eliminate E in the expressions for j_A and J_B and to obtain the total molecular flux j as

$$j = \frac{j_A}{\alpha} = \frac{j_B}{\beta} = \frac{-L_A L_B}{\beta^2 L_A + \alpha^2 L_B}\frac{d\mu}{dx} = -L\frac{d\mu}{dx} \tag{3.39}$$

where $\mu = \alpha\mu_A + \beta\mu_B$ is the chemical potential of the molecular species and L is the phenomenological coefficient relating it to the total molecular flux. Using (3.23) and (3.27) and restricting consideration to the ideal case so that $\gamma = 1$, we can put $L = cD/RT$, $L_A = c_A D_A/RT$ and $L_B = c_B D_B/RT$, where c, c_A, c_B and D, D_A, D_B are the amount-of-substance concentrations and the diffusion coefficients of the molecular species and the ions A, B, respectively; also we can put $c = c_A/\alpha = c_B/\beta$. Using these relations in (3.39), then leads to the expression

$$D = \frac{D_A D_B}{\beta D_A + \alpha D_B} \tag{3.40}$$

for the diffusion coefficient for the molecular species in terms of the diffusion coefficients of the constituent ions. This expression applies to the ideal case, which includes the tracer diffusion and self-diffusion cases, D_A and D_B being measured independently as tracer diffusion coefficients by, say, radioactive tracer methods. More generally, the values of D_i for the species i in (3.40) have to be divided by the appropriate factors $(1 + d\ln\gamma_i/d\ln c_i)$ from (3.27). From (3.40), it follows that when one ion tends to diffuse much more slowly than the other, the diffusion coefficient for the molecular species is essentially equal to that for the slower ion.

An attempt has been made to calculate the diffusion coefficient from assumed atomic interaction potentials and Eyring rate process theory (Miyamoto and Takeda 1983). In spite of assuming fixed atom positions and thus neglecting the important relaxations that can be expected as the diffusing atom passes, rough agreement with observed values in olivine was obtained, encouraging the suggestion that diffusion coefficients too low to be measured, as for example, in pyroxenes, could be estimated in this way.

3.5.5 Polycrystal Diffusion and High Diffusivity

So far we have considered diffusion in a 3D homogeneous body. When the body is a polyphase composite, the average or bulk diffusivity will depend on the diffusivities in the individual phases, on their volume fractions and on their shape and connectivity (Crank 1975, Chap. 12). Also in polycrystalline bodies there are potentially important high-diffusivity or short-circuit paths which are in effect heterogeneities of relatively small volume fraction but high diffusivity. These paths may be interfaces (phase or grain boundaries or free surface) or linear regions (pipes, including dislocations and liquid-filled triple-grain junctions, which can be viewed as distinct narrow regions with a certain effective thickness or diameter δ and a characteristic diffusion coefficient D_{sc} much higher than the diffusion coefficient D_V in the volume of the grains.

When the diffusion distance is much greater than the spacing of interfaces or pipes, the apparent or bulk diffusion coefficient D will be

$$D = D_V(1 - x) + D_{SC}$$
$$\approx D_V\left(1 + \frac{D_{SC}}{D_V}x\right) \quad \text{for } x \ll 1 \tag{3.41}$$

(Le Claire 1976) where x is the mole ratio of the amount of the diffusing species in the high-diffusivity regions to that in the remaining volume (this ratio may be very different from the volume fraction of the high-diffusivity region in the case of a chemically different species on account of segregation). If d is the spacing of the interfaces or pipes, then x is of the order of δ/d for interfaces and $(\delta/d)^2$ for pipes. Thus, since δ is likely to be ~ 1 nm or somewhat less for grain boundaries or dislocations, for example, δ for NiO was found to be 0.7 nm (Atkinson and Taylor 1981), we have $x \sim 10^{-6}$ for a grain size of 1 mm or a dislocation spacing of 1 μm; similarly, $x \sim 10^{-6}$ for triple-grain junctions if $\delta \sim 1$ μm and $d \sim 1$ mm. In these cases, we need $D_{SC}/D_V \sim 10^6$ in order that the short-circuit diffusion and volume diffusion give equal contributions to the bulk diffusion; such a ratio D_{SC}/D_V is typically observed for both dislocations and grain boundaries (for example Atkinson and Taylor 1981).

When the diffusion distance is not large compared with the spacing of interfaces or pipes, the analysis is more complicated because of having to take into account the lateral diffusion from the high-diffusivity regions into the grains. However, such analysis is important in the practical determination of the short-circuit diffusion coefficients, although generally only the value of the product $D_{SC}\delta$ can be directly obtained. For details and further references, see Adda and Philibert (1966, Chaps. 12, 13), Le Claire (1976), Martin and Perraillon (1980), Peterson (1980, 1983).

Owing to the greater degree of disorder in interfaces or pipes, the jump frequency can be expected to be much higher than within the grains due to a lower value of E. This observation is consistent with the observation of lower values of Q, typically about one-half to two-thirds that for volume diffusion in case of grain boundary diffusion in metals and a similar ratio may be expected on nonmetals, for example, 0.7 in case of NiO (Atkinson and Taylor 1981). Consequently, in a given polycrystal, while the volume diffusion through the grains tends to dominate the bulk diffusion at high temperatures, short-circuit diffusion through grain boundaries and dislocations becomes relatively more important as the temperature is decreased and may even become predominant at relatively low temperatures (or when there is a connected network of liquid-filled triple-grain junctions). There is also some suggestion that diffusion is faster in moving than in static grain boundaries; see Peterson (1983) for references and comment. For the influence of the segregation of solute atoms, see Gupta (1977), Bernardini et al. (1982), Cabané-Brouty and Bernadini (1982) and Guiraldenq (1982). The space charge associated with the impurities may also play an important role (Yan et al. 1977).

3.6 Fluid Permeation

The transport of a fluid through a porous solid in response to a pressure gradient in the fluid has some formal similarity to diffusion but it involves the relative movement of two phases rather than the relative movement of components within a single phase. The formal analogy lies in the basic law of Darcy (1856), commonly expressed as

$$q = Ki \tag{3.42}$$

where q is the flow rate, that is, the volume of fluid passing through unit cross-sectional area of the porous body in unit time, i is the hydraulic gradient ($\rho g i = -\mathrm{d}p/\mathrm{d}x$ where ρ is the density of the fluid, g the acceleration of gravity, and p the pressure in the fluid as a function of the coordinate x in the direction normal to the defined area), and K is a constant called, in engineering usage, the hydraulic conductivity or coefficient of permeability and having dimensions length/time (Bear 1972; Hubbert 1956). For our purposes, it is more convenient to express (3.42) in a form free of gravitational connotation that corresponds more obviously to Ficks' first law, namely

$$q = -K' \frac{\mathrm{d}p}{\mathrm{d}x} \qquad (3.43)$$

where $K' = K/\rho g$ is another conductivity constant expressing the rate of permeation per unit pressure gradient. The relations (3.42) and (3.43) only apply for relatively low fluid velocities (Hubbert 1956).

The different roles of the solid and fluid phases are commonly distinguished, under the assumption that they are independent of each other, by writing the conductivity constant K' as the product of the fluidity $1/\eta$ of the fluid (η is the dynamic viscosity) and a constant k characterizing the solid, called the permeability or the intrinsic permeability. Thus we have

$$k = K'\eta = \frac{K\eta}{\rho g} \qquad (3.44)$$

The permeability k has dimensions (length)^2 and hence the SI unit m^2. Another widely used unit is the darcy: 1 darcy $\approx 10^{-12}$ m^2. The permeability k is thus a geometric property of a rock and it can be measured in the laboratory or in the field; see, for example, Brace (1980, 1984). The ranges of typical values for laboratory specimens of various types of rock are given in Fig. 3.4. It should be noted, however, that at pressures below a few megapascals in a gas, when the mean free path becomes comparable to the dimensions of the connected pores or cracks, the distribution of flow within the pores is no longer the same as at high pressures and the values of K and K' will vary with pressure if Darcy's equation is used (von Engelhardt 1960, p. 127). Also the permeability is reduced, becoming a function of fluid content, when the pores are not fully saturated with the fluid, as in the case of soils partially saturated with water (Bear 1972, Chap. 9).

Various models have been proposed as bases for calculating the permeability of a porous solid (Bear 1972, Chap. 5; Scheidegger 1960, Chap. 6). In one widely accepted and intuitively evocative model, the fluid movement is viewed as a flow through an equivalent channel or parallel set of channels of mean length l_e per length l of porous solid traversed (Fig. 3.5). In analogy with the Poiseuille formula for flow in a pipe of diameter $4R$,

$$q = -\frac{1}{2} \frac{R^2}{\eta} \frac{\mathrm{d}p}{\mathrm{d}x},$$

one can then write for the flow in a porous solid

$$q = -C \frac{\phi}{(l_e/l)^2} \frac{R^2}{\eta} \frac{\mathrm{d}p}{\mathrm{d}x} \qquad (3.45)$$

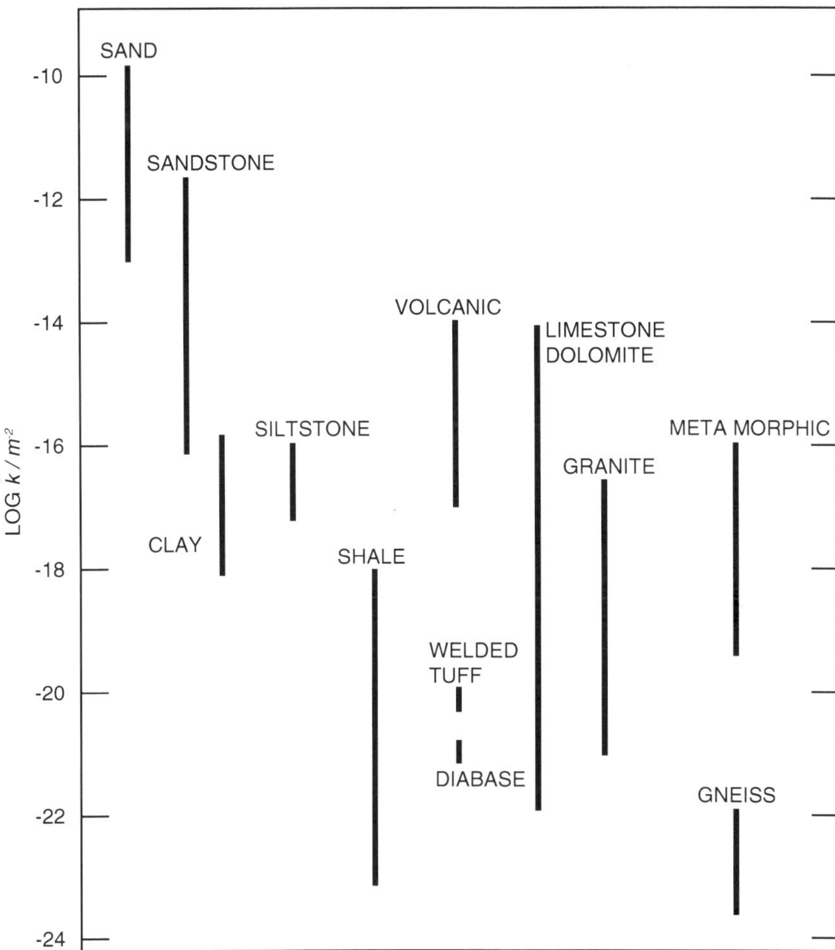

Fig. 3.4   Typical values of permeability k in laboratory specimens of various rocks. The data in this figure are derived from Brace (1980)

where R is now generalized to be the "hydraulic radius" of the channel (i.e., the ratio of the pore volume to the solid–fluid interfacial area, which serves to a first approximation as a measure of the equivalent channel cross-sectional dimension; it is one quarter of the diameter for a uniform circular cross-section, one-half of the narrow dimension for a slot-shaped channel, and $\phi d/6(1 - \phi)$ for an assemblage of spheres of diameter d having porosity ϕ); C is a dimensionless "shape factor" which is somewhat less than the value 1/2 that applies for a circular cross-section in the Poiseuille formula and which in effect allows for the error involved in the use of the hydraulic radius as the equivalent channel cross-sectional dimension in the Poiseuille analogy; the porosity ϕ enters to allow for the fact that the equivalent channel does not occupy the whole of the cross-sectional area of porous solid to which q is related; and (l_e/l) enters twice in order to take into account, first, that the local velocity in the channel is increased by the ratio (l_e/l) relative to that in a channel parallel to the direction of bulk flow and, second, that the pressure gradient along the equivalent channel is less than in the direction of bulk flow (Sullivan and Hertel 1942). The derivation of Chapman (1981, Chap. 3), which leads to $(l_e/l)^3$ in the above equation appears to be in error because of using $\phi(l_e/l)$ instead ϕ of for the cross-sectional area factor since this factor relates to flow in the direction for which q is defined, not the local flow direction. From (3.45), one therefore obtains the following expression for the permeability,

$$k = \frac{C\phi R^2}{(l_e/l)^2} \tag{3.46}$$

The quantity $(l_e/l)^2$ is often called the tortuosity and designated by T, although this name and symbol are variously used also for (l_e/l) and $(l/l_e)^2$ in the literature. The relative length (l_e/l) will depend on the porosity and can be expected to increase as the porosity decreases and the paths followed by the fluid become more tortuous, but it is difficult to obtain a direct measure of it. It is therefore common to approach its evaluation indirectly by invoking an analogy with electrical conductivity and introducing the "formation resistivity factor" F, which is the ratio of the electrical resistivity of the saturated porous body to that of the pure fluid, it being assumed that the fluid is electrically conducting and the solid parts of the body are not, and that the electric current and the fluid flow will follow identical paths (Archie 1942). From Ohm's law, the electrical equivalent of Poiseuille's law, we have

$$j = -\kappa\frac{\mathrm{d}V}{\mathrm{d}x} = -\frac{\kappa}{\kappa_f}\kappa_f\frac{\mathrm{d}V}{\mathrm{d}x} = -\frac{1}{F}\kappa_f\frac{\mathrm{d}V}{\mathrm{d}x} \tag{3.47}$$

where j is the macroscopic current density, $\mathrm{d}V/\mathrm{d}x$ the macroscopic voltage gradient, κ the conductivity of the saturated porous body, and κ_f the conductivity of the fluid. Then, by applying to the analogous model of an equivalent electrical conduit similar arguments to those used to generalize Poiseuille's law to (3.45), we obtain $1/F = \phi/(l_e/l)^2$ which contains a factor $\phi/(l_e/l)$ relating to the cross-sectional area of the conduit and a factor $1/(l_e/l)$ relating to its length. This relationship can then be used in (3.46) to give

$$k = \frac{CR^2}{F} \tag{3.48}$$

The same result can be obtained by noting, in comparing (3.45) and (3.47) for the equivalent channel model, that the conductivity κ_f in the electrical case is the analog of CR^2/η in the fluid flow case. If we now insert in (3.48) the Archie empirical relation $F = \phi^{-m}$, we obtain

$$k = C\phi^m R^2 \tag{3.49}$$

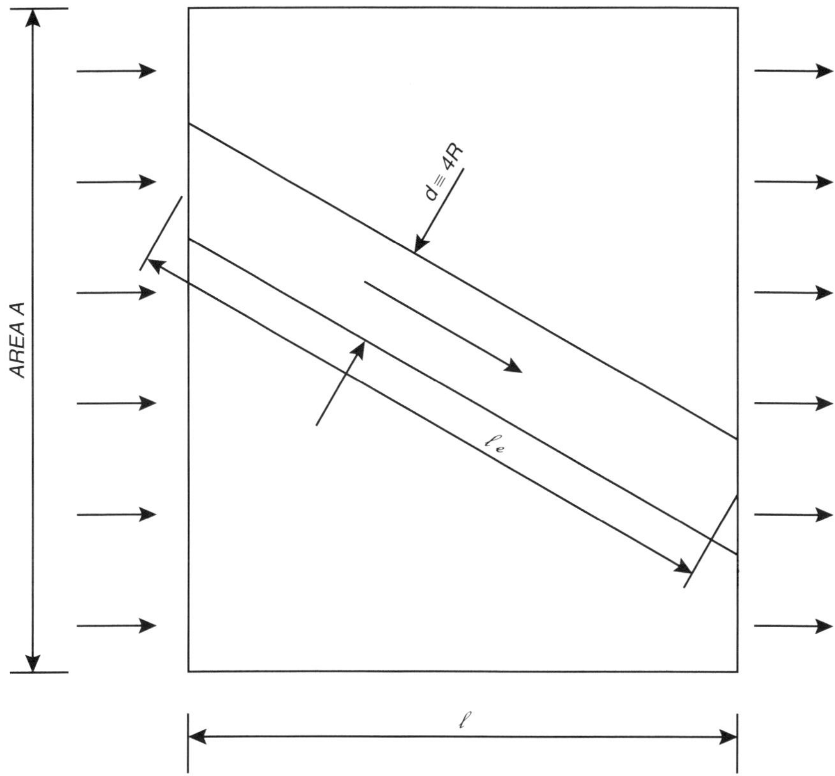

Fig. 3.5 Equivalent channel model for permeability

In the Archie relation, $m = 1.3$ for sand-like granular media and $m = 2$ for a wide range of rocks with porosities 0.001–0.1 (Brace et al. 1965; Gonten and Whiting 1967).

The equivalent channel model on which (3.45) to (3.49) are based is obviously oversimplified and does not take fully into account the variations in cross-section of the pore space along the fluid paths or branching or any lack of connectivity in these paths, and the electrical analog may not be an exact one. However, the relations (3.48) and (3.49) can be useful for practical purposes. Thus, they are found to give a reasonably good fit to observations on fairly porous rocks (see Paterson 1983 for a reappraisal of earlier observations). They probably also apply at least approximately to microcracked rocks such as granite; thus, analysis of measurements by Brace et al. (1968) and Brace (1977) in these terms indicates that increase in confining pressure produces a decrease in hydraulic radius, presumably reflecting a decrease in the mean crack opening width (Paterson 1983). Also, the empirical observation that $k \propto F^{-1.5}$ (Brace et al. 1968) indicates that $R \propto 1/\sqrt{F}$ for granite under varying confining pressure. In the case of granular media, it has been found empirically that the permeability is well described by $k = 0.0006d^2$

where d is the grain size (Bear 1972, p. 133; Hubbert 1956; Rumer 1969). Comparing this expression with (3.49) indicates that the hydraulic radius in granular media is approximately proportional to the grain size since the factor $C\phi^m$ does not vary markedly; putting $C = 0.4$, $\phi = 0.4$ and $m = 1.3$ shows that the expression corresponds to taking the hydraulic radius to be approximately $d/14$.

It is possible that in general the permeability can be expressed in the form

$$k = Cf(\phi)R^2 \tag{3.50}$$

where $f(\phi)$ is a function of ϕ that becomes more sensitive to variation in ϕ as the value of ϕ decreases, especially when connectivity begins to decrease significantly (Bernabé et al. 1982). The fall in permeability toward zero as connectivity is lost is related to similar connectivity effects in electrical conductivity in rocks (see, for example, Waff 1974 for the case of partial melting). This limit is analogous to the percolation limit in resistor networks (Chelidze 1982; Kirkpatrick 1973) or to the critical point in phase transitions. It has been treated by the theory of renormalization groups (Allègre et al. 1982; Madden 1983).

The approach to fluid permeation through Darcy's law with a constant permeability does not apply when local flow velocities become high enough for inertial or turbulent effects to become significant (Scheidegger 1960, Chap. 7). Limitations due to molecular effects in gasses at low pressures have already been mentioned but complications can also arise from ionic effects in the flow of electrolyte solutions, especially if clays are present (Scheidegger 1960, Chap. 7). Finally, external stress applied to the porous medium influences its permeability but the effect depends on both confining pressure and pore pressure and can be related to an effective pressure (see Walsh 1981 for the case of fracture permeability).

For nonsteady-state flow in the Darcy-law regime, the governing equation, analogous to Fick's second law and the heat flow equation, is

$$\frac{\partial p}{\partial t} = \frac{\partial}{\partial x}\left(D\frac{\partial p}{\partial x}\right) \quad or \quad \frac{\partial p}{\partial t} = D\frac{\partial^2 p}{\partial x^2} \tag{3.51}$$

where the second equation applies when D is independent of p. In analogy with the thermal diffusivity, the fluid diffusivity D has the form

$$D = \frac{K'}{\beta'} = \frac{k}{\eta\beta'} \tag{3.52}$$

where β' is the fluid storage capacity per unit volume of the porous body, given by

$$\beta' = \phi\beta + \beta_e - (1 + \phi)\beta_s \tag{3.53}$$

β being the compressibility of the fluid, β_s the compressibility of the solid parts of the body, and β_e the macroscopic compressibility of the porous body with zero or constant pore pressure (Brace et al. 1968); for a summary of the elasticity of porous bodies, see Paterson and Wong {2005, Sect. 7.3.1, where $K = 1/\beta_e$ and $K_s = 1/\beta_s$}.

From the solution of (3.51) for a semi-infinite medium with constant D and a given fluid pressure applied at the surface, the penetration distance in time t is of the order of $\sqrt{Dt}$ or the time to penetrate a distance x is of the order of x^2/D. In the case of a gas such as argon or water vapor as pore fluid, $\beta \sim 10^{-5}$ Pa^{-1} near atmospheric pressure and $\beta \sim 10^{-9}$ Pa^{-1} at pressures of the order of 100 MPa, while $\eta \sim 30.10^{-6}$ Pa s (this value is not very sensitive to pressure in macroscopic flow, although where the pressure is low, η may be effectively higher because of the mean free path effect mentioned earlier). Using the approximation $\beta' = \phi\beta$, the time needed for a small pressure pulse to penetrate 10 mm into a rock having $k = 10^{-19}$ m^2 and $\phi=0.01$ would be of the order of one hour when the fluid pressure is near atmospheric pressure or one second when the fluid pressure is near 100 MPa, respectively; these intervals suggest timescales for the re-equilibration of pore pressure in a laboratory specimen of such a rock after a small perturbation. In the case of large pressure changes, D can no longer be taken as constant because of the large change in β with pressure and a solution for variable D must be sought. For numerical solution of (3.51) in this case, see Lin (1982).

Another effect that may sometimes be of interest is the hydrodynamic dispersion, represented by the spreading or mixing of a tracer solute introduced at a point in the system (Bear 1972, Chap. 10). The effect can be described in terms of a coefficient of hydrodynamic dispersion which is a function of the velocity of the fluid, the molecular diffusion coefficient of the solute in the fluid, and a geometrical property of the porous solid known as its dispersivity, related to but not solely determined by the permeability.

References

Adda Y, Philibert J (1966) La diffusion dans les solides. Press University, Paris, 1285 pp

Allègre CJ, Le Mouel JL, Provost A (1982) Scaling rules in rock fracture and possible implications for earthquake prediction. Nature 297:47–49

Allnatt AR, Lidiard AB (1993) Atomic transport in solids. Cambridge University Press, Cambridge 572 pp

Anderson DE (1981) Diffusion in electrolyte mixtures. In: Kinetics of geochemical processes. Mineral Soc Am, pp 211–260

Anderson DE, Graf DL (1976) Multicomponent electrolyte diffusion. Annual Rev Earth Planet Sci 4:95–121

Archie GE (1942) The electrical resistivity log as an aid in determining some reservoir characteristics. Trans AIME 146:54–62

Arrhenius S (1889) Über die Reaktionsgeschwindigkeit bei der Inversion von Rohzucker durch Saüren. Zeitschrift für physikalische Chemie 4:226–248

Atkins PW (1978) Physical chemistry. Oxford University Press, Oxford, 1018 pp

Atkins PW (1986) Physical chemistry, 3rd edn. Oxford University Press, Oxford, 857 pp

Atkinson A, Taylor RI (1981) The diffusion of 63Ni along grain boundaries in nickel oxide. Phil Mag A43:979–998

Bailey JE, Hirsch PB (1962) The recrystallization process in some polycrystalline metals. Proc Roy Soc (London) A267:11–30

Bear J (1972) Dynamics of fluids in porous media. American Elsevier, New York 764 pp

Bernabé Y, Brace WF, Evans B (1982) Permeability, porosity and pore geometry of hot-pressed calcite. Mech Materials 1(173):183

Bernadini J, Gas P, Hondros ED, Seah MP (1982) The role of solute segregation in grain boundary diffusion. Proc Roy Soc (London) A379:159–178

Berner, R A, 1981, Kinetics of weathering and diagenesis. In: Kinetics in geochemical processes. Min Soc Am 111–134

Brace WF (1977) Permeability from resistivity and pore shape. J Geophys Res 82:3343–3349

Brace WF (1980) Permeability of crystalline and argillaceous rocks. Int J Rock Mech Min Sci 17:241–251

Brace WF (1984) Permeability of crystalline rocks: new in situ measurements. J Geophys Res 89:4327–4330

Brace WF, Orange AS, Madden TR (1965) The effect of pressure on the electrical resisitivity of water-saturated crystalline rocks. J Geophys Res 70:5669–5678

Brace WF, Walsh JB, Frangos WT (1968) Permeability of granite under high pressure. J Geophys Res 73:2225–2236

Brady JB (1975a) Reference frames and diffusion coefficients. Am J Sci 275:954–983

Brady JB (1975b) Chemical components and diffusion. Am J Sci 275:1073–1088

Burton WK, Cabrera N, Frank FC (1951) The growth of crystals and the equilibrium structure of their surfaces. Phil Trans Roy Soc London Series A 243:299–358

Cabrané-Brouty F, Bernadini J (1982) Segregation and diffusion. J de Phys 43 Colloq C6: C6-163–C166-171

Cahn RW (1983) Recovery and recrystallization. In: Physical metallurgy, 3rd edn. North-Holland Publ Co, Amsterdam, pp 1595–1671

Carslaw HS, Jaeger JC (1959) Conduction of heat in solids, 2nd edn. Clarendon Press, Oxford, 510 pp

Cashman KV (1990) Textural constraints on the kinetics of crystallization of igneous rocks. Rev Mineral Geochem 24:259–314

Chai BHT (1974) Mass transfer of calcite during hydrothermal recrystallization. In: Geochemical transport and kinetics pp 205–218

Chapman RE (1981) Geology and water. an introduction to fluid mechanics for geologists. Martinus Nijhoff/Dr W Junk Publishers, The Hague, 228 pp

Chelidze TL (1982) Percolation and fracture. Phys Earth Planet Int 28:93–101

Christian JW (1965) The theory of transformations in metals and alloys. Pergamon, Oxford 973 pp

Christian JW (1975) Transformations in metals and alloys part I : equilibrium and general kinetic theory, 2nd edn. Pergamon Press, Oxford 586 pp

Cooper AR (1974) Vector space treatment of multicomponent diffusion. In: Geochemical transport and kinetics. Carnegie Institution of Washington, Washington, pp 15–30

Crank J (1975) The mathematics of diffusion, 2nd edn. Clarendon Press, Oxford, 414 pp

Cussler EL (2009) Diffusion—Mass transfer in fluid systems, 3rd edn. Cambridge University Press, Cambridge, 654 pp

Darcy H (1856) Les Fontaines Publiques de la Ville de Dijon. Victor Dalmont, Paris

De Groot SR, Mazur P (1962) Non-equilibrium thermodynamics. North Holland, Amsterdam 510 pp

Dohmen R, Chakraborty S (2007) Fe-Mg diffusion in olivine II: point defect chemistry, change of diffusion mechanisms and a model for calculation of diffusion coefficients in natural olivine. Phys Chem Min 34:409–430, errata 597–598

Fick A (1855) On liquid diffusion. Phil Mag 10:30–39

Fisher GW, Lasaga AC (1981) Irreversible thermodynamics in petrology. In: Lasaga AC, Kirkpatrick RJ (eds) Geochemical processes. Reviews in Mineralogy, vol. 8, pp 171–209

Flynn CP (1972) Point defects and diffusion. Clarendon Press, Oxford 826 pp

Glasstone S, Laidler KJ, Eyring EH (1941) The theory of rate processes. Mc-Graw Hill, New York 611 pp

Gonten D, Whiting RL (1967) Correlations of physical properties of porous media. J Soc Petrol Eng 7:266–272

Guillopé M, Poirier J-P (1979) Dynamic recrystallization during creep of single-crystalline halite: an experimental study. J Geophys Res 84:5557–5567

Guiraldenq P (1982) Diffusion intergranulaire et largeur des joints de grains. J de Phys 43 Colloq C6:C6-137–C136-145

Gupta D (1977) Influence of solute segregation on grain-boundary energy and self-diffusion. Met Trans 84:1431–1438

Hartley J, Crank J (1949) Some fundamental definitions and concepts in diffusion processes. Trans Faraday Soc 45:801–818

Hobbs BE (1968) Recrystallization of single crystals of quartz. Tectonophysics 6:353–401

Hofmann AW, Giletti BJ, Yoder HS, Yund RA eds. (1974) Geochemical transport and kinetics. Conference at Warrenton, Virginia, June 1973, Carnegie Institution of Washington Publ No. 634, 353 pp

Hooyman GJ, Holtan H, Mazur P, De Groot SR (1953) Thermodynamics of irreversible processes in rotating systems. Physica 19:1095–1108

Howard RE, Lidiard AB (1964) Matter transport in solids. Reports Progress Phys 27:161–240

Hubbert MK (1956) Darcy's law and the field equations of the flow of underground fluids. Trans AIME 207:222–239

Kirkaldy JS, Young DJ (1987) Diffusion in the condensed state. The Institute of Metals, London 527 pp

Kirkpatrick S (1973) Percolation and conduction. Rev Mod Phys 45:574–588

Kirkpatrick RJ (1981) Kinetics of crystallization of igneous rocks. In: Kinetics of geochemical processes. Reviews in mineralogy. Min Soc Amer 8:321–398

Lasaga AC (1979) Multicomponent exchange and diffusion in silicates. Geochim Cosmochim Acta 43:455–469

Lasaga AC (1981) Transition state theory. In: Kinetics of geochemical processes. Reviews in mineralogy, vol. 8, Min Soc Amer 135–169

Lasaga AC, Kirkpatrick RJ (1981) Kinetics of geochemical processes. Reviews in mineralogy, vol. 8, Min Soc Amer 398 pp

Lasaga AC, Richardson SM, Holland HR (1977) The mathematics of cation diffusion and exchange between silicate minerals during retrograde metamorphism. In: Energetics of geological processes, Springer, New York, pp 353–388

Le Claire AD (1976) Diffusion. In: Treatise on solid state chemistry. Reactivity of solids, vol 4. Plenum Press, New York, pp 1–59

Lehner F (1990) Thermodynamics of rock deformation by pressure solution. In Deformation processes in minerals, Ceramics and Rocks. Unwin Hyman, London, pp 296–333

Lücke K, Stüwe H-P (1971) On the theory of impurity controlled grain boundary motion. Acta Metall 19:1087–1099

Lin W (1982) Parametric analysis of the transient method of measuring permeability. J Geophys Res 87:1055–1060

Madden TR (1983) Microcrack connectivity in rocks: a renormalization group approach to the critical phenomenon of conduction and failure in crystalline rocks. J Geophys Res 88:585–592

Manning JR (1968) Diffusion kinetics for atoms in crystals. Van Nostrand, Princeton 257 pp

Manning JR (1974) Diffusion kinetics and mechanisms in simple crystals. In: Geochemical transport and kinetics, vol 634. Carnegie Institute of Washington Publication, Washington, pp 3–13

Mantina M, Wang Y, Arroyave R, Chen LQ, Liu ZK, Wolverton C (2008) First-principles calculation of self-diffusion coefficients. Phys Rev Lett 100:5901–5904

Mantina M, Wang Y, Chen LQ, Liu ZK, Wolverton C (2009) First principles impurity diffusion coefficients. Acta Mater 57:4102–4108

Martin G, Perraillon B (1980) Measurements of grain boundary diffusion. In: Grain boundary structure and kinetics. 1979 ASM materials science seminar, Milwaukee, Metals Park, Ohio, Am Soc for Metals, pp 239–295

Mehrer H (2007) Diffusion in solids: fundamentals methods materials, diffusion-controlled processes. Springer, New York, 654 pp

Miyamoto M, Takeda H (1983) Atomic diffusion coefficients calculated for transition metals in olivine. Nature 303:602–603

Onsager L (1945) Theories and problems in liquid diffusion. Ann N Y Acad Sci 46:241–265

Paterson MS (1973) Nonhydrostatic thermodynamics and its geologic applications. Rev Geophys Space Phys 11:355–389

Paterson MS (1983) The equivalent-channel model for permeability and resistivity in fluid-saturated rock—a re-appraisal. Mech Mater 2:345–352

Peterson NL (1980) Grain-boundary diffusion-structural effects, models, and mechanisms. In: Grain-boundary structure and kinetics. 1979 ASM materials science seminar, Milwaukee, Metals Park, Ohio, Am Soc for Metals, pp 209–237

Peterson NL (1983) Grain-boundary diffusion in metals. Int Met Rev 28:65–91

Philibert J (1991) Atom movements, diffusion and mass transport in solids, English edition: (trans: Rothman SJ). EDP Sciences, 580 pp

Poirier J-P (1985) Creep of crystals. High-temperature deformation processes in metals, ceramics and minerals, Cambridge Univ Press, New York 260 pp

Renard F, Schmittbuhl J, Gratier J-P, Meakin P, Merino E (2004) Three-dimensional roughness of stylolites in limestones. J Geophys Res 109(B03209):002512. doi: 03210.01029/02003JB002555

Rumer R, (1969) Resistance to flow through porous media. In: Flow through porous media, Academic Press, New York, pp 91–108

Rutter EH (1983) Pressure solution in nature, theory and experiment. J Geol Soc London 140:725–740

Sakai T (1989) Dynamic recrystallization of metallic materials. In: Rheology of solids and of the earth, Oxford University Press, Oxford, pp 284–307

Scheidegger AE (1960) The physics of flow through porous media, 2nd edn. University of Toronto Press, Toronto, 313 pp

Schmid SM, Paterson MS, Boland JN (1980) High-temperature flow and dynamic recrystallization in Carrara marble. Tectonophysics 65:245–280

Shewman PG (1963) Diffusion in solids. McGraw Hill, New York 203 pp

Shewman P (1989) Diffusion in solids, vol 2. TMS Publications, 246 pp

Shimizu I (1998) Stress and temperature dependence of recrystallized grain size: a subgrain misorientation model. Geophys Res Lett 25:4237–4240

Shimizu I (1999) A stochastic model of grain size distribution during dynamic recrystallization. Phil Mag A 79:1217–1231

Sullivan RR, Hertel KL (1942) The permeability method for determining specific surface of fibers and powders. In: Advances in colloid science, vol 1. Interscience Publishers, New York, pp 37–80

Thompson J (1962) On crystallization and liquefaction. As influenced by stresses tending to change of form in the crystals, Proc Roy Soc (London) A11:473–481

von Engelhardt W (1960) Der Porenraum der Sedimente, Springer, 207 pp

Waff HS (1974) Theoretical considerations of electrical conductivity in a partially molten mantle and implications for geothermometry. J Geoph Res 79(4003):4010

Walsh JB (1981) Effect of pore pressure and confining pressure on fracture permeability. Int J Rock Mech Min Sci 18:429–435

Watson EB, Baxter EF (2007) Diffusion in solid-earth systems. Earth Planet Sci Lett 253:307–327

Wilkinson DS (2000) Mass transport in solids and fuids. Cambridge University Press, Cambridge 292 pp

Yan MF, Cannon RM, Bowen HK, Coble RL (1977) Space-charge contribution to grain-boundary diffusion. J Am Ceram Soc 60:120–127

Zhang M-X, Kelly PM (2009) Crystallographic features of phase transformations in solids. Prog Mater Sci 54:1101–1170

Chapter 4
Mechanical Fundamentals

4.1 Introduction

There are some very broad distinctions that can usefully be made in classifying types of mechanical behavior and the approaches to their study. The first of these distinctions is between *brittle* and *ductile* behavior. We can define brittleness as the liability to gross fracturing without substantial permanent change of shape in response to loading beyond the elastic range. Conversely, ductility is the capacity for substantial permanent change of shape without gross fracturing. In this context, "gross" means on the scale of the whole body or region under consideration and the use of the terms brittle and ductile is only meaningful with proper reference to scale. For the study of brittle behavior, see Jaeger (1969), Paterson and Wong (2005), and Jaeger et al. (2007). In this chapter, we are mainly concerned with ductile behavior or plastic deformation.

A second distinction is that between *athermal* and *thermal* regimes of behavior. These terms refer to the degree of sensitivity of the flow behavior to change in temperature and are to be regarded as relative only. Thus, in an athermal regime, the behavior is relatively insensitive to change in temperature and can often be treated as being independent of temperature, although temperature effects may not be absent altogether. Also, athermal behavior cannot be associated uniquely with relatively low temperatures and thermal behavior with relatively high temperatures. Rather, there is a tendency for there to be three regimes of behavior:

1. Low-temperature thermal
2. Athermal
3. High-temperature thermal

These regimes are depicted schematically in Fig. 4.1 in a plot of the flow stress against temperature (the flow stress is the stress needed to bring about plastic flow). The scheme of Fig. 4.1 derives originally from the observed behavior of body-centered cubic metals, but it seems to have a wider applicability (Di Persio and

M. S. Paterson, *Materials Science for Structural Geology*,
Springer Geochemistry/Mineralogy, DOI: 10.1007/978-94-007-5545-1_4,
© Springer Science+Business Media Dordrecht 2013

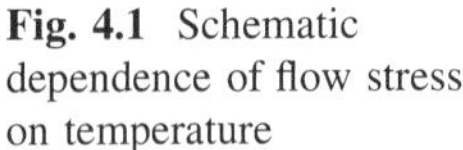

Fig. 4.1 Schematic dependence of flow stress on temperature

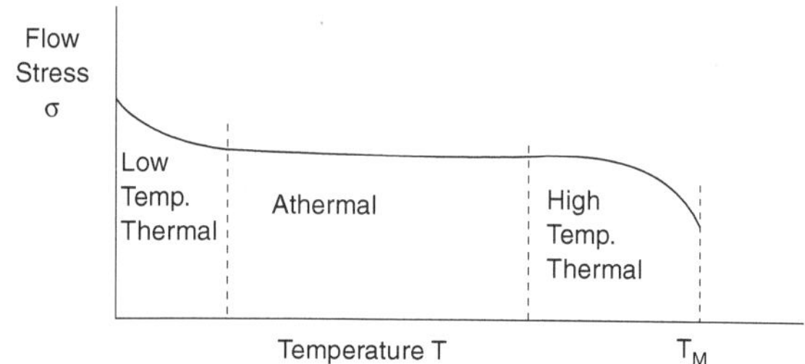

Escaig 1984). However, all three regimes are not necessarily found in a particular material. For example, in copper, the athermal regime seems to extend down to absolute zero (no significant low-temperature thermal regime), while in pure dry quartz, the low-temperature thermal regime seems to extend more or less to the melting point. In other materials such as germanium and silicon, low-and high-temperature thermal regimes appear to overlap in such a way that no significant athermal regime is defined. In pure metals, the high-temperature regime can often be usefully defined as that above about half the absolute melting point but no such general rule can be applied to the wider spectrum of materials, including minerals and rocks. However, within given chemical classes of materials, it may be possible to define similar homologous temperature rules, the homologous temperature being taken as relative to some reference temperature at which a change of state occurs that involves the same components as are taking part in the mechanical process of interest, this reference temperature not necessarily being the melting point, especially where incongruent melting is concerned.

Third, a broad distinction can be made between phenomenological and mechanistic aspects of deformation. The *phenomenological view* is a global one, dealing with a body of material as a whole or as a continuum. It is only concerned with macroscopic variables such as stress, strain and temperature, and with establishing empirical relationships among them. The nature of the material is represented by empirical constants or parameters (elastic modulus, activation energy, etc.) that appear in these relationships. Variations in the state of the material may be represented by additional parameters (internal variables), such as grain size, in the empirical relationships but the material is still treated as a continuum from the mechanical point of view. In contrast, *the mechanistic view* attempts to recognize the local processes contributing to the global behavior. It therefore tends to concentrate on the microscopic rather than the macroscopic scale. The aim is to describe the processes or mechanisms in terms of re-arrangements of fundamental structural units and to develop models for the global behavior in terms of these processes. Ideally, this approach eventually gives a rationalization for the empirical relationships established in a phenomenological approach, and further provides a basis for establishing the range of validity of such relationships in the light of structural observations.

In this chapter, we mainly survey the phenomenological aspects of deformation, those aspects that can be represented as the properties of a homogeneous continuum, taking no account of what might be happening on the microscopic scale. The microscopic aspect of deformation will be touched only in the final section, forming a transition to the three Chaps. 5–7 on mechanisms of deformation.

4.2 Phenomenological Approach

In the phenomenological approach to plastic deformation, the primary mechanical variables are the stress σ and the strain ε. These are tensorial quantities, normally fully representable by six independent components (see texts such as Jaeger 1962; Means 1976 for simple introductions to stress and strain; also see texts such as Nye 1957; Reid 1973).

In general, stress is a measure of the intensity of force acting on a 3-dimensional (3-D) element of the body. Its complete description requires nine components, six of which can be shown to be independent. The full specification of stress is therefore in the form of a symmetrical second-rank tensor $\sigma = \sigma_{ij}$, $i, j = 1, 2, 3$.

The stress σ can often usefully be viewed as consisting of two parts, one representing the hydrostatic aspect of behavior and relevant to changes in volume, and the other representing the non-hydrostatic aspect and relevant to changes in shape. Formally, the stress tensor σ_{ij} can always be written as the sum of a hydrostatic component $\frac{1}{3}\sigma_{ii}$ and a deviatoric component $\sigma_{ij} - \frac{1}{3}\sigma_{ii}$. However, in experimental studies and in geology, it is often convenient to use a slightly different resolution of the stress tensor, which we shall exemplify here in the case of an *axisymmetric* stress state, the one most commonly involved in experimental work. The principal components of the *axisymmetric* total stress tensor, σ_1, σ_2, σ_3 are then written as

$$\sigma_1 = p + \sigma$$
$$\sigma_2 = p \qquad\qquad (4.1)$$
$$\sigma_3 = p$$

where p is the "confining pressure" and σ is the axial stress difference or "differential stress". The hydrostatic component of the stress is now $\left(p + \frac{\sigma}{3},\ p + \frac{\sigma}{3},\ p + \frac{\sigma}{3}\right)$ and the deviatoric component is $\left(\frac{2\sigma}{3},\ -\frac{\sigma}{3},\ -\frac{\sigma}{3}\right)$. In experimental work with axisymmetric stress, the components σ_2 and σ_3 are normally generated by applying a hydrostatic pressure p to the faces of the specimen parallel to the 1 axis through the agency of a fluid or weak solid. This pressure is properly called the confining pressure and is to be distinguished from the hydrostatic component of the stress. It is conventional in rock mechanics to designate compressive stresses as positive, in contrast to engineering convention where tensile stress is positive.

In general, strain consists of the displacement of points in a 3-D space relative to a 3-D reference frame. Its complete description therefore requires nine components, six of which can be shown to be independent. Therefore, the complete specification of strain is in the form of a symmetric second-rank tensor $\varepsilon =$

ε_{ij}, $i, j = 1, 2, 3$. "Engineering" components of strain are commonly used in reporting experimental work. Thus, normal strains e are given as relative elongation or shortening based on the original length l_0 ($e = \frac{-\Delta l}{l_0}$, using the rock mechanics convention that shortening strain is positive). Shear strains are given by $\gamma = \tan \theta = \frac{s}{l_0}$ where s is the shearing displacement and l_0 is the reference length normal to the shearing displacement. However, when specifying finite deformations, the engineering definition of normal strain has the disadvantage that a given increment of strain has different physical significance as the actual length changes. Therefore, for theoretical discussion, it is better to define the strain by integration of strain increments each of which is the relative elongation based on the current length. This leads to the definition of the so-called natural strain $\varepsilon = -\ln \frac{l}{l_0}$ (in the convention of shortening strain being positive). Similarly, in finite deformations, the reference area for normal stresses changes with the deformation and it is desirable to specify the normal stress components as "true stress" calculated on current rather than original cross-sectional area.

The primary environmental variable is normally the temperature T. However, it is also often convenient, when relating the stress difference σ to the deformation in axisymmetric deformation, to treat the "pressure" as a distinct environmental variable, specifying it as the confining pressure, p, or the hydrostatic component of the total stress, $p + \frac{\sigma}{3}$, as appropriate. Other environmental variables that may be relevant at times are the pore pressure and the activities of chemical components in reservoirs available to the specimen.

The other quantities that enter into any relationship describing the mechanical behavior are the material parameters representing the characteristic properties of the particular material. In practice, they are defined in a largely empirical way, dictated by the forms of relationship between the mechanical variables that are found best to describe the observed behavior. Also, particular parameters may be needed to describe different structural states of a given type of material (for example, grain size).

We now consider some general relationships between the mechanical variables that apply in the intermediate athermal and high-temperature thermal fields depicted in Fig. 4.1. In practice, the boundary between these fields may be rather diffused but for highly ductile materials such as metals it can commonly be taken as being roughly one-half of the absolute melting point for materials that melt simply.

4.3 The Athermal Field

4.3.1 The Stress–Strain Curve

For simplicity, we initially only consider behavior in a simple compression or tensile test, with or without superposed confining pressure p. The state of stress can be then described by the stress difference σ and the strain by the relative change of length e or the natural strain ε, as defined in Sect. 4.2.

Athermal or temperature-insensitive ductile behavior is characterized by a stress–strain relationship $\sigma = \sigma(\varepsilon;\ \dot{\varepsilon}, T, \ldots)$ where the dependences on strain rate $\dot{\varepsilon}$, temperature T, etc., are relatively weak. The main aspects of athermal behavior are therefore represented in the stress–strain curve, where it is convenient to treat the strain as the independent variable, following the usual testing procedure. There is usually a fairly distinct initial elastic region, terminated at a more or less well defined level of stress called *the initial yield stress* σ_y. Unloading beyond this point will reveal that the specimen has undergone permanent or plastic strain. On reloading, a stress of approximately the same level as previously reached must be applied before yielding again occurs and plastic flow continues; hence, the stress at any point on the stress–strain curve beyond the initial yield stress can be called the *flow stress*. Commonly, the flow stress increases as straining progresses, a phenomenon called *strain hardening* or work hardening, characterized by the slope $\gamma = \partial\sigma/\partial\varepsilon$ of the stress–strain curve. If $\gamma = 0$ or $\gamma < 0$, the material is said to be *perfectly plastic* or *strain softening*, respectively. The existence of strain hardening or softening indicates that there is continual change within the material during plastic straining whereby its ability to support stress is changed.

The initial states are commonly not at thermodynamic equilibrium, nor is the strain-hardened state, and so there is very limited scope for thermodynamic discussion of plastic behavior in the temperature-insensitive field. If the deformed material is subsequently heated sufficiently, that is, annealed, the strain hardening is removed, partially in the case of recovery (Sect. 3.3.2) or more or less fully in the case of recrystallization (Sect. 3.3.3).

4.3.2 Low Temperature Creep

Although the main interest in the athermal field centers on the stress–strain curve and time or rate effects play a secondary role, the latter effects are sometimes of interest, as in the case of creep in bodies held under stress for long periods of time. In the creep test, the strain ε is measured as a function of elapsed time t at constant stress. In the relatively low-temperature field, the creep relation $\varepsilon = \varepsilon(t)$ is widely found to have the form

$$\varepsilon = \varepsilon_0 + \alpha \ln(1 + vt) \tag{4.2}$$

where ε_0 is the initial strain and α, v are constants that depend on stress and temperature. Such behavior is called *logarithmic creep*. As indicated by measurements on various rocks by Misra and Murrell (1965) under conditions for logarithmic creep, typical values of α are 10^{-4} to 10^{-6} and of v are 10^{-2} to $1\,\mathrm{s}^{-1}$ and sometimes larger. Thus over most of the time span of creep experiments, at least after a few minutes, $vt \gg 1$ and the relation (4.2) can be written in the form

$$\varepsilon = \varepsilon_0 + \alpha \ln t \tag{4.3}$$

where the $\alpha \ln v$ term has been subsumed into the constant ($\alpha \ln v$, in fact, tends to be negligible compared with the elastic strain). This is the form of logarithmic law originally proposed by Phillips (1905) and widely used (for example, Griggs 1939). On the other hand, the relation of Lomnitz (1956), often quoted by seismologists, is equivalent to (4.2). Other relationships for low-temperature creep have been proposed with more parameters or more terms but need not be elaborated here (Benioff 1951; Griggs 1939; Jeffreys 1958; Michelson 1917, 1920).

Time-dependent strain can also occur in elastic vibrations, the theoretical analysis of which in turn also implies specific forms of creep relation that may be relevant at very small stresses and low temperatures. Thus, the theory of the "standard linear solid" (Zener 1948, p. 43) leads to a relation of the form

$$\varepsilon = \varepsilon_0 + \varepsilon_t \left(1 - e^{-t/t_0}\right) \tag{4.4}$$

where t_0 and ε_t are constants (in this linear theory, $\varepsilon_0 = \sigma/M_u$ and $\varepsilon_t = (\sigma/M_R) - \varepsilon_0$ where σ is the stress and M_u, M_R are the unrelaxed and relaxed elastic moduli, respectively).

4.4 The Thermal Field

4.4.1 Stress-Strain-Time Relationships

In temperature-sensitive mechanical behavior, the variables elapsed time t and temperature T assume the same degree of importance as the stress σ and strain ε. The basic relationship to be sought is then of the form

$$\varepsilon = \varepsilon(\sigma, t, T, y, \ldots)$$

involving the four variables just mentioned as well as independent structural parameters (or internal variables) y, such as initial grain size or concentrations of impurities. Experimentally, this multivariable situation is normally approached by observing the relationship only between two of the variables while the others are held at fixed, known values.

The simplest practical test from the point of view of interpretation is the creep test in which one measures $\varepsilon = \varepsilon(t)$ at constant σ, T, y. The approach, normal at low temperatures, of measuring the stress–strain curve at constant $\dot{\varepsilon}$, $T \ldots$ is very useful in exploratory work when it is not known what stress levels are of most interest but interpretation may be more difficult since the role of time is not fully represented in the strain rate (for example, static recovery effects may be occurring). Therefore, our discussion here will be in terms of the creep test (note also that, unless otherwise specified, we shall take σ to be the "true stress", based on

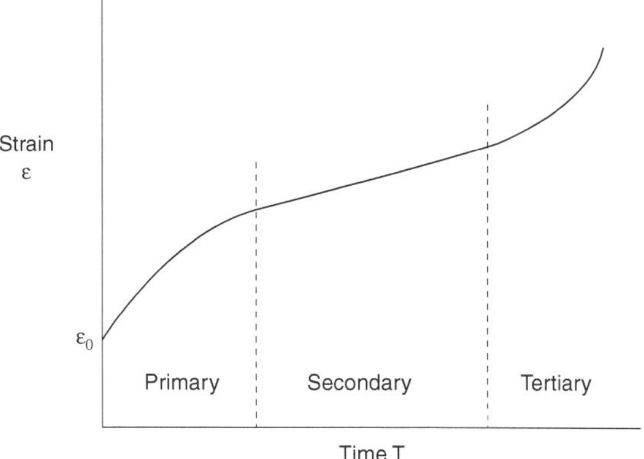

Fig. 4.2 Stages of a creep test

current cross-sectional area, and ε to be natural strain, Sect. 4.2). In a creep test, it is convenient to distinguish four stages (Fig. 4.2), characterized as follows:

1. The instantaneous strain ε_0. This is the strain that occurs during the application of the stress and in the interval before practical strain measurements as a function of time are begun. It corresponds more or less to the strain determined in a short-term stress–strain test but of course it depends on the rate at which the stress is applied and it includes some strain that might be counted as creep strain in the initial stages of an ideal creep test in which the stress were applied strictly instantaneously and strain recording begun immediately.
2. Primary creep. The first stage of the recorded creep curve in which the strain rate normally decreases continuously but at a diminishing rate, as shown in Fig. 4.2. However, in some cases, there may be an incubation period of little or no creep followed by an acceleration in strain rate before the stage of diminishing strain rate is entered, giving an overall sigmoidal shape to the creep curve in the primary creep stage.
3. Secondary creep. A stage in which the strain rate reaches its minimum value and during which the rate of change of strain rate is relatively small or sensibly constant.
4. Tertiary creep. A final stage of accelerating creep, representing the onset of some sort of instability leading to failure; less commonly met in compression tests than in extension tests.

Leaving aside the consideration of tertiary creep and of any initial incubation stage, it is commonly assumed that the creep curve can be represented as the sum of three terms, comprising the instantaneous strain ε_0 and two time-dependent forms, a transient creep term $B(t)$, and a steady-state term $\dot{\varepsilon}_s t$ ($\dot{\varepsilon}_s$ a constant), as follows:

$$\varepsilon = \varepsilon_0 + B(t) + \dot{\varepsilon}_s t \tag{4.5}$$

The *transient creep term $B(t)$* tends to predominate in the primary stage of the creep test. A number of empirical forms have been used for $B(t)$. These include:

1. $B(t) = \alpha \ln(1 + vt)$ or $\alpha \ln t$, analogous to the form mentioned above for representing the whole of the time-dependent strain in the athermal field. This form has not found much application at high temperatures.

2. $B(t) = \beta t^m$, where β and m are constants ($m < 1$). The transient creep behavior of a wide variety of materials can be fitted to a rough approximation with $m = 1/3$. The term Andrade creep is often then applied to this form, after Andrade (1910), although Andrade actually proposed $B(t) = \ln(1 + \beta t^{1/3})$, which is equivalent to $\beta t^{1/3}$ for small transient creep strains (i.e. $\ll 1$). By allowing both parameters β and m to vary, better fits can, of course, be achieved, in which case values of m varying from 0.03 to nearly 1 are found (Garofalo 1965, p. 16). The variation with stress can often be expressed by putting $\beta \propto \sigma^{n'}$, so that $B(t) = C'\sigma^{n'}t^m$ where n' and C' are constants. Differentiation then leads to a primary creep rate of

$$\dot{\varepsilon} = C\sigma^n \varepsilon^{-p}$$

where $C = m(C')^{1/m}$, $n = n'm$ and $p = 1/m - 1$. These expressions are commonly used in engineering design, being also treated as covering the whole of the creep strain when there is no marked secondary or tertiary stages. Both the logarithmic and βt^m forms of $B(t)$ have the property that the "transient" creep component of the strain increases indefinitely as t increases.

3. $B(t) = \varepsilon_t(1 - e^{-t/t_0})$, where ε_t and t_0 are constants. This is identical to the anelastic form (4.4), but now applied to substantial plastic strains. It often gives better fits than the βt^m form (Garofalo 1965, p. 16). It also implies that there is an upper limit to the contribution of the transient creep term, represented by the constant ε_t. The constant t_0 can then be viewed as an empirical relaxation time for the transient creep. Studies on high-temperature transient creep in metals (Amin et al. 1970) suggest that, in the absence of significant grain boundary sliding, the influence of stress and temperature is mainly reflected in the parameter t_0 and that the product $t_0\dot{\varepsilon}_s$ is a constant of the order of 100, not varying greatly from one material to another.

The *steady-state creep* term $\dot{\varepsilon}_s t$ in (4.5) tends to predominate in the secondary stage of the creep rate. It should be emphasized that the observed minimum creep rate is not necessarily equal to the steady-state creep rate $\dot{\varepsilon}_s$ but only represents an upper limit to it since there may still be some contribution from the transient creep term up to the onset of tertiary creep; however, in the absence of a tertiary instability, the expression (4.5) represents a total strain rate that decreases continuously with increasing time and asymptotically approaches the steady-state creep rate $\dot{\varepsilon}_s$.

Although it is of practical and conceptual convenience to identify linearly additive instantaneous, transient and steady-state terms in the creep function (4.5), this procedure may eventually prove to be an artificial, and even possibly invalid, one from a theoretical or mechanistic point of view. Thus, as Amin et al. (1970), Mukherjee (1975), and Poirier (1976) point out, the primary creep stage, the behavior in which is normally treated as being dominated by the transient creep term $B(t)$ in (4.5), should probably be regarded more accurately as a transitional stage of structural evolution in which the mechanisms of deformation are being established. The nature of these mechanisms may well not change fundamentally during the course of the deformation, but the strain rate that they contribute may be expected to change as the structural details evolve until some sort of dynamic equilibrium or saturation is reached at the steady state. Moreover, the nonelastic strain occurring prior to the recorded primary stage is somewhat artificially divorced from it by experimental limitations and probably should be treated separately from the elastic strain calculated using a dynamically determined, unrelaxed elastic modulus (cf. Eq. 4.4). Any fully developed theory is therefore more likely to yield, in addition to the elastic strain, a single unified creep function, probably of complex form and representing within itself the several aspects of preliminary nonelastic strain, the measured primary stage and the asymptotic approach to a steady state to be expected in the secondary stage if a tertiary stage or instability is sufficiently delayed.

The concept of steady-state creep calls for brief comment at this point. At the present experimental or empirical level, it embodies the notion of behavior that is independent of the strain. Care should be taken in any attempt to attribute deeper significance to it than this, for two reasons. On the one hand, the experiments are often limited to moderate amounts of strain and the demonstration of the constancy of creep rate is an approximate one, described as such for convenience in analysis but which might not be justifiably so designated if a larger strain interval were explored (as in torsion experiments, Paterson and Olgaard 2000). On the other hand, one may be tempted to infer from the existence of a steady-state creep rate that, under the given values of the independent macroscopic variables, the specimen is now in a stationary state in the sense used in the thermodynamics of irreversible processes (for example, Prigogine 1967, Chap. 6) and that this state would correspond to the optimization of a dissipation potential or entropy production rate (for example, Ziegler 1977, Chap. 15). However, Rice (1970) has concluded that, except in special cases, "no firm basis... is available for a stationary creep potential", such as would give the steady-state strain rate when differentiated with respect to the stress; that is, in general, the steady-state strain rate cannot be expected to be a function of state that is determined completely by the instantaneous values of the macroscopic variables, independently of history, but that additional internal variables (such as uncontrolled structure parameters) will be needed to define fully the state of the specimen.

In spite of a lack of fundamental significance for a steady state in creep deformation, the concept has been widely used in discussions of the rheology of rocks in geological contexts. It simplifies theoretical analysis by eliminating one variable, enabling one to concentrate on $\dot{\varepsilon}_s = \dot{\varepsilon}_s(\sigma, T, y, \ldots)$ instead of $\varepsilon = \varepsilon(\sigma, t, T, y, \ldots)$ in the thermal field. Consequently, experimental work has been commonly concentrated on determining the steady-state strain rate as a function of stress and temperature, that is, establishing a "flow law". Often this aim has been pursued, not through the creep test itself, but through experiments at constant strain rate, taken to strains sufficient for the rate of strain hardening to become sufficiently small, so that the specimen can be regarded as deforming at a stress and a strain rate that are both sensibly constant; it is then assumed that this combination represents a steady state, independent of the testing history and therefore equivalent to what would eventually be reached in a creep test, in the absence of a tertiary stage.

4.4.2 Analysis of "Steady State" Deformation

We now consider the relationship commonly sought in triaxial tests for the steady-state creep rate in the form

$$\dot{\varepsilon}_s = \dot{\varepsilon}_s(\sigma, T, p, y),$$

where an explicit distinction is made between the stress difference σ and the pressure p, as explained in Sect. 4.2; y serves as an internal variable covering any structural factors to be taken into explicit account. In a first approximation, it greatly facilitates analysis to assume that the dependence of $\dot{\varepsilon}_s$ on each variable can be considered separately.

The *temperature dependence* of the steady-state creep rate is normally well represented over limited ranges of temperature by an Arrhenius form $\dot{\varepsilon}_s \propto \exp(-Q/RT)$, where R is the gas constant (8.32 J K^{-1} mol^{-1}), T is the absolute temperature, and $Q = -R\partial \ln \dot{\varepsilon}_s / \partial(1/T)$ is a measure of the degree of temperature sensitivity, called the apparent or empirical activation energy (see Sect. 3.1.2). It should be recognized that in fitting the experimental results, in this way, all aspects of the temperature dependence are lumped together and Q may not represent an actual activation energy for a particular elementary process. For example, a part of the temperature dependence may arise from variations in elastic modulus with temperature (which may eventually prove not even to be of exponential form) and another part from the temperature dependence of a diffusivity (Bird et al. 1969; Poirier 1976, p. 47), in which case the apparent activation energy for steady-state creep could not, on the above analysis, be identified exactly with the activation energy for diffusion. It may also be remarked that, at this level of discussion, the apparent activation energy Q cannot be given a specific thermodynamic designation, such as enthalpy or Gibbs free energy, until the thermodynamic system has been adequately defined (Poirier 1976, pp. 37–38 and 98–106).

The stress dependence of $\dot{\varepsilon}_s$ can generally be represented over limited ranges of stress either by $\dot{\varepsilon}_s \propto \exp(\sigma/\sigma_0)$ or by $\dot{\varepsilon}_s \propto \sigma^n$, where σ_0 and n are constants. The exponential form tends to be found at relatively high stresses (where it can alternatively be represented as $\dot{\varepsilon}_s \propto \sinh(\sigma/\sigma_0)$), while the power law tends to give better fits at lower stresses. A two parameter representation, $\dot{\varepsilon}_s \propto [\sinh(\sigma/\sigma_0)]^n$, has been proposed to give a wider range of fit with one expression, covering both exponential and power law behavior (Garofalo 1965), but this in turn can be inadequate over the full range of interest, and so it would seem better for empirical representation to use several single parameter expressions that give best fits in limited ranges of stress, especially if these regimes can be correlated on microstructural evidence with changes in mechanism. Sometimes an exponential stress dependence is represented in the form $\dot{\varepsilon}_s \propto \exp(-\sigma V'/RT)$ with a view to representing the role of stress in assisting thermally activated processes as contributing an energy $\sigma V'$, where the constant V' has the dimensions of volume.

For the *pressure dependence* of $\dot{\varepsilon}_s$, the form $\dot{\varepsilon}_s \propto \exp(-pV^*/RT)$ is used, with an implication that its role is through influencing thermally activated processes, as in diffusion. The degree of pressure dependence is thus measured by the constant V^*, called the *activation volume* and defined as $V^* = -RT\partial \ln \dot{\varepsilon}_s/\partial p$. There is potentially some confusion in terminology here since the quantity V' representing the stress dependence of $\dot{\varepsilon}_s$, introduced in the previous paragraph, is also commonly called an activation volume, especially in the metallurgical literature. Sometimes this ambiguity is resolved by writing $V' = Ab$ (b being the Burgers vector for given dislocations in the crystals) and calling A the activation area. However, it is simpler to distinguish V' and V^* as activation volumes for stress and pressure dependence, respectively.

The concept of structure dependence, represented by a term in y, calls for some preliminary comment. Rheological behavior clearly involves microstructural aspects such as mobility of crystal defects, processes at grain boundaries and the presence of impurities, and creep rates may be influenced by the changes in concentration of these entities, which are covered by the generic term "structure". However, in discussing the role of structural factors one must distinguish, at least in principle, between those which play the role of independent variables, under the control of the experimenter and those which are dependent or uncontrolled. Quantities specifying structural features that persist unmodified through the range of strain under consideration can be treated as *independent* variables; examples are: the initial grain size in cases where recrystallization is absent or minor, a concentration of crystal defects that is continuously equilibrated with respect to an externally controlled environment (for example, a fixed oxygen fugacity), the concentration of an impurity in solid solution or forming a separate phase, and the relative amounts of the phases in a multiphase material. In contrast, variables such as mobile dislocation density, subgrain size, and recrystallized grain size in situations where the dislocations, subgrains, and grains have been generated during the deformation are, in general, *dependent* variables. Only those quantities that can be considered as independent variables in a given test are covered by the symbol y in this section. An example of a structure dependence is given by the dependence of the steady-state creep rate on grain size d which is observed at high temperatures where relative grain sliding is significant, in the so-called superplastic regime; it is found that $\dot{\varepsilon}_s \propto d^x$ and the sensitivity to grain size is then specified by the constant $x = \partial \ln \dot{\varepsilon}_s / \partial \ln d$. However, when attention is not specifically directed to structural, or any other, variables, their influence is subsumed in a "constant" factor A included as a multiplier in the empirical steady-state creep law.

Recapitulating, in the simplest or first analysis, the steady-state creep behavior is commonly and conveniently represented in the form

$$\dot{\varepsilon}_s = A f(\sigma) \exp(-Q/RT) \tag{4.6}$$

where $f(\sigma) = \exp(\sigma/\sigma_0)$ or σ^n according to which gives the better fit to the observations, and A is a multiplier to be treated as a constant when only variations of σ and T are concerned or to be expressed as an appropriate function $A(p,y)$ when independent variations of pressure or structure are considered. When no expression of form (4.6) with a single set of constants can be found to fit the observations over the whole range of the variables, it is usual to consider this range in several parts or regimes within each of which the observations can be fitted to an expression of form (4.6) with a single set of constants. Such a definition of distinct steady-state rheological regimes is, of course, an empirical procedure and may initially appear rather arbitrary but it can take on more physical significance if the change from one regime to another can be shown also to involve microstructural changes suggestive of change in the deformation mechanism or in factors controlling the deformation. The microstructural study may, alternatively, lead to the postulation of models for the deformation mechanism that imply rheological laws different in form from (4.6) which it will then be desired to test for fit.

4.5 Instabilities and Localization

So far we have taken it for granted that a specimen will undergo a uniform deformation when loaded under conditions of uniform stress. However, this is not always the case and rather restrictive conditions may have to be satisfied before uniformity of deformation can be expected. When these conditions are not met, heterogeneity of deformation can develop, giving rise to structural patterns on various scales, apparently analogous in many cases to the geological features studied by structural geologists.

In considering instability, it is often convenient to distinguish between shape or external geometrical instability and material or internal instability (Drucker 1960); (Biot 1965, pp. 192–204) (Paterson and Weiss 1968), although the distinction is not always a clearcut one, however, and in some cases may simply refer to two different aspects of the same unstable behavior. The first category is typified by the necking instability in a tensile test specimen or by the Euler buckling of a slender column in compression, while the second is typified by the development of Lüders' bands in the deformation of mild steel or of kink bands in crystals compressed parallel to a unique slip plane. The latter effects can be described without reference to the boundaries of the specimen or system while the boundaries are involved in an essential way in describing the former.

In mechanical testing, one can generally regard the combination of specimen and loading rams as a system subject to prescribed displacements or loads applied at certain boundaries. The precise behavior within this system may involve either of the types of instability just distinguished. Thus, in compression tests there is, on the one hand, the possibility of a shape instability in which the initial alignment of loading ram and specimen is replaced by a buckled configuration (for creep buckling, see Hoff 1958) and, on the other hand, even if the gross alignment is preserved, the distribution of deformation within the specimen itself may become heterogeneous due to material instability, as in the formation of localized shear zones. In extension testing, there is no question of buckling instability but necking can occur when the loss of strength due to a virtual reduction in cross-sectional area locally is less than the gain due to strain hardening associated with the locally increased strain or strain rate. The corresponding shape instability in compression is local bulging (Jonas and Luton 1978), (Jonas et al. 1976) but it tends to be less important than buckling, especially when the deformation is strain-rate sensitive, and it is less frequently observed than shearing instability in the material itself. The analysis of the necking or bulging instability has been developed by many writers since the classical paper of Considère (1885), especially in recent years in connection with "superplasticity", the phenomenon that is defined macroscopically by the ability of a specimen to undergo exceptionally large elongations without failure resulting from excessive necking (see, for example, Hart et al. 1995); (Jonas et al. 1976; Reid 1973, pp. 38–43). Of particular, interest in the latter connection is the role of the strain rate dependence in determining the rate at which necking develops; superplasticity depends more on the slowness of development of a neck than on inherent instability against any necking at all.

Shearing instability within a deforming body is of potential importance in all types of mechanical testing as well as in natural situations involving complex stresses. It involves the localization of strain within a zone parallel to a surface of pure shear or zero extensional strain, that is, to a characteristic or slip surface in the sense of the mathematical theory of plasticity (Hill 1950); (Prager and Hodge 1951); (Hoffman and Sachs 1953). There have been many analyses of instability based on the work done on a deforming specimen (Argon 1973; Backhofen 1972; Evans and Wong 1985; Poirier 1980). However, such analyses do not take into account the development of local heterogeneity of deformation during the onset of instability and can fail to account for observations in torsion tests (Paterson 2007). A full analysis needs to be based on continuum mechanics and perturbation theory, such as set out by Fressengeas and Molinari (1987) and Shawki and Clifton (1989).

4.6 Failure Criteria and Flow Laws for General Stress States

So far, we have discussed deformation behavior in situations involving a simple homogeneous state of stress, usually specified by a single component of stress (uniaxial stress state), with possibly the superposition of a hydrostatic component to be taken into account. However, in practical applications, it is frequently necessary to deal with general states of stress (complex, triaxial, or multiaxial stresses) in which all three principal components and their orientations have to be taken into account, as well as their variation from point to point. The problem then arises of how to generalize the stress–strain–strain rate relationships so far discussed, that is, to establish general constitutive relations; see Malvern (1969, Chap. 6), Kocks (1975), Rice (1970), 1975), and Kestin and Bataille (1980) for general considerations on constitutive relations and their extension to time-dependent plasticity, including the question of memory effects and the use of internal variables. There are two distinct approaches to the problem, one being a generalization of the treatment of viscous fluids and the other an extension of the mathematical theory of plasticity, on which we shall now comment briefly.

The simplest approach through the theory of viscous flow is to retain the normal treatment based on isotropic Newtonian or linear viscosity but to assign to the dynamic viscosity an effective or equivalent value, given by $\eta = \sigma/3\dot{\varepsilon}$ where σ, $\dot{\varepsilon}$ are the normal stress and strain rate, respectively, measured in a uniaxial experiment carried out to steady state within the ranges of stresses that is of interest (Griggs 1939; the factor three arises naturally from the constitutive equations for isotropic linear viscous flow with zero bulk viscosity). However, this procedure is only suitable for an approximate treatment of the broad aspects of the flow and it automatically suppresses any characteristic features arising from the actual nonlinear nature of the viscosity. The alternative is to solve the equations of viscous flow with a viscosity that is no longer a constant but is itself a function of stress. Difficulties associated both with establishing the form of this function and of solving the nonlinear equations have largely inhibited development along these lines.

The theory of plasticity, as developed in its simplest form for isotropic metals at low temperatures, generalizes the uniaxial stress–strain relations in terms of "effective" stresses and strain increments, these being scalar functions of the actual stress and strain increment components that serve to describe the mechanical behavior in the same form as in the uniaxial case (not to be confused with the "effective stress" in situations involving pore pressure, Paterson and Wong 2005, p. 148, although the general concept is the same in that we seek to specify that aspect of the stress state that is effective in the particular situation). Following von Mises (Hill 1950, p. 26), the effective stress σ^* is taken to be

$$\sigma^* = \left(1/\sqrt{2}\right)\left[(\sigma_1 - \sigma_2)^2 + (\sigma_2 - \sigma_3)^2 + (\sigma_3 - \sigma_1)^2\right]^{1/2}$$

where σ_1, σ_2, σ_3 are the principal stresses. The corresponding effective plastic strain increment $d\varepsilon^*$ is taken to be

$$d\varepsilon^* = \left(\sqrt{2}/3\right)\left[(d\varepsilon_1 - d\varepsilon_2)^2 + (d\varepsilon_2 - d\varepsilon_3)^2 + (d\varepsilon_3 - d\varepsilon_1)^2\right]^{1/2}$$

where ε_1, ε_2, ε_3 are the principal plastic strain increments. Von Mises' yield criterion for a perfectly plastic material then states that yielding occurs whenever the effective stress σ^* reaches a certain value k, a constant for the material. This criterion is observed to be widely applicable for metals when the hydrostatic component of the stress is not large compared with the effective stress, although some metals follow more closely the Tresca criterion according to which yielding occurs when the maximum shear stress $(\sigma_1 - \sigma_3)/2$ reaches a fixed value for a given material. It seems likely that the von Mises criterion could also be useful for rocks where nondilatant ductile behavior is involved at low temperature and moderate pressure, and that the strain-hardening extensions of the theory could even be useful at higher temperatures where steady state is not reached. However, there are few observational data for testing this application (see, for example Robertson 1955, for marble).

In extending the approach of the theory of plasticity to creep, it is often assumed, following Odqvist (1935), that creep under general stress states can also be expressed in terms of relationships between the effective stress σ^* and the effective plastic strain rate $\dot{\varepsilon}^*$,

$$\dot{\varepsilon}^* = \left(\sqrt{2}/3\right)\left[(\dot{\varepsilon}_1 - \dot{\varepsilon}_2)^2 + (\dot{\varepsilon}_2 - \dot{\varepsilon}_3)^2 + (\dot{\varepsilon}_3 - \dot{\varepsilon}_1)^2\right]^{1/2}$$

($\dot{\varepsilon}_1$, $\dot{\varepsilon}_2$, $\dot{\varepsilon}_3$ are the principal plastic strain rates), that are similar in form to the relationships found between σ and $\dot{\varepsilon}$ in uniaxial tests; however, in some cases the maximum shear stress and maximum shear-strain rate are found to be more suitable as "effective" stress and strain rate, and in other cases more elaborate forms have been proposed (see brief review by Finnie and Abo el Ata, 1971). An example of this approach is that of Nye (1953) for the case of high temperature steady-state creep of ice, which may also be applicable to rocks in particular circumstances, such as where the range of pressure is not great and solution transfer effects are not involved. Nye assumed that there is a universal steady-state relationship $\dot{\varepsilon}^* = f(\sigma^*)$ of the same form as that established in a uniaxial experiment, namely, $\dot{\varepsilon}^* = A(\sigma^*)^n$, where A and n are constants (note that the effective strain rate and stress used by Nye differ by numerical factors from those given above). This universal relationship is then incorporated in an application of the theory of plasticity analogous to its extension from perfectly plastic material to one with isotropic strain hardening for which a universal stress–strain curve is introduced (Malvern 1969, p. 364); the position- and time-dependent scalar multiplier $d\lambda$ of the theory of plasticity now becomes $d\lambda = 3f(\sigma^*)/2\sigma^*$. It has also been suggested that such a procedure can be extended to cover transient creep as well, and general relationships such as $\varepsilon^* = C(\sigma^*)^{n'}(\varepsilon^*)^{-p}$, $\dot{\varepsilon}^* = C'(\sigma^*)^n\phi(t)$, and $\dot{\varepsilon}_{ij} = \sigma_{ij}F(J_2)\phi(t)$ have been proposed, where C, C', n, n', p are constants, $\phi(t)$ is a function of time, equal to unity for steady-state creep, $F(J_2)$ is a function of

$J_2 = (1/3)(\sigma^*)^2$ and $\dot{\varepsilon}_{ij}$, σ_{ij} are components of the general plastic strain rate and stress tensors, respectively (see, for example, Johnson et al. 1962; Smith and Nicolson 1971). However, little has been done to extend these approaches to anisotropic materials.

The procedures just described, which assert in general that the flow behavior is determined entirely by the shear or deviatoric stress components, are based on many observations showing that the mean stress or hydrostatic component has little influence in plastic deformation. These observations have been mainly on metals and mainly at low temperatures. However, even with metals it is known that there is a small increase in flow stress with increase in mean stress, which becomes more obvious when the mean stress is large compared with the shear stresses. Further, it has been observed that a significant influence of mean stress can appear in high-temperature creep of metals when cavity growth occurs (Dyson et al. 1981; Lonsdale and Flewitt 1977; Needham and Greenwood 1975). Similar pressure effects have been observed in polymers and glasses, where they are often rationalized in terms of a concept of "free volume" in the structure. Finally, in granular flow of soils and cataclastic flow of rocks, it is always necessary to take account of the role of the pressure component, this is being covered in the simplest case by the Coulomb failure criterion (Jaeger and Cook 1976, p. 95; Paterson and Wong 2005, p. 24). The factor common to all these cases is that effects associated with volume change during the flow can no longer be neglected. This situation tends to occur in a number of situations with rocks, so that the hydrostatic component of the stress will often have to be taken into account in treating flow under general states of stress, as in: (1) low-temperature cases where some cataclastic component of flow is involved, as already indicated; (2) cases involving changes in porosity which may include generalized types of high-temperature cavitation, even under predominantly compressive conditions when fluids are present; (3) situations in the lower crust and mantle of the earth where the hydrostatic component of the stress becomes very large relative to the deviatoric components and influences the elastic constants significantly.

The role of pressure is commonly incorporated in the failure criteria and flow laws for general states of stress at high temperature by assuming that it can be expressed through an exponential multiplying factor $\exp(-pV^*/RT)$ as in uniaxial tests (see Sect. 4.4.2). This multiplying factor can be applied respectively to the equivalent viscosity and to the universal strain rate versus stress relationship in the two simple approaches for general stress states, mentioned above. More generally, in the plasticity theory approach, the pressure dependence can be allowed for by postulating a form of yield surface which, instead of being a cylinder parallel to an axis equally inclined to the three principal stress axes in stress space, is a cone that opens in the direction of increasing compressive stresses in a way expressing the appropriate linear (Coulomb), exponential, or other form of pressure dependence.

4.7 Mechanistic Approach

In the following chapters, we turn from the macroscopic or continuum view of deformation to the microscopic or discontinuum view. We shall now consider the elementary processes whereby the deformation takes place physically, that is, the deformation mechanisms. The study of mechanisms is involved with the real structure of the material, that is, with the nature of the building blocks or elementary structural units of which the material is made, the way in which these units are assembled and interact, and the way in which the assemblage is affected by the deformation.

In practice, there is a hierarchy of structural units at different scales to be considered but for a particular deformation mechanism we can usually distinguish structural units at a particular scale that can be regarded as the fundamental flow units for that mechanism. Depending on the scale, there are three general categories of flow units (Paterson 1979):

1. Atomic: individual atoms or molecules.
2. Intragranular: especially the intracrystalline blocks or glide packets that slide over each other in crystallographic slip.
3. Granular: individual grains, clastic fragments or particles.

The identification of the principal flow units and their pattern or relative movement defines the *microgeometry* of the deformation.

The *microdynamics* of the deformation involves the dynamical interactions between the flow units. These interactions can generally be analyzed in terms of processes in the zones of contact between the flow units, for which it is sometimes useful to introduce finer-scale structural entities, such as dislocations or other crystal defects. Thus, the microdynamical treatment may variously concentrate on such effects as friction at particle contacts, the dynamics of dislocation multiplication and propagation, the accommodation of geometrical impediment between flow units, or the kinetics of diffusion or solution and redeposition, and it will entail establishing which is the rate controlling process in a particular situation.

The variety in types of flow units and of interaction between them naturally leads to there being a wide variety of possible deformation mechanisms. However, in dealing with rocks and minerals, this range can be usefully divided into three categories corresponding to the categories of flow units listed above:

1. Atomic transfer flow: This may involve solid-state diffusional flow as in change of shape by material transfer by inter- or intra-crystalline diffusion, or viscous flow in amorphous material; point defects may be important for the diffusion process. Alternatively, it may involve solution-transfer processes, involving change of shape by material transfer via a fluid phase, at high or at low temperature.
2. Crystal plasticity: Intragranular deformation by slip and twinning, much studied in metals and other ductile materials over a wide range of temperatures; involves crystal dislocations in a variety of ways.

3. Granular flow: This may occur with low cohesion between the flow units as typically in soils and poorly cemented sedimentary rocks at low temperatures and includes cataclastic deformation of rocks; inter-particle interference and friction, fracturing of particles and dilatancy are important factors. Alternatively, it may occur where there is relatively high cohesion between the units, typically at high temperature, and including "superplastic" behavior and, at least in a geometric sense, some materials with small amounts of partial melt; accommodation mechanisms that satisfy intergranular strain compatibility are important.

The following chapters deal in turn with these three categories.

References

Amin KE, Mukherjee AK, Dorn JE (1970) A universal law for high-temperature diffusion controlled transient creep. J Mech Phys Solids 18:413–426

Andrade ENdaC (1910) On the viscous flow of metals and allied phenomena. Proc Roy Soc (London) A84:1–12

Argon AS (1973) Stability of plastic deformation. In: The Inhomogeneity of Plastic Deformation. American Society for Metals, Metals Park, Ohio, 161–189

Backhofen WA (1972) Deformation processing. Addison Wesley, London 326 pp

Benioff H (1951) Earthquakes and rock creep, part 1. Bull Seismol Soc Amer 41:31–62

Biot MA (1965) Mechanics of incremental deformations. Wiley, New York, 504 pp

Bird JE, Mukherjee AK, Dorn JE (1969) Correlations between high-temperature creep behaviour and structure, In: Quantitative relation between properties and microstructure; international conference, July 27–Aug 4, Haifa. Israel Univ Press, Haifa, 255–342

Considère A (1885) L'emploi du fer et de l'acier dans les constructions. Ann des Ponts et Chaussées 9:574–775

Di Persio J, Escaig B (1984) Dislocation cores in molecular crystals. Dislocations, Paris, Editions du CNRS, 267–282

Drucker DC (1960) Plasticity. In: Proceedings 1st symposium on naval structural mechanics, Pergamon, 407–448

Dyson BF, Verma AK, Szkopiak ZC (1981) The influence of stress state on creep resistance: experiments and modelling. Acta Metall 29:1573–1580

Evans B, Wong T-F (1985) Shear localization in rocks induced by tectonic deformation. In: Bazant Z (ed) Mechanics of geomaterials. rocks, concretes, soils, John Wiley, 189–210

Finnie I, Abo El Ata MM (1971) Creep and creep rupture of copper tubes under multiaxial stress. In: Smith AI, Nicolsoned AM (eds) Advances in creep design. Applied Science Publishers, Ltd, London, 329–352

Fressengeas C, Molinari A (1987) Instability and localization of plastic flow in shear at high strain rates. J Mech Phys Solids 35:185–211

Garofalo F (1965) Fundamentals of creep and creep-rupture in metals. Macmillan, New York, 258 pp

Griggs DT (1939) Creep of rocks. J Geol 47:225–251

Hart BS, Flemings PB, Deshpanda A (1995) Porosity and pressure: role of compaction disequilibrium in the development of gas pressures in a Gulf Coast Pleistocene basin. Geology 23:45–48

Hill R (1950) The mathematical theory of plasticity. Clarendon Press, Oxford, 355 pp

Hoff NJ (1958) A survey of the theories of creep buckling. In: Proceedings of the 3rd US national congress of applied mechanics, held at Providence RI, June 11–14, New York, ASME, 29–49

Hoffman O, Sachs G (1953) Introduction to the theory of plasticity for engineers. McGraw-Hill Book Co, New York, 276 pp

Jaeger JC (1962) Elasticity, fracture and flow, 2nd edn. Methuen, London, 208 pp

Jaeger JC (1969) Elasticity, fracture and flow (3rd ed), Methuen, London, 268 pp

Jaeger JC, Cook NGW (1976) Fundamentals of rock mechanics (2nd ed), Chapman and Hall, London, 585 pp

Jaeger JC, Cook NGW, Zimmerman RW (2007) Fundamentals of rock mechanics, 4th edn. Blackwell Publishing, Oxford 475

Jeffreys H (1958) A modification of Lomnitz's law of creep in rocks. Geophys J R Astr Soc 1:92–95

Johnson AE, Henderson J, Khan B (1962) Complex-stress creep, relaxation and fracture of metallic alloys. HMSO, Edinburgh, 76 pp

Jonas JJ, Holt RA, Coleman CE (1976) Plastic stability in tension and compression. Acta Metall 24:911–918

Jonas JJ, Luton MJ (1978) Flow softening at elevated temperature. In: Advances in deformation processing. Proceedings of 21st sagamore army materials research conference Aug 1974, Plenum Press, New York, pp 215–243

Kestin J, Bataille J (1980) Materials with memory. An essay in thermodynamics. J Non-Equilib Thermodynamics 5:19–33

Kocks UF (1975) Constitutive relations for slip. In: Argon AS (ed) Constitutive equations in plasticity. MIT Press, Cambridge, Mass, pp 81–115

Lomnitz C (1956) Creep measurements in igneous rocks. J Geol 64:473–479

Lonsdale D, Flewitt PEJ (1977) Effect of hydrostatic-pressure on microstructural changes which occur during creep in alpha-iron. Scripta Met 11:1119–1125

Malvern LE (1969) Introduction to the mechanics of a continuous medium. Englewood Cliffs, Prentice-Hall, New Jersy, 713 pp

Means WD (1976) Stress and strain. Basic concepts of continuum mechanics for geologists. Springer, New York, 339 pp

Michelson AS (1917) The laws of elastic-viscous flow, part 1. J Geol 25:405–410

Michelson AS (1920) The laws of elastic-viscous flow. part 2. J Geol 28:18–24

Misra AK, Murrell SAF (1965) An experimental study of the effect of temperature and stress on the creep of rocks. Geophys J R Astr Soc 9:509–535

Mukherjee AK (1975) High-temperature creep. In: Herman H (ed) Plastic deformation of materials. Treatise on materials science and technology, vol 6. Academic Press, New York, pp 163–224

Needham NG, Greenwood GW (1975) The creep of copper under superimposed hydrostatic pressure. Metal Sci 9:258–262

Nye JF (1953) The flow law of ice from measurements in glacier tunnels, laboratory experiments and the Jungfraufirn borehole experiment. Proc Roy Soc (London) A219:477–489

Nye JF (1957) Physical properties of crystals. Clarendon Press, Oxford, 322 pp

Odqvist FKG (1935) Creep stresses in a rotating disc. In: Proceedings of the 4th international congress applied mechanics, Cambridge University Press, Cambridge, 228–229

Paterson MS, Weiss LE (1968) Folding and boudinage of quartz-rich layers in experimentally deformed phyllite. Geol Soc Am Bull 79:795–812

Paterson MS (1979) The mechanical behaviour of rock under crustal and mantle conditions. In: McElhinny MN (ed) The Earth: its origin, structure and evolution. Academic Press, London, pp 469–489

Paterson MS, Olgaard DL (2000) Rock deformation tests to large shear strains in torsion. J Structural Geol 22:1341–1358

Paterson MS, Wong T-f (2005) Experimental rock deformation—the brittle field 2nd ed. Springer, Berlin, 347 pp

Paterson MS (2007) Localization in rate-dependent shearing deformation, with application to torsion testing. Tectonophysics 445:273–280

Phillips P (1905) The slow stretch in India rubber, glass, and metal wires when subjected to a constant pull. Phil Mag (6th ser), 9:513–531

Poirier J-P (1976) Plasticité à Haute Température des Solides Cristallins, Paris, Editions Eyrolles, 320 pp

Poirier JP (1980) Shear localization and shear instability in materials in the ductile field. J Structural Geol 2:135–142

Prager W, Hodge PG (1951) Theory of perfectly plastic solids. Wiley, New York

Prigogine I (1967) Introduction to thermodynamics of irreversible processes (3rd edn). Interscience Publishers, New York, 147 pp

Reid CN (1973) Deformation geometry for material scientists. Pergamon Press, Oxford, 211 pp

Rice JR (1970) On the structure of stress-strain relations for time-dependent plastic deformation of metals. J Appl Mech, Trans ASME 37:728–733

Rice JR (1975) Continuum mechanics and thermodynamics of plasticity in relation to microscale deformation mechanisms. In: Argon AS (ed) Constitutive equations in plasticity. The MIT Press, Cambridge Mass, 23–79

Robertson EC (1955) Experimental study of the strength of rocks. Geol Soc Am Bull 66:1275–1314

Shawki TG, Clifton RJ (1989) Shear band formation in thermal viscoplastic materials. Mech Mater 8:13–43

Smith AI, Nicolson AM (eds) (1971) Advances in creep design, Applied Science Publisher, London, 485 pp

Zener C (1948) Elasticity and anelasticity of metals. University of Chicago Press, Chicago, 170 pp

Ziegler H (1977) An introduction to thermomechanics. North-Holland Publishing Co, Amsterdam, 308 pp

Chapter 5
Deformation Mechanisms: Atomic Transfer Flow

5.1 A General Model

In this category of deformation mechanisms we are concerned with processes in which individual atoms or small groups of associated atoms are removed from certain interfaces or discontinuities within the structure of the body (sources) and are transferred to other interfaces or discontinuities (sinks) in such a way that the overall shape of the body is changed, that is, the body undergoes macroscopic strain. The sources and sinks may be dislocation cores, planar crystal defects, grain boundaries or free internal or external surfaces, and the transfer may take place by a variety of mechanisms, including solid state diffusion (intra- or intergranular) and transfer via a fluid phase (in case of a porous or partially melted body). The overall kinetics may be controlled by the kinetics of the transfer process or by the kinetics of the detachment and re-attachment processes. We have used the term "atomic transfer flow" in introducing this class of mechanisms in order to emphasize that the transfer occurs more or less atom by atom rather than by the movement of relatively large blocks of atoms; however, the term "diffusion creep" is commonly used in the same sense, especially when it is wished to emphasize diffusion as the transfer process or as being rate controlling.

Atomic transfer flow is essentially a solid state process in that the atoms detached or re-attached belong to solid grains, even though the transfer may in some cases occur via a fluid medium. This type of flow is therefore to be distinguished fundamentally from viscous flow in fluids where, although individual atom movement is again involved, the atoms do not have defined sites in sources and sinks and the resistance to flow derives from the process of momentum transfer as the atoms move about. In Newtonian viscous flow in a fluid, the flux of stream momentum is proportional to the gradient in stream velocity of the particles and so the stress, deriving from the rate of change of momentum, is proportional to the strain rate (Cottrell 1964, p. 26). The viscosity associated with momentum transfer in the diffusive movement of atoms in a solid would be enormous and so

M. S. Paterson, *Materials Science for Structural Geology*,
Springer Geochemistry/Mineralogy, DOI: 10.1007/978-94-007-5545-1_5,
© Springer Science+Business Media Dordrecht 2013

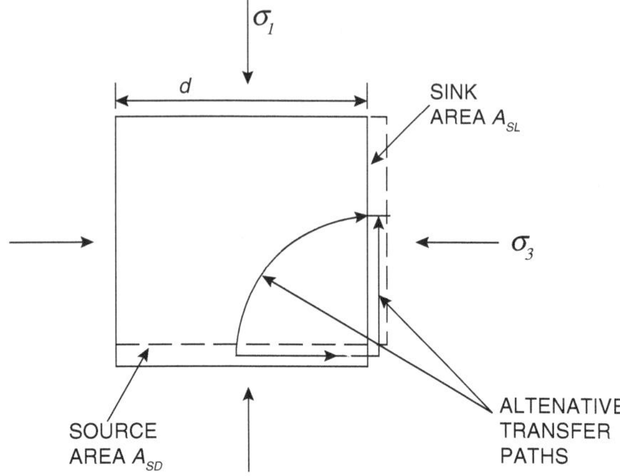

Fig. 5.1 Schematic representation of a single domain consisting of one source-sink pair undergoing shape change by atomic transfer flow

flow in a solid by a process analogous to that in a viscous fluid is normally negligible.

As might be expected for a process that involves the relative displacement of individual atoms rather than large groups of atoms, atomic transfer flow tends to be more important at relatively low stresses, insufficient to initiate the displacement of the large groups but sufficient to bias the motion of the individual atoms. It will be seen later that atomic transfer flow can be of practical importance at high temperatures when solid-state diffusion is the transport mechanism and at moderate or low temperatures when fluid transfer is involved.

The various types of atomic transfer flow can be classified according to the nature of the sources and sinks and of the transfer process. Several that may be relevant for rocks are set out in subsequent sections. However, underlying all is a common theoretical framework that we first set out.

We consider the macroscopic strain of a body that result from the transfer of material from source to sink via a well-defined path under uniaxial stress (Fig. 5.1). The model is defined in terms of the following elements and symbols:

1. The stress-supporting part of the body is considered, at least for preliminary discussion, to consist of a single component of molar volume V_m.
2. The sources and sinks, regarded as surfaces of area A_s, are characterized by a rate constant k_s governing the rate of any reaction required to effect detachment or re-attachment of material and a numerical parameter α_s representing the fraction of surface sites in source or sink at which the detachment or re-attachment reaction can occur. (The suffixes "*so*" and "*si*" are used to distinguish source and sink, respectively).
3. The transfer path, of cross-sectional area A_t, is assumed to consist of a medium in which the species being transferred has a diffusion coefficient D and solubility c ($c = 1/V_m$ when no other substance exists in the path).

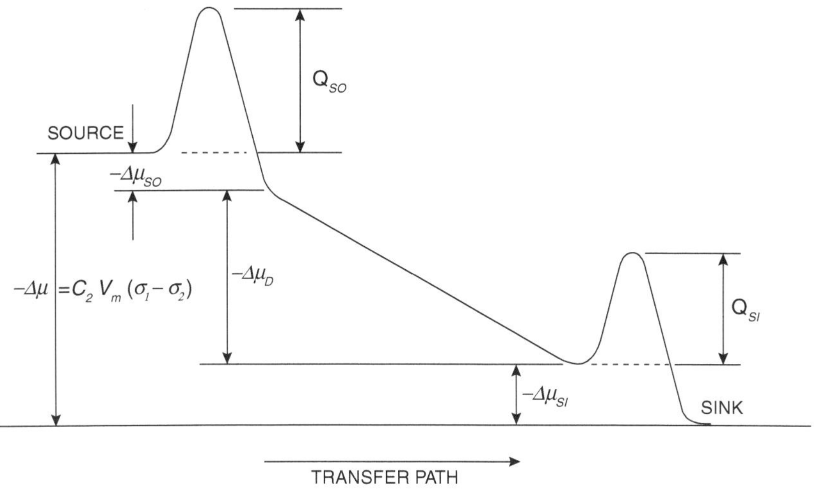

Fig. 5.2 Components of thermodynamic potential difference driving transfer of material from source to sink

The relative shortening of the body, represented by that of the domain in Fig. 5-1, is of the order of nV_m/V where n is the amount of substance transferred (in moles) and V is the volume of the domain. The macroscopic strain rate $\dot{\varepsilon}$ in the direction of σ_1 is then given by

$$\dot{\varepsilon} = C_1 \frac{V_m}{V}\frac{dn}{dt} \tag{5.1}$$

where C_1 is a dimensionless geometrical factor, generally >1, that takes into account the actual orientations of sources and sinks and the actual relative movements of the domains required to fit them together again after transferring the material.

The overall thermodynamic potential difference $-\Delta\mu$ driving the transfer of material is given by $-\Delta\mu = C_2 V_m(\sigma_1 - \sigma_3)$ where μ is the chemical potential and C_2 is another dimensionless geometrical factor, generally <1, taking into account the actual orientations of the sources and sinks (see Sect. 3.2 on sign of $-\Delta\mu$), and $\sigma_1 - \sigma_3$ is the applied stress difference. The quantity $-\Delta\mu$ is made up of three parts (Fig. 5.2), the potential differences $-\Delta\mu_{so}$ and $-\Delta\mu_{si}$ required to drive any reactions involved in releasing or attaching material at source and sink, respectively, and the potential difference $-\Delta\mu_D$ required to drive the diffusion in the transfer path. Thus, we can write

$$(-\Delta\mu_{so}) + (-\Delta\mu_{si}) + (-\Delta\mu_D) = C_2 V_m(\sigma_1 - \sigma_3) \tag{5.2}$$

We now relate dn/dt separately to the reactions at source and sink and to the diffusion in the transfer path. In the case of the reactions a formulation analogous to (3.5) can be used under the assumption that the reaction will be first order, with a rate proportional to the number of sites in source or sink at which atoms or molecules can potentially be readily detached or attached, respectively; that is, we assume

$$\frac{\mathrm{d}n}{\mathrm{d}t} = \frac{\alpha_s A_s b}{V_m} k_s = \frac{A_s k}{V_m} \tag{5.3}$$

where b is the thickness of a monolayer, and k_s is the rate coefficient at the potential site. We can write $\alpha_s b k_s = k$ since it is only this product that is directly measured. Tsai and Raj (1982b) introduce a rate proportional to the molar concentration of the diffusing species instead of to the number of reaction sites but, as they point out, only the product of this term with the rate coefficient is measurable and so empirically the two models cannot be distinguished from deformation experiments alone.

Returning to (5.3) and using an argument analogous to that used in deriving (3.12), we can now write

$$k_s = \frac{\bar{k}_s(-\Delta\mu_s)}{RT}$$

under the assumption that $-\Delta\mu_s \ll RT$; here the $\bar{k}_s$ are of the form $A\exp(-Q/RT)$ where the A are constants with dimensions of frequency and the Q's are activation energies for the reactions. Then the (5.3) become

$$\frac{\mathrm{d}n}{\mathrm{d}t} = \frac{b\alpha_{so}A_{so}\bar{k}_{so}}{V_m RT}(-\Delta\mu_{so}) \tag{5.4a}$$

$$\frac{\mathrm{d}n}{\mathrm{d}t} = \frac{b\alpha_{si}A_{si}\bar{k}_{si}}{V_m RT}(-\Delta\mu_{si}) \tag{5.4b}$$

at source and sink, respectively. In the case of diffusion, (3.28) can be used to give, in the ideal case,

$$\frac{\mathrm{d}n}{\mathrm{d}t} = jA_t = A_t\frac{cD}{RT}\frac{(-\Delta\mu_D)}{l}$$

or, using (5.2),

$$\frac{\mathrm{d}n}{\mathrm{d}t} = cD\frac{A_t}{l}\frac{[C_2 V_m(\sigma_1 - \sigma_3) - (-\Delta\mu_{so}) - (-\Delta\mu_{si})]}{RT} \tag{5.5}$$

Two situations must now be considered:

1. If $\Delta\mu_{so}$ and $\Delta\mu_{si}$ can be treated as dependent variables, they can be eliminated from (5.4) to (5.5) and the resulting value of $\mathrm{d}n/\mathrm{d}t$ substituted in (5.1) to give

$$\dot{\varepsilon} = \frac{C' V_m cD}{1 + V_m cD\frac{A_t}{bl}\left(\frac{1}{\alpha_{so}A_{so}k_{so}} + \frac{1}{\alpha_{si}A_{si}k_{si}}\right)}\frac{A_t}{lV}\frac{V_m(\sigma_1 - \sigma_3)}{RT} \tag{5.6}$$

where $C' = C_1 C_2$ is a dimensionless constant. In order to bring out the physical content of (5.6) it is simplest to consider the two limiting cases.

a. *Diffusion-controlled case.* When the rate coefficients k_s are sufficiently large relative to the diffusion coefficient D that the bracketed term in the denominator sum can be neglected, the diffusion in the transfer path is rate limiting and the creep rate is

$$\dot{\varepsilon} = C' V_m c \frac{D}{lV/A_t} \frac{V_m(\sigma_1 - \sigma_3)}{RT} \tag{5.7}$$

Expressing l, V and A_t in terms of grain and grain boundary dimensions leads to Nabarro-Herring, Coble and fluid transfer diffusion creep laws, as will be shown later.

b. *Reaction-controlled case.* When the diffusion coefficient is sufficiently large relative to the rate coefficients so that the bracketed term in the denominator sum of (5.6) is much greater than unity, the reactions are rate controlling and the creep rate is

$$\dot{\varepsilon} = C' \frac{bk}{V/A_s} \frac{V_m(\sigma_1 - \sigma_3)}{RT} \tag{5.8}$$

where we have, for illustration, assumed that the reaction at one of the source-sink pair is much slower than at the other and we have written $k = \alpha_s \bar{k}_s$, an effective rate coefficient with the dimensions of frequency.

2. If the $\Delta\mu_s$ have independently fixed values, (5.4) and (5.5) must be independently substituted in (5.1) to calculate virtual strain rates, the minimum of which will be the realizable strain rate. It immediately follows that a positive strain rate can only be obtained if

$$C_2 V_m(\sigma_1 - \sigma_3) > (-\Delta\mu_{so}) + (-\Delta\mu_{si})$$

that is, that there will be a threshold stress for creep since $(-\Delta\mu_{so})$ and $(-\Delta\mu_{si})$ must both be positive for source and sink to function as such (cf. Ashby 1969). At stresses immediately above the threshold stress the diffusion will be rate controlling because of the relatively small value of $-\Delta\mu_D$ but at stresses above a certain higher level the reaction rate at source or sink will become rate controlling and the strain rate will become independent of further increase in stress.

Before proceeding to consider more specific models of atomic transfer flow, attention should be drawn to some general aspects of this class of deformation mechanisms:

1. In the case of ionic substances or multicomponent bodies the individual ions or components may possibly travel along different transfer paths and tend to have

different transfer rates. However, if no compositional change or segregation occurs, then from a macroscopic or thermodynamic point of view only the molecular species or an average substance need be considered, to which an effective diffusion coefficient can be attributed. If, on the other hand, the deformation accompanies a segregation or differentiation of the body, as may well happen in geological deformations, then the diffusion of the individual components has to be considered separately and the flow equations adapted appropriately.

2. In the expressions (5.6–5.8) the stress dependence is linear or Newtonian as long as there is no stress dependence in the other parameters. However, non-linear stress dependence could, in principle, arise if any of the dimensional parameters l, A_t, α, A_s associated with the diffusion path or reaction zones were to be stress dependent. Also steady state behaviour, or constancy of strain rate at constant stress, will only appear to the extent that these dimensional parameters remain constant over the duration of the flow.

3. It is insufficient to discuss only the transfer of material from sources to sinks without considering how the parts of the body fit together after the transfer, as has already been indicated in defining the constant C_1 in (5.1). We must therefore now consider the nature and implications of these compatibility constraints.

5.2 Compatibility Considerations

In general, the sources and sinks will need to constitute a more or less continuous system of interfaces that will define the parts or domains of the body (commonly the grains of a polycrystal) the fitting together of which will have to be maintained during the deformation if constancy of volume is to be maintained. In addition, both a correlation in the relative rates of material removal or emplacement at different points in the source/sink system and a correlation in the relative motion of the parts of the body that accompany the material transfer are required to maintain the fitting together of the parts. These constraints on the geometry of the source/sink system and on the correlations in the transfer and relative displacement constitute the compatibility conditions for the deformation mechanism. The general framing of appropriate compatibility conditions does not yet seem to have been set out formally for diffusion creep although the qualitative requirements are fairly obvious. In the formal framing, generalized dislocation notions may be useful, involving Burgers vectors with components normal to the source or sink interfaces, representing the material transfer, and components parallel to the interfaces, representing sliding at the interfaces required to satisfy the compatibility conditions. The details of the consequences of the compatibility constraints will depend on the specific models but where grain boundaries are the source/sink system some sliding at grain boundaries will be entailed.

Compatibility conditions will normally contain a presumption of constancy of volume apart from elastic effects. Failure to meet the compatibility conditions will tend to lead to the formation or elimination of voids, that is, to volume changes and hence to a much stronger pressure dependence in the flow law than might otherwise appear (the diffusion coefficient would normally introduce only slight pressure dependence, although appreciable pressure dependence may arise from the reaction rate in case of reaction-controlled flow where, for example, solubility in a fluid is a parameter in the reaction kinetics).

Unless the geometry of the source/sink system is suitably specialized, compatibility conditions will require local variability in the rate at which material is transferred; for example, where an asperity or change in orientation tends to obstruct geometrically necessary sliding on a grain boundary, the rate of transfer will need to be accelerated locally in order to accommodate the sliding. Such variability is unlikely to be consistent with a homogeneous stress distribution, in which the only variation in normal stress component across given sources and sinks arises from variations in orientation, and therefore the stress distribution will necessarily be heterogeneous on the domain or grain scale and on finer scales. If the local strain rate is linear in the stress difference, a theoretical treatment in terms of a homogenous stress, such as given in this and subsequent sections may represent a suitable averaging procedure for many purposes but it could contain more serious error in non-linear cases or in describing specific local aspects of linear cases.

It is also to be emphasized that the maintenance of fit between domains through the compatibility conditions imposes such an intimate interdependence between the shape changes of the individual domains and the relative displacements of the domains that these two effects must be regarded as two aspects of the same basic process. The sliding at interfaces is thus an integral part of the atomic transfer deformation process, which can be viewed equally as based on the material transfer or the relative domain displacement aspects. This equivalence has been stated particularly clearly by Lifshitz (1963) and by Raj and Ashby (1971).

5.3 Nabarro-Herring Creep

The model for atomic transfer or diffusion creep first put forward was that based on grain boundaries as the sources and sinks and on volume diffusion through the grains as the transfer mechanism, with diffusion control, propounded by Nabarro (1948) and Herring (1950). The Nabarro-Herring creep law can be obtained from (5.7) by taking the volume of the grains, V, to be or order d^3 and the cross-sectional area and length of the diffusion path, A_t and l, to be of order d^2 and d, respectively, so that $lV/A_t \sim d^2$. Then, with $c = 1/V_m$, (5.7) becomes

$$\dot{\varepsilon} = C_{\mathrm{NH}} \frac{V_m D_v}{RT} \frac{\sigma_1 - \sigma_3}{d^2} \tag{5.9}$$

where the numerical constant C_{NH} now contains also the proportionality constants relating V, A_t and l to the powers of d and D_V is the bulk or volume diffusion coefficient. Values of C_{NH} of around 12–14 have been calculated for particular grain geometrics by Herring (1950), Raj and Ashby (1971) and others; for a summary, see Poirier (1976, 1985) who also discusses the alternative presentation of the theory in terms of shear strain rate and shear stress, in which case C_{NH} is multiplied by 3.

As an illustration of the typical application of (5.9), $\dot{\varepsilon} \approx D_V/d^2$ if we assume $V_m = 10^{-4}$ m^3, $T = 1200$ K and $\sigma_1 - \sigma_3 = 10$ MPa. If, further, $D_V = 10^{-18}$ m^{2}s^{-1}, then $\dot{\varepsilon} = 10^{-6}$s^{-1} for $d = 1\,\mu$m and $\dot{\varepsilon} = 10^{-12}$s^{-1} for $d = 1$ mm.

For metals it is common to formulate the theory in terms of vacancy diffusion (Poirier 1976, 1985), which presumes a vacancy diffusion mechanism for the atomic transfer process. In the case of compound substances, however, as in minerals and rocks, it would seem more appropriate to discuss the transfer process directly in terms of the compound itself. Several atomic species may be involved but, except where chemical segregation or differentiation is occurring, the creep rate is to be related to the total molecular flux. The relevant diffusion coefficient and molar volume in (5.9) will then be those of the molecular species of which the material consists. In the case of ionic compounds of form $A_\alpha B_\beta$, and since for a pure substance it is the self-diffusion that is involved, the diffusion coefficient of the molecular species will be given by the expression (3.40), that is, by

$$D = \frac{D_A^* D_B^*}{\beta D_A^* + \alpha D_B^*} \tag{5.10}$$

where D_A^* and D_B^* are the self-diffusion coefficients of the cation and anion, respectively; thus, if $D_B^* < < D_A^*$, then $D \approx D_B^*/\beta$ and the creep rate will be determined essentially by the self-diffusion coefficient of the anion. In the case of ceramics, such as beryllium oxide, it has been deduced that the anion diffusion tends, in fact, to be rate-controlling in Nabarro-Herring creep (Gordon 1973, 1975).

In more complex systems, where more than one component is to be recognized, it is possible that a certain amount of creep can occur by selective diffusion of the more mobile components, leading to a differentiation or segregation. However, more complicated compatibility requirements will arise in such cases in connection with accommodating the immobility of the more sluggish components and the selective transport of the others. Such segregation effects have long been used to aid in recognizing diffusion creep, the classic case being that of magnesium $-1/2$ % zirconium alloy in which the re-deposited material is relatively devoid of the slower-diffusing zirconium, as shown by the distribution of zirconium hydride precipitates after hydrogen treatment (Squires et al 1963); also see Poirier (1976, 1985) for this and other examples.

5.4 Coble Creep

If the grain boundaries act as the transfer paths as well as being the sources and sinks for diffusion creep in a polycrystalline material, with the diffusion rate-controlling, the mechanism is referred to as Coble creep (Coble 1963). The Coble creep law can be obtained from (5.7) in a similar way to the Nabarro-Herring law (5.9), with $V \sim d^3$ and $l \sim d$ but now with the cross-sectional area of the order $d\delta$ where δ is the effective width of the grain boundary for grain boundary diffusion, so that $lV/A_t \sim d^3/\delta$. Then with $c = 1/V_m$, (5.7) becomes

$$\dot{\varepsilon} = C_{\mathrm{CO}} \frac{V_m D_{\mathrm{GB}} \delta}{RT} \frac{\sigma_1 - \sigma_3}{d^3} \tag{5.11}$$

where D_{GB} is the grain boundary diffusion coefficient and C_{CO} is a numerical constant differing somewhat from C_{NH} in (5.9) because of the different geometry of the transfer path, C_{CO} being about three times greater that C_{NH}. Values of C_{CO} of around 45 have been calculated by Coble (1963) and by Raj and Ashby (1971) for particular assumed geometries meeting compatibility requirements.

As an illustrative example for comparison with the previous Nabarro-Herring example, again taking $V_m = 10^{-4} \mathrm{m}^3$, $T = 1200\,\mathrm{K}$ and $\sigma_1 - \sigma_3 = 10\,\mathrm{MPa}$ and assuming $\delta = 1\mathrm{nm}$, a value of $D_{\mathrm{GB}} = 10^{-12}\,\mathrm{m}^2\mathrm{s}^{-1}$ would give $\dot{\varepsilon} \approx 10^{-3}\mathrm{s}^{-1}$ for $d = 1\,\mu\mathrm{m}$ and $\dot{\varepsilon} \approx 10^{-15}\mathrm{s}^{-1}$ for $d = 1\mathrm{mm}$. Thus laboratory experiment on fine-grained materials will tend to emphasize Coble creep even where Nabarro-Herring creep might be more favoured with coarser grain sizes geologically.

Since both the material transfer and the relative grain movements associated with the two types of diffusion creep are additive, the total creep rate can be obtained by summing (5.9) and (5.11) in cases where both are contributing significantly. For the contributions of Nabarro-Herring and Coble creep to be similar, D_{GB} has to be about three orders of magnitude greater than D_V in the case of 1 μm grain size, or six orders of magnitude greater in the case of 1 mm grain size, if δ is of the order of 1 nm. Diffusion measurements (Sect. 3.5.5) indicate that a ratio of six orders of magnitude between D_{GB} and D_V is realistic but the ratio will depend on temperature, diminishing at higher temperatures because of lower activation energies for grain boundary diffusion. It has been noted that the cation diffusion is often rate controlling in grain boundary diffusion and hence in Coble creep, in contrast to the control by the anions in Nabarro-Herring creep (Gordon 1973, 1975).

5.5 Fluid-Transfer Diffusion Creep

If material is transferred from sources to sinks by diffusion through a fluid phase existing in intergranular cavities or cracks, and if the transfer process is rate controlling, the creep relation should be of the same form as for Coble creep,

(5.11), except that now δ becomes the mean thickness of the intergranular film (or δ/d is replaced by the volume fraction ϕ of fluid) and D_{GB} is replaced by $V_m c D_F$ where c and D_F are the molar concentration and diffusion coefficient, respectively, of the material in the fluid ($V_m c$ is the volume fraction of the fluid phase occupied by the diffusing material, here V_m is, strictly, the molar volume of the material in solution rather than in the solid phase but in the present approximation we shall ignore the distinction). Thus for diffusion-limited fluid transfer creep we have the relation

$$\dot{\varepsilon} = C_{FT} \frac{V_m^2 c D_F \delta}{RT} \frac{\sigma_1 - \sigma_3}{d^3} \tag{5.12}$$

where the value of the numerical constant C_{FT} is similar to that for Coble creep. Relations of this form or equivalent to it have been derived by Weyl (1959), Stocker and Ashby (1973), Rutter (1976, 1983), Elliott (1973), Raj and Chung (1981), Raj (1982).

Two limitations on the applicability of (5.12) can be foreseen. First, if the fluid itself is moving through the porous body, the material transfer rate may be modified in a way that depends on the direction and rate of the fluid flow relative to the principal stress directions. Second, because of the more rapid diffusion in liquids or, particularly, the more rapid transport in a moving fluid, the transfer kinetics are less likely to be rate controlling in fluid transfer creep than are the interface kinetics at source and sink. We now consider the situation where the latter are rate controlling.

5.6 Reaction-Controlled Creep

Each of the specific models in the previous three sections has been treated under the assumption that the diffusion from source to sink is the rate-controlling step. We now consider the parallel cases in which the creep rate is controlled by the rate of reaction involved in the release and/or absorption the diffusing species at the source and/or sink, respectively, as first discussed by Ashby (1969) and Greenwood (1970) and subsequently by Raj and Chyung (1981) and others. If the chemical potential difference driving the reaction is determined by the flux from which the strain rate derives, then the general expression (5.8) can be applied. The diffusion path no longer enters into consideration; only the source-sink geometry and the reaction rate are involved.

Taking the sources and sinks to be grain boundaries in all three cases so that $V \sim d^3$ and $A_s \sim d^2$, then $V/A_s \sim d$ in (5.8) and we obtain the creep equation

$$\dot{\varepsilon} = C_R \frac{V_m b k}{RT} \frac{\sigma_1 - \sigma_3}{d} \tag{5.13}$$

In order to use such a relationship the rate coefficient k must be independently known (see Tsai and Raj 1982b for a calculation of a creep rate using an independently determined rate coefficient). It will be noted that, provided k and the grain size d are independent of the stress, the creep rate is linear or Newtonian in the stress and is inversely proportional to the grain size, in contrast to the d^{-2} and d^{-3} dependence of Nabarro-Herring and Coble creep, respectively.

The relationship (5.13) has been developed further by Burton (1972, 1983) under the assumption that, in the absence of fluid phases and chemical reactions, the rate coefficient k is determined by a spiral dislocation growth mechanism within the grain boundaries, according to which $k \propto D(\sigma_1 - \sigma_3)/E_d$ where D is the diffusion coefficient, and E_d the energy per unit length of the grain boundary dislocations. The equation (5.13) then becomes

$$\dot{\varepsilon} = C_B \frac{V_m b D}{E_d RT} \frac{(\sigma_1 - \sigma_3)^2}{d} \tag{5.14}$$

where C_B is a new dimensionless constant incorporating C_R and the proportionality constant in k. This equation illustrates how a non-Newtonian flow law can arise from stress dependence in the kinetic factors.

The numerical constant C_R might be expected to be of similar order to the constants for the Nabarro-Herring and Coble formulae because of similar compatibility requirements, although the formulae of Raj and Chyung (1981) implies that it is of order unity; the most appropriate value does not seem to have been investigated closely. In Burton's model the constant C_B incorporates a factor specifying the density of jogs on the grain boundary dislocation, Burton's estimate of which leads to $C_B \sim 1/10$; Burton also estimates E_d to be one-tenth that for intragranular dislocations, giving $E_d \sim 10^{-9}\,\mathrm{Jm}^{-1}$.

Using $C_R = 10$, $V_m = 10^{-4}\,\mathrm{m^3 mol}^{-1}$ and $RT = 10^4\,\mathrm{Jmol}^{-1}$, (5.13) becomes $\dot{\varepsilon} = 10^{-6}\mathrm{s}^{-1}$ at $\sigma_1 - \sigma_3 = 10\,\mathrm{MPa}$ and $d = 1\,\mu\mathrm{m}$ if $k = 10^{-12}\,\mathrm{ms}^{-1}$ (a measurable rate, falling within the range of those determined in a ceramic system by Tsai and Raj (1982a, 1982b); such a creep rate is comparable to that for Nabarro-Herring creep under these conditions. From (5.6) and using the assumptions underlying (5.9) and (5.13), the cross-over from reaction control at relatively small grain sizes to diffusion control at larger grain sizes is, in fact, to be expected at $d \sim D/k$, that is, at $d \sim 1\,\mu\mathrm{m}$ when $D = 10^{-18}\,\mathrm{m^2 s}^{-1}$ and $k = 10^{-12}\,\mathrm{ms}^{-1}$. Alternatively, if Burton's model is applied, with $C_B = 1/10$, $b = 1\,\mathrm{nm}$ and $E_d = 10^{-9}\,\mathrm{Jm}^{-1}$, and hence $k = 10^{-18}(\sigma_1 - \sigma_3)\,\mathrm{m^2 s}^{-1}$, the cross-over is to be expected at $d = 100/(\sigma_1 - \sigma_3)\,\mathrm{m}$ when $D = 10^{-18}\,\mathrm{m^2 s}^{-1}$, that is, at $d = 10\,\mu\mathrm{m}$ if $(\sigma_1 - \sigma_3) = 10\,\mathrm{MPa}$; below this stress $\dot{\varepsilon} \propto (\sigma_1 - \sigma_3)^2$ and above it $\dot{\varepsilon} \propto (\sigma_1 - \sigma_3)$ until dislocation creep with high stress exponent (Sect. 6.6.6) takes over, thus giving one possible explanation of the sigmoidal shaped log strain rate versus log stress plots sometimes observed around the "superplastic" regime (Burton 1972).

The numerical illustrations given in the previous paragraph are most likely to be relevant to single phase systems without a fluid phase and indicate a marginal importance for reaction control at finer grain sizes and lower stresses. However, in geological systems with fluid phases present, multiple phases and potentially more complicated reactions, reaction control may be relatively more important.

5.7 Dislocation Climb Creep

Dislocations (Chap. 6) can also, in principle, serve as sources and sinks for atomic transfer creep. The addition or removal of material at the dislocation cores corresponds to climb of the dislocations, which can be viewed as the removal of material from, or addition to, the "extra half plane" associated with the edge component of the dislocation. However, dislocations acting as sources must have different Burgers vectors from those acting as sinks in order that the material transfer be effective as a strain mechanism, so there must be a multiplicity of Burgers vectors present. Also the dislocation lines must be of such orientations that they have at least some edge character. Dislocation arrays meeting these requirements will, in general, be three dimensional networks, with the dislocations either more or less randomly distributed or organized into subgrain boundaries.

Theories of dislocation climb creep have, in general, assumed diffusion control. In applying (5.7), the transfer path length l can be taken as being of the order of the average mesh dimension of the dislocation network, that is, $\rho^{-\frac{1}{2}}$ where ρ is the dislocation density, the domain volume V can be taken as of the order of the average mesh volume $\rho^{-\frac{3}{2}}$, and the transfer path cross-sectional area A_t as of the order of ρ^{-1} if volume diffusion is involved. Then, with $c = 1/V_m$, (5.7) becomes

$$\dot{\varepsilon} = C_{DC} \frac{V_m D_V \rho}{RT} (\sigma_1 - \sigma_3) \tag{5.15}$$

where D_V is the volume diffusion coefficient. The numerical coefficient C_{DC} will depend on mean values of the angle between the dislocation lines and the stress direction as well as on other geometric factors similar to those arising for Nabarro-Herring creep. However, the compatibility requirements are somewhat different in character from where grain boundaries are involved as sources and sinks, and grain boundary sliding will no longer contribute to the strain. To achieve an arbitrary strain within a crystal by dislocation climb at least six Burgers vectors must be independently involved (Groves and Kelly 1969, Sect. 6.8.2) and this requirement therefore must probably also be met in polycrystalline deformation, although there will tend to be some heterogeneity in behaviour from grain to grain and within grains. However, the independence of the activity of the climb systems of different Burgers vectors may be put in question, firstly, by any tendency for long range stress fields to build up and lead to interaction between the different groups of dislocations and, secondly and more seriously, by the pinning of dislocations at the nodes of the network.

The last effect will tend to limit severely the amount of strain that can be produced at the initial rate given by (5.15), making the dislocation climb creep a transient effect. However, if a certain threshold stress is exceeded the dislocation segments between the pinning points can act as Bardeen-Herring climb sources for dislocation multiplication (Bardeen and Herring 1952) and unlimited amounts of climb can be produced. The dislocation spacing will still be limited by the need for the externally applied stress to balance the internal stress arising from mutual repulsion of like dislocations. These internal stresses are proportional to the spacing between the dislocations and hence inversely proportional to $\rho^{\frac{1}{2}}$, leading to the dislocation density being of the order of $(\sigma_1 - \sigma_3)^2/(Gb)^2$. Using this relationship in (5.15) leads to

$$\dot{\varepsilon} = C_N \frac{V_m D_V}{RT} \frac{(\sigma_1 - \sigma_3)^3}{(Gb)^2} \tag{5.16a}$$

or, with $V_m \approx Lb^3$,

$$\dot{\varepsilon} = C_N' \frac{b D_V}{G^2 RT} (\sigma_1 - \sigma_3)^3 \tag{5.16b}$$

which is the formula of Nabarro (1967); G is the shear modulus, b the Burgers vector, L the Avagadro constant and C_N, C_N' numerical constants of order 0.01 and 0.01/L respectively.

If the transfer process is pipe diffusion along the dislocation cores, we have $A_t \sim b^2$ instead of ρ^{-1} in (5.7), where b is the Burgers vector, and hence instead of (5.15) we have the initial strain rate

$$\dot{\varepsilon} = C_{\mathrm{DC}/P} \frac{V_m D_P b^2 \rho^2}{RT} (\sigma_1 - \sigma_3) \tag{5.17}$$

where D_P is the pipe diffusion coefficient and $C_{\mathrm{DC}/P}$ is a numerical constant, again probably of the order of magnitude of unity to ten. Dislocation multiplication leading to a steady state dislocation density of the order of $(\sigma_1 - \sigma_3)^2/(Gb)^2$ would correspondingly yield the further formula of Nabarro (1967) for the case of pipe diffusion,

$$\dot{\varepsilon} = C_{N/P}' \frac{b D_P}{G^4 RT} (\sigma_1 - \sigma_3)^5 \tag{5.18}$$

again with $C_{N/P}'$ of the order 0.01.

If we use the parameters $C_N = C_{N/P} = 0.01$, $V_m = 10^{-4}\,\mathrm{m}^3$, $T = 1{,}200$ K, $G = 50\,\mathrm{GPa}$ and $b = 0.5\,\mathrm{mm}$, and we consider a stress $\sigma_1 - \sigma_3 = 10\,\mathrm{MPa}$, the Nabarro formula (5.16b) gives a strain rate of the order of $10^{-10}\,\mathrm{s}^{-1}$ for dislocation climb creep sustained by Bardeen-Herring dislocation multiplication if we assume material transfer by volume diffusion with $D_V = 10^{-18}\,\mathrm{m}^2\mathrm{s}^{-1}$. Alternatively, (5.18) gives a strain rate of the order of $10^{-11}\,\mathrm{s}^{-1}$ if we assume transfer by dislocation core diffusion with $D_P = 10^{-12}\,\mathrm{m^2 s^{-1}}$. Comparison with cases of Nabarro-Herring creep calculated for similar conditions indicates that the two types of diffusion creep give comparable strain rates at 10 MPa stress when the grain size is of the order of 100 μm. However, owing to its stronger stress dependence, the dislocation climb creep can be expected to predominate down to smaller grain sizes at higher stresses. Also, for a given D_P/D_V ratio, the core diffusion mechanism can be expected to be relatively more important as the stress is increased. However, these predictions rest on the presumption that the compatibility requirements are met and that the climb multiplication process is effective. In practice, with minerals, the number of Burgers vectors available may often be insufficient for the dislocation climb mechanism to operate exclusively and so climb creep may more commonly appear as a contributing mechanism, complementary to others.

Harper-Dorn creep (1957) is another type of creep process that is often thought to involve dislocation climb at very low stresses but with dislocation density that is independent of stress, giving rise to linear stress dependence (Nabarro 2000; Poirier 1985, pp. 114–117).

References

Ashby MF (1969) On interface-reaction control of Nabarro-Herring creep and sintering. Scripta Metall 3:837–842

Bardeen J, Herring C (1952) Diffusion in alloys and the Kirkendall effect. In: Imperfections in nearly perfect crystals: Symposium held at Pocono Manor, 12–14 Oct 1950, New York,Wiley, 261–288

Burton B (1972) Interface reaction controlled diffusional creep: a consideration of grain boundary dislocation climb sources. Mater Sci Eng 10:9–14

Burton B (1983) The characteristic equation for superplastic flow. Phil Mag A48:L9–L113

Coble RL (1963) A model for boundary diffusion controlled creep in polycrystalline materials. J Appl Phys 34:1679–1682

Cottrell AH (1964) The mechanical properties of matter. Wiley, New York, 430 pp

Elliott D (1973) Diffusion flow laws in metamorphic rocks. Bull Geol Soc Am 84:2645–2664

Gordon RS (1973) Mass transport in the diffusional creep in ionic solids. J Am Ceram Soc 56:147–152

Gordon RS (1975) Ambipolar diffusion and its application to diffusion creep. In: Mass transport phenomena in ceramics. Materials science research, Vol. 9, New York, Plenum Press, 445–464

Greenwood GW (1970) The possible effects of diffusion creep of some limitation of grain boundaries as vacancy sources or sinks. Scripta Metall 4:171–174

Groves GW, Kelly A (1969) Change of shape due to dislocation climb. Phil Mag 19:977–986

Harper JG, Dorn JE (1957) Viscous creep of aluminum near its melting temperature. Acta Metall 5:654–665

Herring C (1950) Diffusional viscosity of a polycrystalline solid. J Appl Phys 21:437–445

Lifshitz IM (1963) On the theory of diffusion-viscous flow of polycrystalline bodies. J. Exptl. Theoret. Phys. (U.S.S.R.), 44, 1349–1367; tr. in Soviet Physics JETP 1317(1344), 1909–1920

Nabarro FRN (1948) Deformation of crystals by the motion of single ions. In: Report of a conference on strength of solids held at H H Wills Laboratory (ed), Univ of Bristol, 7–9 July 1947, London, The Physical Society, 75–90

Nabarro FRN (1967) Steady-state diffusional creep. Phil Mag 16:231–237

Nabarro FRN (2000) Harper-Dorn creep—a legend attenuated? Phys stat sol (a) 182:627–629

Poirier J-P (1976) Plasticité à Haute Température des Solides Cristallins. Editions Eyrolles, Paris, 320 pp

Poirier J-P (1985) Creep of Crystals. High-temperature Deformation Processes in Metals, Ceramics and Minerals, New York, Cambridge Univ Press, 260 pp

Raj R, Ashby MF (1971) On grain boundary sliding and diffusional creep. Metall Trans 2:1113–1127

Raj R, Chyung CK (1981) Solution-precipitation creep in glass ceramics. Acta Metall 29:159–166

Raj R (1982) Creep in polycrystalline aggregates by matter transport through a liquid phase. J Geophys Res 87:4731–4739

Rutter EH (1976) The kinetics of rock-deformation by pressure-solution. Philos Trans R Soc London, Ser A 283:203–219

Rutter EH (1983) Pressure solution in nature, theory and experiment. J Geol Soc London 140:725–740

Squires RL, Weiner RT, Phillips M (1963) Grain boundary denuded zones in a magnesium ½ wt % Zirconium alloy. J Nucl Materials 8:77–80

Stocker RL, Ashby MF (1973) On the rheology of the upper mantle. Rev Geophys Space Phys 11:391–426

Tsai RL, Raj R (1982a) A theoretical estimate of solution-precipitation creep in MgO-fluxed Si_3N_4. J Am Ceram Soc 65:C88–C90

Tsai RL, Raj R (1982b) Creep fracture in ceramics containing small amounts of a liquid phase. Acta Met 30:1043–1058

Weyl PK (1959) Pressure solution and force of crystallization—a phenomenological theory. J Geophys Res 64:2001–2025

Chapter 6
Deformation Mechanisms: Crystal Plasticity

6.1 Basic Geometry of Slip and Twinning

The deformation mechanisms of the greatest importance in the intragranular plastic deformation of crystalline materials are slip and twinning. In these mechanisms, the strain or change of shape is achieved by the relative movement of blocks of atoms rather than by the more or less independent movement of individual atoms that characterizes the atomic transfer mechanisms considered in the previous chapter. Deformation of means of the slip and twinning mechanisms is commonly referred to as crystal plasticity. It can be effective at all temperatures but is of overriding importance in the athermal regime (Sect. 6.6.1).

Macroscopically, the basic process in both slip and twinning consists of a more or less uniform simple shear (in the continuum mechanics sense, Means 1976, p. 146). The shearing occurs parallel to a well-defined crystallographic direction and, in most cases, on a well-defined crystallographic plane (Fig. 6.1). When the crystallographic orientation remains unchanged relative to the direction and plane of shearing, the process is known as crystallographic slip or translation gliding. When the orientation changes to a crystallographic twin orientation, the process is known as mechanical twinning or twin gliding. For brevity, we refer to the two processes as slip and twinning, respectively.

Slip. Close examination of the slip process shows that it is not homogeneous on the microscopic scale. The blocks that slide over each other can often be distinguished through the steps produced in a bounding surface at the loci of sliding. These steps may be visible microscopically as fine lines if the surface has been previously polished (Fig. 6.2). Such lines, which can be seen in the optical microscope, are known as "slip bands". The electron microscope (Heidenreich and Shockley 1948) reveals that they tend to be of multiple structure, consisting at the near atomic scale of finer steps called "slip lines", although this term is

M. S. Paterson, *Materials Science for Structural Geology,*
Springer Geochemistry/Mineralogy, DOI: 10.1007/978-94-007-5545-1_6,
© Springer Science+Business Media Dordrecht 2013

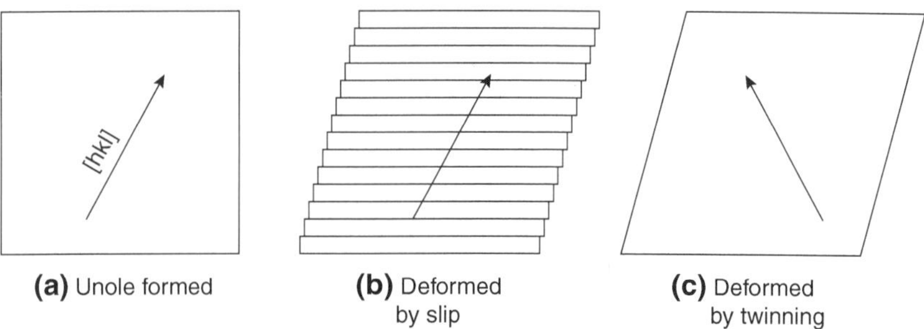

Fig. 6.1 Representation of the distinction between deformation by gliding and deformation by twinning. The *arrow* represents a crystallographic direction [*hkl*]

Fig. 6.2 Slip bands resulting from glide on a {100} plane in a <001> direction in a galena crystal compressed vertically normal to an {011} plane (width of specimen 5 mm). The concentration of slip bands constitutes a "deformation band" (a copy of Fig. 27 from Lyall 1965)

commonly used more comprehensively also to cover the features seen optically; these "slip lines" should not be confused with the trajectories of maximum shear stress of the mathematical theory of plasticity, which are also known as "slip lines" (see Jaeger and Cook 1979, p. 120). In any case, it is clear that, fundamentally, the slip process consists of the translation of one part of a crystal relative to the remainder at a surface that is more or less atomically sharp, and without loss of cohesion at that surface.

Macroscopically, slip is characterized by a slip direction, a slip plane (or other surface), and an amount of shear. Taken together, the combination of direction and plane is known as the slip system or glide system (the term "glide" is used interchangeably with slip in this context). When there is no clearly defined slip plane, as in "pencil glide", the process can be resolved into two or more simple shears with a common slip direction, the slip direction always being well defined. The slip direction and, when well defined, the slip plane usually have low crystallographic indices. An important property of the slip process is that more than one slip system may operate in a given crystal; the slip systems then intersect each other and the combination is referred to as "multiple slip". The intersecting slip systems may or may not be crystallographically equivalent.

The mechanics of the slip process are most simply described with reference to the slip system. If a crystal is subjected to a uniaxial normal stress σ in a direction inclined at angles λ to the slip direction and χ to the slip plane normal (Fig. 6.3) then the shear stress τ acting on the slip plane in the slip direction, the so-called "resolved shear stress", is given by

$$\tau = \sigma \cos \chi \cos \lambda = S\sigma \tag{6.1}$$

where $S = \cos \chi \cos \lambda$ is known as the Schmid factor. Similarly, the increment of resolved shear strain $d\gamma$ is related to the corresponding normal strain increment $d\varepsilon$ by

$$d\gamma = \frac{d\varepsilon}{\cos \chi \cos \lambda} = \frac{d\varepsilon}{S} \tag{6.2}$$

The relations $\tau = S\sigma$ and $\gamma = \int S^{-1} d\varepsilon$ can then be used to obtain the values of resolved shear stress τ and resolved shear strain γ from measurements of σ and ε. The quantities τ and γ are the most important quantities for describing the operation of the slip system.

Schmid's law states that slip will occur when a certain critical resolved shear stress is exceeded (Schmid 1924; Schmid and Boas 1936, 1950). Schmid's law is applicable mainly in the athermal regime (Sect. 6.6.1) but it is sometimes extended to steady-state creep in the high-temperature thermal regime in the sense that a characteristic resolved shear stress will be required to produce a specified resolved shear-strain rate. Although Schmid's law is very useful as a first approximation and is widely invoked, it should be borne in mind that in practice it does not always apply strictly and at times can break down seriously, for example, in body-centered cubic (b.c.c.) metals (Louchet 1979; Puls 1981). An important factor

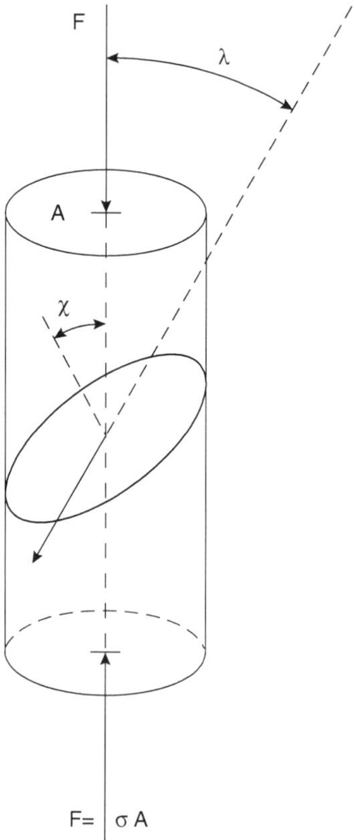

Fig. 6.3 Depicting a single crystal under an axial force $F = \sigma A$ (σ = axial stress; A = cross-sectional area), with a slip plane whose normal is at an angle χ to the axis of the stress and whose slip direction is at an angle λ to the axis of the stress

causing departure from Schmid's law is latent hardening, which is the increase in critical resolved shear stress on a given slip system due to the activity of an intersecting slip system.

Twinning. Mechanical twinning is distinguished from slip in that:

1. It is essentially homogeneous down to the unit cell scale, each equivalent atomic layer being translated by the same amount with respect to the one immediately adjacent to it (Fig. 6.4).
2. The amount of this translation is fixed and such that the crystal structure is transformed into a structure identical to the original.

In practice, especially in the grains in polycrystalline aggregates, twinning tends to occur in discrete bands due to the fortuity of its nucleation and to the constraints of strain compatibility between grains. Under given constraints, the twin bands tend to be narrower when the twinning shear is higher.

The characteristic geometric elements of mechanical twinning (Fig. 6.5) are the two circular sections K_1 and K_2 of the strain ellipsoid and the two directions η_1 and

Fig. 6.4 Artificial twinning
of calcite parallel to plane
$(01\bar{1}2)$ produced by forcing a
knife-edge into a cleavage
rhomb at point *a* (derived
from the sketch in Fig. 516,
p. 210, from Dana 1932)

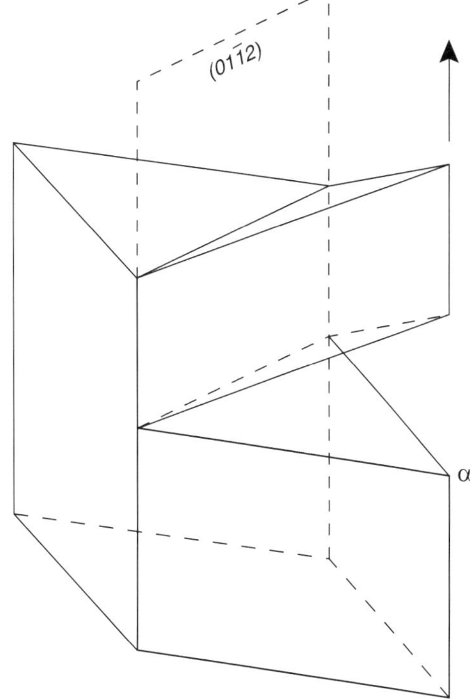

Fig. 6.5 Section of strain
ellipsoid for mechanical
twinning. Section is normal to
intermediate principal strain
axis, a direction of no strain.
The plane of the drawing is
known as the "shear plane".
K_1 and K_2 are the circular
sections or planes of no
distortion

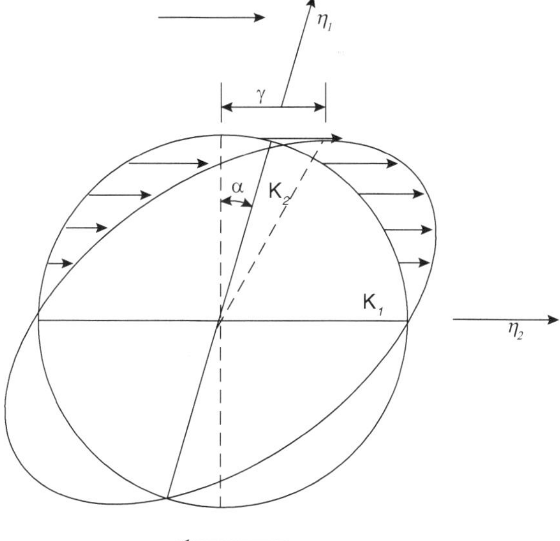

η_2 in which the shear plane intersects K_1 and K_2, the shear plane being defined as the plane containing the normals of K_1 and K_2 and itself being normal to the intermediate principal strain axis, that is, the plane of the circular section most nearly parallel to the longest dimension of the deformation twin in the shear plane; it is, in general, parallel to the composition plane, the plane across which the two twin individuals would join with best fit or least interfacial energy to form a so-called coherent boundary.

In crystals of relatively low symmetry, either K_1 and η_2 have rational crystallographic indices and K_2, η_1 are irrational (defining type I twins), or K_2, η_1 are rational and K_1, η_2 irrational (type II twins). In the case of type I twinning, the twin individuals are related by a reflection with respect to K_1 or by a twofold rotation about the normal to K_1. In the case of type II twinning, the symmetry relationship is a twofold rotation about η_2 or a reflection with respect to the plane normal to η_2. However, in higher symmetry crystals, all four elements K_1, K_2, η_1, η_2 can be rational and the four types of symmetry relationship just mentioned become equivalent, giving rise to so-called compound or degenerate twins, in which the composition plane can then also be always referred to as the twinning plane. In all cases, the twinning is fully specified by either K_1 and η_2 or K_2 and η_1, as appropriate. However, a convenient alternative, often used, is to specify the twinning in terms of the plane K_1, the direction η_1 and the magnitude of the shear $\gamma = 2\cot\phi = 2\cos\phi$ (Fig. 6.5).

For further discussion of the geometry and crystallography of mechanical twinning, see Cahn (1953, 1954), Hall (1954), Pabst (1955), Klassen-Neklyndova (1964), Christian (1965, Chap. 20) and Hirth and Lothe (1982, Chap. 23). It should be borne in mind that some treatments in the metallurgical literature refer only to relatively high symmetry crystals and do not make some of the distinctions in terminology that are appropriate for lower symmetry crystals.

Martensitic or displacive phase transformations constitute a more general class of shear deformation processes related to mechanical twinning. In these, instead of the crystal structure being restored as in twinning, albeit with a new orientation, a new structure is produced by the shearing, namely, that of a polymorphous phase. Mechanical twinning is thus a special case of this more general class of transformations. Ideal kinking (Orowan 1942; Paterson and Weiss 1966) can also be regarded as having some analogy to twinning and it has even at times been described as "irrational twinning" but it is better viewed as a special type of deformation band, that is, a particular heterogeneous distribution of slip (see review by Cahn 1953).

At the atomic scale, slip and twinning and displacive transformation take place by the movement of dislocations, the properties of which we now consider. Later in the chapter (Sects. 6 and 7) we consider slip and twinning in terms of dislocation motion.

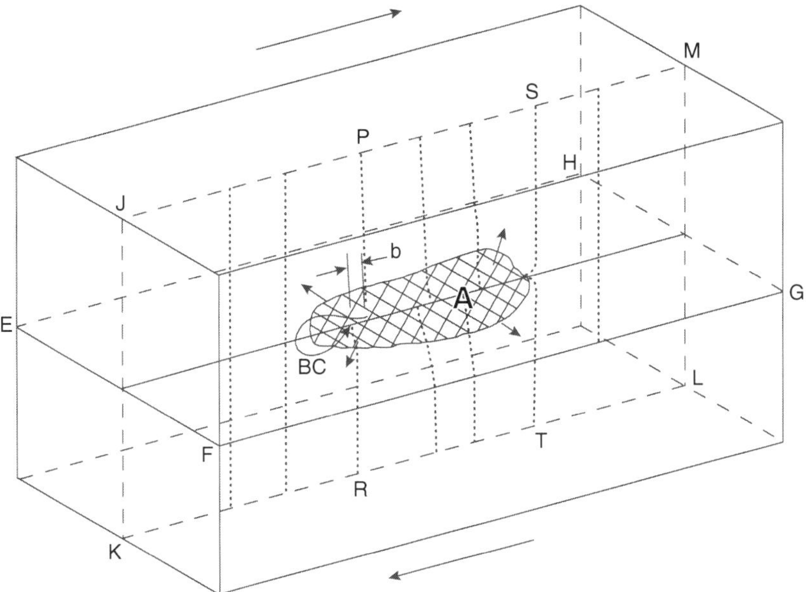

Fig. 6.6 Depicting a dislocation as the boundary of a slipped patch *A* in the slip plane *EFGH*, in which the columns of atoms crossing the slip plane are displaced by the amount of the Burgers vector *b*, as shown by the Burgers circuit *BC*

6.2 Properties of Dislocations in Crystals

6.2.1 Theoretical Shear Strength and the Concept of a Dislocation

The simple translation of one layer of atoms of a crystal over another with the simultaneous breaking and restoration of all the interlayer bonding would, theoretically, require a very high shear stress parallel to the layer. This stress, known as the theoretical shear strength, has been estimated as being around 1/15–1/10th of the elastic shear modulus of a crystal, within perhaps a factor of two (Hirth and Lothe 1982, p. 5; Kelly 1966, p. 12). In practice, plastic flow is commonly observed at stresses that may be several orders of magnitude below this level. The discrepancy can be explained by supposing that the breaking and restoration of bonding is done sequentially at a "front" that moves over the plane on which the translation is occurring; that is, at any given instant, bonds are only being broken along this "front" and so a much smaller total force and, hence, smaller macroscopic stress need be applied. This "front" or linear crystal defect is known as a dislocation. A dislocation can thus be regarded as the boundary between the slipped and unslipped parts of the plane or surface on which translation is occurring and its motion in the plane or surface represents the spreading of the slipped patch (Fig. 6.6). The macroscopic plastic strain results from the movement of a large number of dislocations distributed through the crystal. The dislocation is thus the basic entity at the atomic scale underlying the mechanisms of slip and mechanical twinning. Brief histories of the notion of a dislocation are given by Nabarro (1967, Chap. 1), Jouffrey (1979) and Hirth (1985).

The idea of the propagation of dislocations as the mechanism of slip in crystals was first put forward independently in 1934 by Orowan (1934), Polanyi (1934) and Taylor (1934). This picture has been abundantly verified observationally since the 1950s (Hirth and Lothe 1982, p. 9) and a very large body of literature on both observation and theory has appeared. A comprehensive account of the theory of dislocations in crystals cannot be attempted within the scope of the present volume but the principal properties will be summarized in the following sections. For general accounts of dislocation theory, the classic texts are Cottrell (1953), Friedel (1964), Nabarro (1967) and Hirth and Lothe (1982), while succinct accounts will be found in Weertman and Weertman (1964, 1992), Nicolas and Poirier (1976), Haasen (1978) and Poirier (1985) and in many other books and reviews in materials science. Groh et al. (1979) have also published a useful compilation of papers in French on a broad range of dislocation topics.

Figure 6.6 depicts the type of dislocation with which we are concerned here and which has been called by Nabarro (1967) the Burgers type to distinguish it from the more general types associated with the names of Weingarten, Volterra and Somigliana. Such a dislocation is a line that that can be thought of as the boundary of a surface in which the body is imagined to have been cut and across which the material on one side is imagined to have been translated homogeneously relative to the material on the other side before repairing the cut to restore continuity of the material. The virtual translation vector b is known as the Burgers vector. In the case of crystals, b is normally equal to a translation vector of the crystal lattice, in which case the dislocation is called a perfect dislocation; the minimum possible value of b is thus the minimum lattice parameter except when a nonprimitive unit cell is chosen. If b is not a translation vector of the lattice, the dislocation is known as a partial dislocation, which necessarily bounds a surface of faulting in the crystal structure. Where a dislocation line is normal to the Burgers vector it is said to be of pure edge character and, where parallel to the Burgers vector, of pure screw character; otherwise, it is of mixed character.

The essential nature of a dislocation is revealed by considering a Burgers circuit, which is a notional path through a sequence of lattice sites forming a closed loop around the dislocation. If a path is followed through a corresponding sequence of lattice sites in a perfect crystal, this path fails to close and the vector required to close the gap is the Burgers vector of the dislocation. Taking the sense of the Burgers vector to be positive when going from finish to start of the circuit (FS), a sense or sign can then be attributed to a given segment of dislocation line such that the Burgers circuit is seen to be performed in a clockwise sense when looking in the positive direction along the dislocation line (RH). Conversely, if one first chooses a positive sense for the dislocation line, the positive sense of the Burgers vector can be defined by the reverse argument. This FS/RH sign convention is, however, not always followed and an opposite convention is also common in the literature (see discussion and references in Hirth and Lothe 1982, pp. 17–24). The important point to make is that, whatever the sign convention, parallel positive and negative dislocation segments having the same Burgers vector are mutually annihilating if brought together.

It follows from the nature of a dislocation that there is always an internal stress field associated with it in the crystal. In the approximation that the material is taken to be elastically isotropic, with shear modulus G and Poisson ratio v, the stresses around a dislocation in an infinite crystal, expressed in cylindrical coordinates r, θ, z with z parallel to the dislocation line, are as follows for pure screw and pure edge dislocations (Hirth and Lothe 1982, Chap. 3):

Screw dislocation:

$$\sigma_{rr} = \sigma_{\theta\theta} = \sigma_{zz} = 0$$

$$\sigma_{\theta z} = \frac{Gb}{2\pi r}; \quad \sigma_{zr} = \sigma_{r\theta} = 0 \tag{6.3a}$$

Edge dislocation:

$$\sigma_{rr} = \sigma_{\theta\theta} = \frac{Gb \sin \theta}{2\pi(1-v)r}; \quad \sigma_{zz} = 2v\sigma_{rr}$$

$$\sigma_{r\theta} = \frac{Gb \cos \theta}{2\pi(1-v)r}; \quad \sigma_{\theta z} = \sigma_{zr} = 0 \tag{6.3b}$$

(Note that the sign convention of compressive stress being positive has been used, leading to the signs of σ_{rr}, $\sigma_{\theta\theta}$ and σ_{zz} being opposite to those usually given in textbooks; also for edge dislocations, θ is taken to be zero when r coincides with b. For the corresponding elastic displacements, see Hirth and Lothe (1982, Chap. 3).

The importance of dislocations in crystal plasticity lies in the slip or twinning that is brought about when they are moved through the crystal. The direction of the Burgers vector can then be identified with the slip direction and the plane containing the Burgers vector and the dislocation line is the slip plane (corresponding glide elements are defined in the case of twinning). If ρ is the total length of mobile dislocation line per unit volume (that is, the dislocation density) and s is the average distance moved by the dislocation line in the slip plane, then the macroscopic resolved shear strain resulting from the dislocation motion is

$$\gamma = \rho b s \tag{6.4a}$$

where b is the magnitude of the Burgers vector. This relationship or its kinetic equivalent

$$\dot{\gamma} = \rho b v \tag{6.4b}$$

where $\dot{\gamma}$ is the strain rate and v the dislocation velocity, is known as the Orowan equation.

The movement of a dislocation line out of the slip plane is called *climb*. Since climb generally requires the removal or supply of substance along the dislocation path, it is referred to as nonconservative dislocation motion, in contrast to the conservative motion in slip. The volume ΔV of substance added or removed during the climb of a dislocation segment of length l through a distance s normal to the slip plane is given by

$$\Delta V = lbs \sin \theta \qquad (6.5)$$

where b is the magnitude of the Burgers vector and θ its inclination to the slip line in the slip plane. In the case of a pure screw dislocation, for which the directions of Burgers vector and dislocation line are parallel, no unique slip plane is defined and so the dislocation can move conservatively in any plane; this motion is referred to as cross-slip when the plane is not the primary slip plane.

The application of stress to the region containing a dislocation has a mobilizing tendency that is equivalent to applying a force normal to the dislocation line. From work considerations it follows that the component of this force in the slip plane, the *glide force*, F_g, is

$$F_g = \tau bl \qquad (6.6a)$$

where τ is the resolved shear stress in the slip plane, b the magnitude of the Burgers vector and l the length of the dislocation segment concerned. Correspondingly, the component of the force in the plane normal to the slip plane, the *climb force*, F_c, is

$$F_c = \sigma bl \sin \theta \qquad (6.6b)$$

where σ is the deviatoric normal-stress component parallel to the Burgers vector and θ the angle between the slip line and the Burgers vector. Thus, the climb force arising from the stress component σ is σbl for a pure edge dislocation and zero for a pure screw dislocation. For a more general treatment of the force acting on a dislocation, see Weertman and Weertman (1964, pp. 54–61) and Hirth and Lothe (1982, Chap. 3).

6.2.2 *The Energy of a Dislocation*

The energy of a dislocation can be considered in two parts, associated with the core and the long-range elastic stress field, respectively. The core is the cylindrical region of radius r_0 immediately surrounding the dislocation line, within which the crystal structure is disrupted or distorted beyond the limits of linear elasticity; r_0 is usually taken to be several times b in magnitude. The region of the long-range elastic stress field is taken as extending from the cylinder of radius r_0 to the surface of the crystal or, more usually, to a radius R beyond which the internal stress field is to be regarded as belonging to other dislocations ($R \sim \rho^{-1/2}$ where ρ is the dislocation density). Within the region of the long-range stress field, the distortion of the crystal structure is within the linear elastic range and the classical theory of elasticity can be applied, while within the core region the energy can only be calculated using considerations at the atomic level.

The energy E_{el} in the long-range stress field of a segment of length l of a straight dislocation is approximately Gb^2l if the dislocation density is low, where b is the magnitude of the Burgers vector and G is the shear modulus in the approximation of isotropic elasticity (the value of which is probably best obtained by Reuss averaging of the actual anisotropic elastic properties: Hirth and Lothe 1982, p. 424). More exactly, the energy depends somewhat on the character of the dislocation line and on R/r_0, according to

$$\frac{E_{el}}{l} = \frac{\alpha G n^2}{4\pi} \ln\left(\frac{R}{r_0} - 1\right) \tag{6.7}$$

where $\alpha = \cos^2\theta + \frac{\sin^2\theta}{1-v}$, θ being the angle between the dislocation line and the Burgers vector and v the Poisson ratio; the factor -1 arises when the surface at radius R is taken to be stress free (for derivation, see Hirth and Lothe 1982, Chap. 3). When full account is taken of the actual anisotropic elasticity of the crystal (for example, Steeds 1973) (Hirth and Lothe 1982, Chap. 13), a more complicated function K of the elastic constants replaces αG in (6.7). However, the effect on the calculated energy is usually not very great. Thus, Heinisch et al. (1975) showed that for olivine and orthopyroxene the isotropic approximation gives energies that are mostly within 10 % of those calculated on anisotropic elasticity, while in the more anisotropic cases of quartz and calcite the differences do not exceed about 30 % or so. The change in energy when curvature of the dislocation is taken into account is also usually relatively small. Thus, the elastic energy per unit length for a dislocation loop of radius R_l is approximately $(Gb^2/2\pi)\ln(R_l/r_0)$, giving a similar value to (6.7) except for very small loops (Nabarro 1967, p. 75).

In the long-range linear elastic field around any dislocation with an edge component there are regions of volumetric expansion and contraction on opposite of the dislocation. These effects closely compensate each other so that the overall elastic dilatation is close to zero. This also applies for a screw dislocation, which, to the first approximation, has no dilatational components in its long-range elastic field, However, due to anharmonic effects in or near the core, there is actually a small net volume increase, of the order of b^2l for a dislocation segment of length l, for both screw and edge dislocations (Hirth and Lothe 1982, p. 231; Poirier 1985, p. 152; Seeger and Haasen 1958).

The core energy, which depends on the structural arrangement and bonding energies of the atoms in the core region, is more difficult to estimate (Hirth and Lothe 1982, Chap. 8). Earlier calculations based on assumed interatomic potentials of conveniently simple form, such as the well-known Peierls–Nabarro model, can be regarded as little more than illustrative. However, ab initio molecular orbital calculations have made some progress and recent calculations based on more realistic potentials and structural relaxation procedures appear to yield results that are useful approximations, especially in simple ionic crystals (Puls 1981). For example, Bucher (1982) has calculated the core energy to be approximately 0.7×10^{-9} J m^{-1} (1.2 eV per length b) for edge dislocations in sodium chloride, while Heggie and Jones (1986) have obtained approximately 1×10^{-9} J m^{-1} (3 eV per length b) for a basal 60° dislocation in quartz, to be compared with values of Gb^2 of about 2×10^{-9} and 11×10^{-9} J m^{-1}, respectively. The general indication is that, at least for simple structures, the core energy does not exceed a few tenths of Gb^2 per unit length, that is, that rather less energy is associated with the core than with the long-range stress field. Also the core energy may be reduced by structural changes in the core such as the incorporation of impurity atoms.

In summary, the total energy of a dislocation can be taken as being of the order $\frac{1}{2}Gb^2$ to Gb^2 per unit length, the greater part of which is normally associated with the long-range stress field. For a typical mineral with $b \sim 0.5$ nm and $G \sim 50$ GPa, the dislocation energy is therefore of the order of 10^{-8} J m^{-1} (or ~ 30 eV per length b). It follows that, at a dislocation density of 10^{12} m^{-2}, the contribution of the dislocations to the internal energy of the crystal will tend to be rather less than 1 J mol^{-1}, and even at extremely high dislocation densities of 10^{15}–10^{16} m^{-2} the contribution will not exceed the order of 1 kJ mol^{-1} (cf. measurements of stored energy in heavily cold-worked calcite by Gross 1965).

The energy that we have been considering so far is the increment in the Gibbs energy of the crystal due to introducing an individual dislocation segment (it is Gibbs energy because temperature and pressure, or stress, have been taken as the independent variables, Sect. 2.2). This energy will include a $-T\Delta S$ term involving entropy ΔS that can be associated mainly with the thermal vibrations of the atoms, and hence is expressed in the temperature dependence of the elastic constants. When we now consider an assemblage of dislocations, there is an additional configurational entropy to be taken into account, but it can be shown that the corresponding $-T\Delta S$ term is relatively small and so the increment in the Gibbs energy of the crystal due to the presence of an assemblage of dislocations will be slightly smaller than the sum of the contributions of the individual dislocation segments as calculated above (Cottrell 1953, p. 39; Friedel 1964, p. 73; Nabarro 1967, p. 683). If the assemblage is viewed as consisting of a number of dislocation loops (or independent segments) and the equilibrium concentration of loops is calculated by minimizing the Gibbs energy of the assemblage, the concentration is found to be negligibly small at any temperature below the melting point unless the loops are exceedingly small (Friedel 1964, p. 74, 1982). That is, the introduction of dislocations into a crystal normally increases its Gibbs energy and so the dislocation content of a crystal at equilibrium in respect of its defect structure at any given temperature and pressure will be negligibly small. This argument has also been extended to form the basis of a theory of melting, according to which the melting point is the temperature at which there is an abrupt transition from a state in which the dislocation content at equilibrium is zero to a state in which it is very large (Nabarro 1967, p. 688).

Since the energy of a dislocation line is increased or decreased by increasing or decreasing its length, respectively, the line can be regarded as being under tension, the derivative of the energy with respect to the length being known as the *line tension*. The line tension is identical with the energy per unit length, as given by the sum of (6.7) and the core energy per unit length, and hence is of the order of Gb^2, with the dimensions of force. However, analogy with a string under tension is not an exact one because there is interaction between the parts of a dislocation (Sect. 6.3.4) and so the line tension depends on the dislocation configuration (Hirth and Lothe 1982, p. 174).

6.2.3 The Peierls Potential

In considering the mobility of a dislocation line in the absence of the influence of thermal agitation or extrinsic factors such as interaction with other dislocations or the presence of impurities, it is the fluctuation in the dislocation energy with position in the crystal that is of primary importance rather than the absolute value of the energy. The fluctuating component of the energy, which is associated with the crystalline periodicity and has maxima at spacings equal to the Burgers vector or submultiples of it, arises entirely within the core energy and is known as the Peierls energy or Peierls potential since Peierls (1940), at the instigation of Orowan, was the first to attempt to calculate it (Hirth and Lothe 1982, p. 217).

In view of the difficulty of modeling the dislocation core, it is conceptually useful to begin by assuming an empirical expression for the dislocation energy E_d of the form

$$E_d = E_0 + E_p \sin^2 \pi n x \tag{6.8}$$

where E_0 is the non fluctuating part of E_d and includes the energy of the long-range elastic stress field, E_p is the amplitude of the Peierls potential, xb is the displacement of the dislocation core from a position of minimum energy, and n is an integer (usually 1 or 2) that allows for different forms of the Peierls potential (Hirth and Lothe 1982, Chap. 8). The force needed to move a straight segment of dislocation line over the Peierls barrier E_p, the Peierls force, is then given by the maximum slope of the potential E_d, that is, by $(1/b)(dE_d/dx)_{\mathrm{Max}} = \pi n E_p$. From (6.5), the corresponding stress, the Peierls stress τ_p, is $(\pi n/b^2)(E_p/l)$. If we write $\pi n(E_p/l) = \beta Gb^2$, we obtain

$$\tau_p = \beta G \tag{6.9}$$

where β is an empirical numerical constant. If we also write $(E_0/l) = \alpha Gb^2$ where $\alpha \sim \frac{1}{2}$ to 1 (Sect. 6.2.2), then $\beta = \alpha \pi n(E_p/E_0)$ is of the order of the ratio of E_p to the total dislocation energy.

On the basis of interpretation of experimental observations and of a few calculations, it is generally accepted that the value of β varies from the order of 10^{-4} or less for close-packed metals (for example, copper and basal slip in zinc) to a maximum of the order of 10^{-2} for covalent crystals such as germanium and silicon (Haasen 1978, p. 256; Hirth and Lothe 1982, p. 241). The best calculated values for β for NaCl, KCl, and MgO are in the range $(1-3) \times 10^{-3}$ (Hirth and Lothe 1982, p. 232), corresponding to a Peierls stress of around 50 MPa for the alkali halides and 150 MPa or more for MgO. The value of the Peierls stress for dry quartz deduced by Blacic and Christie (1984) from extrapolation of experimental observations, namely, about 3,000 MPa, corresponds to a value of β of about 6×10^{-2} and to a value of E_p somewhat less than 1 eV per length b. Thus, it is seen that the fluctuating component of the dislocation energy is always a small fraction of the total.

Simple theoretical models indicate that the Peierls stress depends sensitively (for example, exponentially) on the width of the dislocation, as defined by the zone within which atoms are displaced by more than a certain fraction of the Burgers vector from an ideal structural site. In close-packed metals the width of the dislocations is many times the Burgers vector, that is, their cores are very smeared out, and the Peierls stress is small, while in covalent crystals the dislocation cores tend to be very narrow and the Peierls stress relatively high. Theoretical models also predict that, in general, the Peierls stress will tend to decrease as the interplanar spacing increases, promoting the occurrence of slip on low-index planes, and that the Peierls stress for edge dislocations will tend to be less than that for screw dislocations (Hirth and Lothe 1982, p. 240). A high value of the Peierls potential is manifested microscopically in a tendency for dislocation lines to be straight and crystallographically well defined.

Two remarks are appropriate at this point. First, it is implicit in relating the Peierls stress to dislocation motion that the additional potential energy acquired by the dislocation at the Peierls ridge is mainly dissipated during movement into the next valley and so is not available to assist in surmounting the following ridge. Second, the Peierls considerations so far refer primarily to the situation at absolute zero temperature and to the properties of straight segments of dislocation lines. Relaxing either of the latter constraints leads to easier dislocation movement and introduces some more complex considerations. Thus, if the dislocation line has a step or kink in it, whereby one segment is in advance of an adjacent segment by one lattice spacing, the kink can behave as a sort of second-order dislocation in the linear structure of the parent dislocation, which can move along the dislocation in displacement increments equal to the lattice spacing in that direction. This motion can be expected to occur at a lower applied stress than the uniform motion of the whole dislocation line. However, in its motion along the dislocation, the kink has to surmount potential barriers analogous to the Peierls barrier for motion of a straight dislocation line through a crystal and so the resistance to kink motion can be expressed as an analog of the Peierls stress. Hence, the terms "secondary Peierls potential" and "secondary Peierls stress" are sometimes used for this

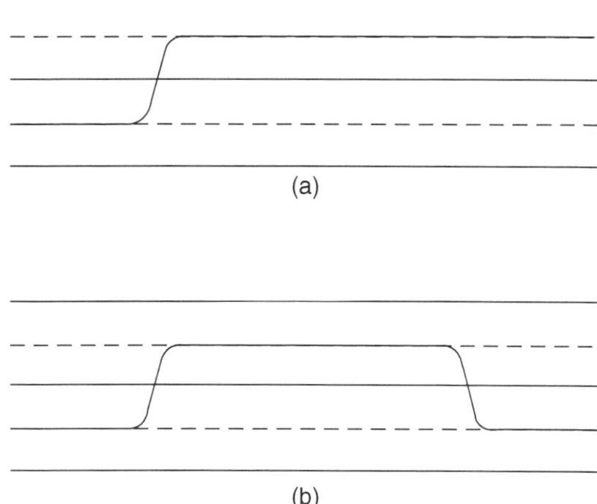

Fig. 6.7 Depicting **a** a single kink and **b** a double kink. The *heavy line* is the configuration of the dislocation; the *light line* represents the Peierls potential peaks and the *dashed lines* represent the Peierls troughs

potential barrier and the stress for motion of a kink, respectively. The secondary Peierls barrier is only likely to be substantial in the case of narrow dislocations, as in covalent solids, but, as long as the crystal is at absolute zero temperature, it will, in any case, be of little importance since, once the kink has traversed the dislocation line, the line will then tend to move forward as a whole at the Peierls stress itself, in the absence of the thermally activated nucleation of kinks at a lower stress.

Raising the temperature above absolute zero will first have the effect of flattening somewhat the profile of the Peierls potential, and hence reducing the Peierls stress. However, this effect is unlikely to be a very marked one (Friedel 1964, p. 67) and it will eventually be overshadowed by the much more important effect of the thermally activated nucleation of kinks on the dislocation, rendering the Peierls considerations for straight segments of dislocations no longer relevant. We therefore now have to consider kinks more seriously, especially in relation to their thermally activated behavior.

6.2.4 Kinks and Jogs

As mentioned in the previous section, if a dislocation line in a given slip plane lies partly in one Peierls valley and partly in the next, the crossover is called a *kink*. If there is a second crossover nearby bringing the dislocation back into the first valley, the combination is called a double kink or, better, a kink pair (Seeger 1984) (Fig. 6.7). Kinks are only sharp or narrow, and usefully distinguished as entities, when the Peierls potential is relatively high.

Although propagating a kink along a dislocation gives the possibility of advancing the dislocation under lower stress than is needed in the absence of kinks, the continuing motion of the dislocation in this way requires the continuing nucleation of new kinks. The new kinks are necessarily nucleated as kink pairs unless the nucleation occurs where the dislocation ends at an interface or junction, or unless the kink is produced by intersection by another dislocation of the same Burgers vector moving on another slip plane. The latter two origins for new kinks are probably, in general, of minor importance for dislocation mobility in materials of high Peierls stress, the dislocation mobility in these being thought rather to depend strongly on the thermal nucleation of kink pairs, at least in certain ranges of temperature.

The energy of a kink, E_k, can be expressed approximately as

$$E_k \approx \frac{2b}{n\pi} \left(2\frac{E_0}{l}\frac{E_p}{l} \right)^{\frac{1}{2}} \tag{6.10}$$

(Hirth and Lothe 1982, p. 254) where n, E_0 and E_p are as defined by (6.8) for a length l of dislocation line. Putting $(E_0/l) = \alpha Gb^2$ and $(E_p/l) = (\beta/n\pi)\,Gb^2$ as in Sect. 6.2.3, we obtain

$$E_k \approx \left(\frac{2\alpha}{n\pi}\right) \beta^{\frac{1}{2}} Gb^3 \tag{6.11}$$

That is, E_k is of the order of $(1/5)\beta^{1/2}Gb^3$ within a factor of two or three, where β is the same numerical parameter as in (6.9). The value of E_k can thus be expected to vary from less than 0.1 eV for the close-packed metals to the order of 1 eV for covalent crystals. A more refined calculation indicates that E_k is several times larger in screw dislocations than in edge dislocations (Hirth 1982, p. 256, where the relationship between E_k and the width of the kink is also discussed). The energy of a kink pair is almost $2E_k$ for all but very small separations of the individual kinks.

When the kink energy is relatively large, it is necessary to raise the temperature quite substantially in order to make possible a significant rate of kink pair nucleation by thermal activation and so facilitate dislocation glide in face of a high Peierls barrier. This situation is presumed to explain why plastic deformation is only achieved at elevated temperatures in materials such as germanium and silicon and it probably also underlies the difficulty of deforming minerals such as quartz and the framework silicates. Raising the temperature in these cases may also facilitate the surmounting of the secondary Peierls barriers along the potential ridge, promoting the mobility of the kinks thus nucleated. For further review of the role of kink nucleation and mobility, see Hirth and Lothe (1982, pp. 532–545), Philibert (1979) and Guyot and Dorn (1967).

A *jog* consists also of a unit step in an otherwise straight dislocation line but, instead of the step lying within the slip plane as for a kink; it now has a component normal to the slip plane. A jog is probably commonly formed in a dislocation as a result of intersection by another dislocation having a Burgers vector inclined to the slip plane of the first dislocation (intersection by a dislocation with Burgers vector lying in the slip plane of the first dislocation would produce a kink in it; see Hull

(1975, Chap. 7) on the geometrical aspects of kinks formed by intersection). A jog is in effect a short segment of dislocation having the same Burgers vector as the main part of the dislocation line but having, in general, a different slip plane. If this second slip plane is not one for easy slip, the presence of the jog will have a dragging effect on the main part of the dislocation. The dragging effect will be especially marked if the jog has to undergo nonconservative motion in being displaced with the main part of the dislocation, as in the case of a jog in a screw dislocation when this displacement of the jog necessary leaves a trail of vacancies or interstitials (in the case of compounds, the "vacancies" or "interstitials" would need to consist of the whole repeating group of atoms, making jog dragging especially difficult in screw dislocations in a crystal of complex structure and suggesting that in such a case a precondition for the easy glide of a segment of screw dislocation would be for it to be cleared of jogs by moving them conservatively along the dislocation). Jogs are also thought to be important in the climb of dislocations, providing sites for more ready attachment or detachment of the diffusing atoms (for example, Nabarro 1967, p. 351). At elevated temperatures the jogs may be formed thermally. For the energetics of jogs, see Hirth and Lothe (1982, pp. 495–497, 569–585).

6.2.5 *Zonal, Extended and Partial Dislocations*

The changes in linkage or interrelationship between neighboring atoms brought about during the passage of a dislocation need not be concentrated on a single structural site or confined to a single structural plane at any instant but can be distributed over a small region. Correspondingly, during the shearing process individual atoms may move in directions not parallel to the slip direction (a movement called "shuffling") and different patterns of movement may occur on adjacent structural layers, the only constraint being that the net bulk displacement of one part of the crystal relative to the remainder corresponds to the Burgers vector and that the original structure be left unchanged after the passage of the dislocation. When the nonparallel atomic movements are confined to a single pair of structural layers, the term "synchro-shear" is applied to the movement pattern (Kronberg 1961) and sometimes, by extension, to the dislocation itself (Amelinckx 1979, p. 393). In more general cases where several layers are simultaneously involved, the term "zonal dislocation" can be used (Amelinckx 1979). In effect, in a zonal dislocation the topological disruptions in the core region can be envisaged as extending over several adjacent structural sites. Thus, in the dislocations envisaged in aluminum oxide (corundum) two adjacent atomic layers are involved in a synchro-shear, while in $[1\bar{1}23](1\bar{1}\bar{2}2)$ slip in zinc zonal dislocations involving three layers are thought to be involved (Amelinckx 1979). Similar enlarged core regions may be expected to be not uncommon in minerals. However, in zonal dislocations the long-range elastic stress field can still be regarded as being determined by a sole singularity represented by the Burgers vector; it is the core energy itself that is minimized by the complex motions or reorganization in the core zone. However, the amplitude of the Peierls potential, and hence the Peierls stress, need not be reduced and in fact may be increased by the core spreading, rendering the dislocation less mobile (Vitek 1985).

If the discrete structural or topological disruptions in the core of a dislocation are distributed over a region of sufficiently large dimensions that the long-range stress field can no longer be regarded as deriving from a single linear singularity, then the dislocation can be called an *extended dislocation*. Such a dislocation can generally be regarded as made up of component *partial dislocations* separated by ribbons of planar defect or fault (Sect. 6.2.1). Any perfect dislocation with Burgers vector $\underline{\underline{b}}$ can be dissociated into a set of j dislocations having Burgers vectors $\underline{\underline{b}}_{=j}$ such that $\underline{\underline{b}} = \sum \underline{\underline{b}}_{=j}$ without affecting the displacements at distances large compared with the spacing of the new dislocations. However, if $\underline{\underline{b}}$ is already a minimum or primitive lattice translation, the components b_j will not generally be lattice translations and so will give rise to partial dislocations. Further, provided that $b_1^2 + b_2^2 + \ldots + b_j^2 < b^2$ and that the spacings of the partial dislocations are distinctly larger than their core diameters, the total elastic energy will be reduced by the dissociation from approximately Gb^2 toward $\sum_j Gb_j$ as the spacing of the partial dislocations becomes large. However, additional energy must be provided for the ribbons of planar defect separating the partial dislocations, in proportion to the area of the ribbons. There will therefore be an equilibrium spacing corresponding to the minimization of the sum of the elastic and planar defect energies. In practice, the equilibrium spacing can be used to estimate the surface energy of the planar defect or fault ribbon; typical values are in the range 0.01–0.1 Jm^{-2} (Amelinckx 1979; Carter 1984). The terms glide dissociation and climb dissociation are often used to distinguish the cases where the partial dislocation lines and fault ribbons lie in a plane that, respectively, contains the Burgers vector of the parent perfect dislocation (Fig. 6.8) and is inclined to their Burgers vector. In the case of screw dislocations, the distinction disappears, at least geometrically, since some sort of distinction may still, in principle, be made dynamically in terms of planes of easy and difficult glide. The glide and climb motion of dissociated dislocations involves a coupling between the individual partials through the energy of the strip of stacking fault joining them. Processes such as cross-slip appear to require a temporary recombination of the partials, while, in the case of climb, the nucleation of jogs is thought sometimes to involve local loop formation (for example Carter 1984; Cherns 1984).

In an ordered structure (Sect. 2.), the actual periodicity is that of the superlattice and so, strictly, the Burgers vector will be a multiple of the Burgers vector for the disordered structure. In practice, dislocations in ordered structures are commonly still described with reference to the lattice defined for the disordered structure. A unit dislocation in the ordered structure is then referred to as a *superdislocation*. The superdislocation can be, and is likely to be, dissociated into a pair of normal

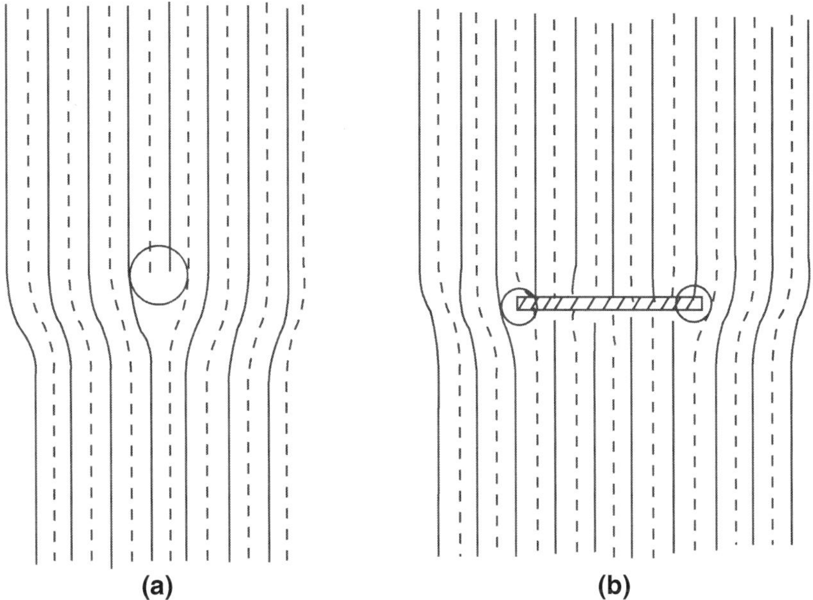

Fig. 6.8 Schematic representation of glide dissociation of an edge dislocation **a**, giving rise to partial dislocations separated by a stacking fault **b**

dislocations separated by a strip of antiphase boundary and, further, the normal dislocations can themselves be dissociated into partial dislocations (Amelinckx 1979).

6.2.6 Dislocation Reactions

The dissociation of a perfect dislocation into partials (Sect. 6.2.5) is one example of a more general class of dislocation reactions which may occur when dislocation lines are brought together or dissociated into separate dislocations (Weertman and Weertman 1964, Chap. 4). In all dislocation reactions, a principle of conservation applies according to which the vector sum of the Burgers vectors must remain the same. This rule applies, in particular, at nodes of the three-dimensional network of dislocations in a real crystal, where it can be expressed in the form that, if the Burgers vectors are given senses according to a consistent rule (Sect. 6.2.1), the vector sum of the Burgers vectors of all the dislocations issuing from a node must be zero (Friedel 1964, Chap. 1).

The potential importance of dislocation reactions can be illustrated with the frequently discussed reaction in face-centered cubic(f.c.c.) crystals which leads to the formation of a sessile dislocation known as a Lomer–Cottrell lock. Consider two perfect dislocations, one of Burgers vector $\frac{1}{2}[01\bar{1}]$ in the (111) plane and one of

Burgers vector $\frac{1}{2}[101]$ in the $(11\bar{1})$ plane (the notation $1/n[uvw]$ denotes the vector in the $[uvw]$ direction having the length 1/nth of the vector $u\underline{a} + v\underline{b} + w\underline{c}$ where $\underline{a}, \underline{b}, \underline{c}$ are the unit vectors defining the unit cell, and $(111)[01\bar{1}]$ is the usual slip system for f.c.c. crystals such as copper). These dislocations can dissociate according to

$$\frac{1}{2}[01\bar{1}] = \frac{1}{6}[\bar{1}2\bar{1}] + \frac{1}{6}[11\bar{2}]$$

$$\frac{1}{2}[101] = \frac{1}{6}[2\bar{1}1] + \frac{1}{6}[112]$$

leading to two pairs of partial dislocations with fault ribbons between each pair. If the two extended dislocations are then moved toward the intersection of their slip planes, the leading partial dislocations can react according to

$$\frac{1}{6}[\bar{1}2\bar{1}] + \frac{1}{6}[2\bar{1}1] = \frac{1}{6}[110]$$

with reduction in total elastic strain energy. The product is a partial dislocation with Burgers vector $\frac{1}{6}[110]$ and line direction $[1\bar{1}0]$, which together define a slip plane (001) that is not a normal slip plane in f.c.c. crystals. The $1/6[110]$ dislocation is also linked by two inclined fault ribbons to the $1/2[11\bar{2}]$ and $1/6[\bar{1}12]$ partial dislocations in the (111) and $(11\bar{1})$ planes, respectively, and hence is sometimes called a stair-rod dislocation. Its sessile character follows from the immobility of these fault ribbons in the (001) slip plane.

6.2.7 Electric Charge on Dislocations

Dislocations in an insulator can, like other crystal defects, be electrically charged; see review by Whitworth (1975), and sections in Hirth and Lothe (1982, Chaps. 12 and 14). This effect has potential implications for the mobility of the dislocations and for the transport of charge during deformation. There are two ways of viewing the charging, depending on whether the covalent or the ionic aspect of the bonding is being emphasized.

In the first case, the charging can be viewed as arising electronically by the transfer of electrons or holes to or from electron energy levels associated with the dislocation or with kinks or jogs in the dislocation. In the case of edge dislocations the extra levels can be associated with "dangling bonds", but additional energy levels can also possibly be associated with distortions in the crystal field in screw dislocations and so these can also, in principle, be charged. This view of dislocation charging has been developed for the semiconductors (Haasen and Schröter 1970; Labusch and Schröter 1975; Mataré 1971).

In the second case, the charging of dislocations can be viewed as arising ionically from a lack of charge balance among ions located within the dislocation core, including at kinks and jogs. It may result from the imbalance of positive and negative ions along the boundary of the "half-slab" that terminates in the core of a dislocation with an edge component in a pure crystal (intrinsic charging) or from the presence of impurity ions of different charge (extrinsic charging). Straight dislocations of orientations that would involve intrinsic charging will tend not to occur because of the high associated electrostatic energy (leading to bowing-out instability) but intrinsic charging at jogs, kinks, and points of emergence may still arise in dislocations that would be intrinsically neutral if perfectly straight (for a detailed discussion of intrinsic charging of dislocations in the NaCl structure see Amelinckx 1979, p. 379). Extrinsic charging, at least in alkali halides, often appears to arise by a "sweeping up" of charge during the movement of the dislocation, in addition to any initial charge from bound point defects (Whitworth 1975). In the alkali halides, in which there tends to be a preponderance of cation vacancies due to their formation energy being lower than that of anion vacancies, the dislocation charge is generally negative in both nominally pure crystals and in those doped with divalent cations, but positive charge can arise in other cases and it is possible that an isoelectric point may exist at which there is a changeover from negative to positive charge with increase in temperature.

The presence of charge on dislocations can give rise to various effects in the electric properties of a crystal, such as transient changes in electric conductivity during plastic deformation because of charge transport. Also the formation of a compensating charge cloud around a charged dislocation will tend to pin the dislocation and lead to hardening but, at least in alkali halides, the magnitude of this effect in the flow stress is thought to be negligible (Whitworth 1975). Very little is known about charge on dislocations in ionic crystals other than the alkali and silver halides, but charging may be expected to be a widespread effect.

6.2.8 Dislocations in Minerals

Since chemical bonding in minerals is generally of a character intermediate between ionic and covalent, it is to dislocations in ionic and covalent compounds that one looks for guidance on the likely characteristics of dislocations in minerals, rather than to metals, except for those properties for which the character of the bonding is not of importance. The latter situation can be expected to hold where the long-range elastic strain energy is the principal determining factor in the behavior of the dislocations, but the nature of the bonding tends to enter importantly where the core properties are playing a determining role.

The following factors tend to be of importance in characterizing the dislocations in minerals (Nabarro 1984a; Paterson 1985):

1. Commonly the unit cell is fairly large and contains many atoms of several kinds, resulting in large Burgers vectors (0.5–1 nm typically) and complex structure of the dislocation cores, with dissociation or zonal spreading expected to be a common feature.

2. Partially covalent character of bonding may be of significance especially where Si–O and Al–O bonds are involved, associated with high Peierls stresses.
3. Chemical substitution and non stoichiometry may be common, involving a rich variety of structural defects.
4. Many mineral structures are of low symmetry, resulting in low multiplicity of slip systems and hence in difficulty of meeting intragranular strain compatibility requirements in rocks.

Requirements deriving from covalent bond character or electrostatic repulsion of ions of unlike charge are factors additional to the long-range elastic strain energy (dependent purely on the Burgers vector and the elastic constants) in determining the most favored slip direction and slip plane in minerals. Thus, the metallurgical rule that slip occurs in the direction of closest packing (shortest Burgers vector) is less strictly followed in minerals. For example, in calcite the most commonly observed slip direction $<\bar{2}021>$ is that of the third-shortest Burgers vector (length 0.81 nm, compared with 0.50 nm for the shortest repeat distance, that in the a axis direction).

Frank (1951) suggested, on the basis of a simplistic thermodynamic argument, that when the Burgers vector exceeds a value of the order of 0.5–1 nm there will be a tendency to form a hollow core in the dislocation of diameter $Gb^2/4\pi^2\gamma$, or somewhat smaller when nonlinear elasticity near the core is taken into account, where G is the shear modulus and γ the surface energy: see Nabarro (1984a, b) for the case of anisotropic crystals. Little direct evidence for such an effect has come so far from electron microscopy of minerals and possibly other types of core response meet the situation. However, the existence of such a tendency may conceivably be a factor favoring such phenomena as the segregation of impurity atoms in the core, the fast diffusion of atoms along the core or the formation of a non crystallographic core as proposed for ice (Di Persio and Escaig 1984, Perez et al. 1975).

6.3 Dislocation Interactions

So far, we have dealt with the properties of an isolated single dislocation in an otherwise perfect crystal. We now consider the interaction of dislocations with other types of crystal defect, including boundaries, and the mutual interaction of dislocations. These interactions, together with the interactions with the crystal itself (as discussed in Sects. 6.2.3–6.2.7), are very important because of their influence on the mobility of dislocations and hence on the plastic deformation of crystals. Such interactions have to be taken into account in developing theories of the flow resistance of crystals under external stress (Sect. 6.6), much of the complexity of which arises from the variety of the interactions. The magnitude of interaction can, in principle, be expressed as the change ΔE in the energy of the dislocation in the presence of the entity with which it is interacting. The force acting on the dislocation will then be $\partial \Delta E / \partial x$ where x is the coordinate in the direction of the motion.

6.3.1 Interaction with Point Defects

Many deformation effects are controlled by the interaction of dislocations with the structural perturbations associated with vacancies, self-interstitials, and impurity or solute atoms, referred to conveniently as point defects (Sect. 1.2.2). The interaction being a mutual one, the local concentration of point defects will also tend to be influenced by the presence of the dislocation. The actual local concentration will depend on whether or not equilibrium has been established and, in case of equilibrium, will be determined by considerations of entropy as well as of the local atomic interaction energies.

It is useful to distinguish three aspects of the interaction of point defects with a dislocation:

1. Atomic bonding interactions at the scale of the dislocation core (chemical aspect).
2. Elastic interactions with the linear elastic stress field of the dislocation (elastic aspect).
3. Electrostatic interactions between charges that may exist on the dislocations and point defects (electrostatic aspect).

In the next three paragraphs, we elaborate a little on the nature of these aspects of interaction (for general references, see Friedel 1964, Chap. 13; Hirth and Lothe 1982, Chap. 14; Nabarro 1967, Chap. 6).

Because of the gross structural distortion in the core of a dislocation it may be expected that site occupancy, compared with the normal topological arrangement of atoms for the perfect crystal, may be seriously modified, for example, by substitution of impurity atoms, by changes in the concentration of solute atoms in solid solutions (such as Mg/Fe ratio in olivine), or by nonoccupancy of normally occupied sites. That is, relative to a local singularity in structure produced simply by reuniting the two surfaces of a hypothetical cut after their relative translation by the Burgers vector, a modification or reconstruction of bonding relationship may often be expected in the core since its energy would be thereby reduced, and segregation or redistribution of impurity or solute atoms may be involved in the modification. The actual core structure can, in principle, be obtained from ab initio or similar quantum mechanical calculations but, in practice, a more rudimentary approach in terms of interaction with vacancies and solute or impurity atoms is usually taken, expressed in terms of binding energies for the interaction, determined more or less empirically (Hirth and Lothe 1982, p. 512). The effect of the segregation of vacancies or solute atoms to the dislocation core will be to modify the Peierls potential, decreasing it in case of attractive interaction; if the Peierls valley is thereby deepened and its walls steepened, the effect will be accompanied by an increase in Peierls stress. If the dislocation is dissociated, segregation to the fault strip between the partial dislocations (Suzuki atmosphere) will have a similar effect. However, unless the Peierls stress is sufficiently high to dominate the dislocation behavior, the role of the binding interaction with the point defect in the core will be less important than the elastic interaction since most of the dislocation energy tends to be in the elastic field.

The interaction of point defects with the long-range elastic stress field is discussed in terms of the elastic distortion associated with the point defect. In the more refined treatments, the point defect is not regarded as a rigid inclusion to be accommodated volumetrically in an elastic matrix but it is viewed as also undergoing elastic distortion itself, with elastic constants different from those of the matrix. The size misfit aspect is then sometimes referred to as a *paraelastic* interaction and the modulus difference aspect as a *dielastic* interaction (Haasen 1983). The elastic interaction energy (negative for attractive interaction) can be shown to be finite for both edge and screw dislocations. In the case of screw dislocations, in which there is no dilatancy to first order in the elastic field, the interaction arises through the dielastic effect involving the shear modulus, as well as to some extent through second-order elastic effects. For the cases in which the temperature is high enough for redistribution of point defects to occur, their equilibrium concentration in the neighborhood of the dislocation can be calculated taking into account the elastic interaction energy per defect and any necessary formation energy for the defect. Thus, an atmosphere of point defects can be established around the dislocation, giving a pinning effect. This atmosphere is known as a *Cottrell atmosphere* in the case of an enhanced concentration of solute atoms. If the total energy of interaction of the atmosphere with the dislocation can be calculated as a function of the displacement of the dislocation, then the interaction force between the dislocation and its atmosphere can be obtained.

Finally, electrostatic interaction will occur between a charged dislocation and charged point defects, a situation that might be expected to be particularly relevant in nonmetallic compounds. However, it is not known, in general, how important this type of interaction is in practice. In the case of alkali halides, at least, it is thought to be of negligible importance mechanically compared with the elastic interactions (Sect. 6.2.7).

The mechanical implications of the interactions between point defects and dislocations depend in general, on the extent to which the distribution of point defects is at equilibrium with the dislocation configuration initially present and on the rate at which redistribution can occur during dislocation motion. The effects are mainly of significance where the defects are solute atoms. Three particular situations may be selected for comment:

1. If, following introduction of the dislocations, the temperature has been sufficiently low that redistribution of point defects in the neighborhood of the dislocations has not occurred, then during deformation the dislocations move through a more or less random but fixed assemblage of centers of interaction which represent local potential barriers to be overcome by the applied stress in

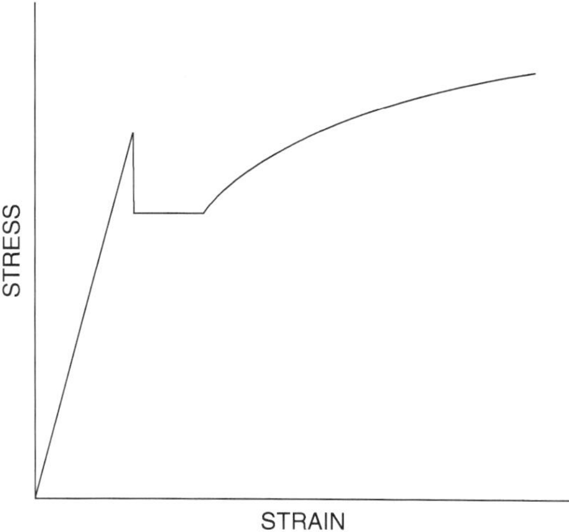

Fig. 6.9 A "yield point" in a stress-strain curve

moving the dislocations. This situation gives rise to an overall resistance to dislocation motion, known as solute or solid solution hardening when the point defects are solute atoms. Theoretical studies suggest that the solute introduces a strengthening proportional to $c^{\frac{1}{2}}$, at low concentration c, to $c^{\frac{2}{3}}$, at medium to high concentration, with a proportionality constant that contains the interaction force or "obstacle strength" with a power somewhat greater than unity (see further Sect. 6.6.3).

2. If, however, solute segregation has occurred, either in the core or nearby, so as to tend to pin or lock the dislocations in their initial positions but the test temperature is too low for diffusion of the solute to occur at an appreciable rate, then in order to initiate plastic deformation a relatively high stress is at first needed, higher than the stress needed to maintain the dislocations in motion once they are freed from their locking atmospheres. In this case, there is a tendency for the stress to fall in the early stages of a stress–strain test, after initiation of plastic deformation, an effect commonly referred to as a "yield point" (Fig. 6.9).

3. Alternatively, if the temperature is high enough that the segregating species can keep pace with the dislocation movement by diffusion, then the interaction between solute and dislocation leads to a viscous drag force on the dislocation. This effect will tend to give rise to linear (Newtonian) viscous behavior in which the velocity of the dislocation is proportional to the resolved shear stress acting in the region of the crystal containing the dislocation. The constant of proportionality will vary directly with the diffusion coefficient of the solute and inversely with the mean solute concentration, as well as depending on the interaction potential governing the distribution of the solute around the dislocation. (See further, Sects. 6.4.1 and 6.6.5).

6.3.2 Interaction with Dispersed Second Phases

Dispersed second phases, commonly formed as fine precipitates, can play a very important role in obstructing dislocation motion. The effects are widely exploited technologically for the strengthening of materials.

The resistance to the motion of a dislocation presented by a distribution of small second phase particles may be expected to resemble in some respects that presented by immobile atoms in solid solution but the individual interactions will, in general, be much stronger. The actual interaction forces will vary widely according to the natures of the particles and of the processes whereby the dislocation cuts through the particles or circumvents them. The latter distinction, of cutting or bypassing, is important in determining the overall nature of the interaction and its consequences for the macroscopic mechanical behavior. The behavior may also involve quite different considerations at low and high temperatures, depending on whether thermal activation is playing a significant role in assisting the cutting or bypassing. (See further, Sects. 6.6.3 and 6.6.7).

The particle size is a very important variable in determining the nature and intensity of the influence of a given dispersed second phase. The respective barriers to a dislocation cutting through particles and circumventing them are commonly such that there is a maximum hardening effect at an intermediate particle size (often submicron) in the range of practical particle sizes (see Sect. 6.6.2). Eventually, at large particle sizes, the second phase can be regarded as simply providing boundaries with which dislocations in the matrix phase interact, as we shall next discuss.

6.3.3 Interaction with Boundaries

The energy of a dislocation, which, as already pointed out, consists mainly of the elastic strain energy in the long-range stress field, is modified when the dislocation is situated sufficiently near a boundary that part of the long-range stress field is located on the other side of the boundary. If the latter region consists of an elastically softer material or of free space, the dislocation near the boundary will have a lower energy than one further away and hence will be subject to an image force attracting it toward the boundary. Conversely, if the second region consists of elastically harder material, the image force will be a repulsive one. For the theory of image forces, Friedel (1964, p. 44), Nabarro (1967, Chap. 5), Haasen (1978, p. 252) and Hirth and Lothe (1982, Chaps. 3 and 5) can be consulted. The image force tends to deflect a dislocation line as it approaches to intersect a surface. Thus, it may influence the configuration of dislocations in the thin foils viewed in transmission electron microscopy if the dislocations are sufficiently freely mobile.

There is also a short-range interaction of a dislocation with a boundary since the boundary has to be cut to produce a step or shear discontinuity when the dislocation crosses the boundary (Friedel 1964, p. 46; Nabarro 1967, p. 282). This interaction does not greatly impede the dislocation emergence at a free surface when only surface energy has to be supplied in forming the step (no effect at all would be expected for a screw dislocation in this case). However, when there is strong material on the other side of the boundary that also has to take part in the step formation, there may be a large resistance to the emergence. This effect could be important in affecting mechanical properties when surface films are present on specimens. In the case of ionic crystals, there may also be charge effects where dislocations emerge at surfaces, and in any material the point of emergence tends to be a preferred site for chemical activity.

Even within an imperfect crystal, planar defects such as faults and antiphase boundaries (surfaces at which there is an ordering discontinuity in an ordered structure) may act as boundaries to interact with a dislocation. There will in general be no image forces involved but some resistance to passage of the dislocation will arise from the creation of additional boundary energy through the formation of a step (in the case that the dislocation has an edge component) or of a shear discontinuity (in case of a screw component).

When a sufficiently large repulsive interaction exists, the dislocation will be arrested at the boundary and so prevent subsequent dislocations on the same slip plane from reaching the boundary. This effect leads to a pile-up of dislocations of like sign in the given slip plane. The dislocations themselves interact repulsively in the pile-up. We therefore now consider the mutual interaction of dislocations.

6.3.4 Mutual Interactions Between Dislocations

Mutual interactions between dislocations are thought to underlie much of the observed plastic behavior of crystals in both athermal and thermal regimes, especially where strain hardening is involved and Peierls stress effects are not dominant. The variety of effects, both observed and envisaged in theories, is very large and has given rise to a large and confusing literature, the most complex in dislocation theory. Here, and in Sects. 6.6.3 and 6.6.6, we can only attempt to indicate the essential elements in this theoretical plethora.

Broadly, the mutual interaction can be a long-range effect arising from the overlap of the long-range elastic stress fields of the dislocations, or a short-range effect involving intersection or reaction. The long-range interactions tend to be more important for dislocations that are more or less parallel to each other and the short-range for those more nearly normal to each other.

Parallel dislocations may either attract or repel each other, depending on their relative positions and Burgers vectors and on their edge or screw character. The attraction or repulsion is determined by whether, when superimposing the stress fields, as given by (6.3a, b) or a combination of these for mixed dislocations, the total elastic strain energy is, respectively, less or greater than the sum of the separate elastic energies. Consideration of specific cases indicates that, for a given Burgers vector and for isotropic elasticity:

1. Parallel screw dislocations always attract or repel each other, depending on whether they are of unlike or like sign, respectively.
2. Parallel edge dislocations of like sign attract or repel each other if the plane joining them is inclined at greater or less than 45° to the slip plane, respectively; these interactions are reversed for unlike signs.

More general rules can be found for mixed and inclined dislocations and for those of different Burgers vector (Hirth and Lothe 1982, Chap. 5; Weertman and Weertman 1964, Chap. 3). The repulsive effect determines the spacing of successive dislocations of like sign in a pile-up in a given slip plane in which the leading dislocation is immobilized. The attractive effect underlies the stability of dipoles, which are pairs of parallel dislocations of opposite sign lying close to each other in parallel slip planes, and of tilt subgrain boundaries, which are arrays, normal to the slip plane, of edge dislocations of like sign.

In the case of dislocations steeply inclined to each other, the long-range elastic interactions may lead to some local distortion of the dislocations but, on the whole, they may be of less importance than short-range interactions and the effects of the actual mutual intersection of the dislocations. Relative to a given dislocation moving in its slip plane, the inclined dislocations crossing the slip plane are commonly described as *forest dislocations* and their intersection by the moving dislocation as forest cutting. Two short-range interaction effects may be specially mentioned. The first is the formation of an *attractive junction* due to the moving and forest dislocations being locally deviated into parallelism and then reacting to form a product dislocation segment of lower energy; the result is to tend to pin the moving dislocation at the junction and to contribute to the building up of a three-dimensional network. The second effect is to introduce an offset in each of the dislocation lines as a result of their intersection, each offset being identical with the Burgers vector of the other dislocation and constituting either a kink or a jog which may influence the future mobility of the dislocation cut. The intersection process may be thermally activated since the formation of the kinks or jogs requires energy which may be in part provided by thermal fluctuation.

6.4 Dislocation Velocity

Since crystal plasticity arises from the motion of dislocations, it is important to understand the factors affecting their velocity, especially in relation to creep. The dislocation velocity is focussed upon as one of the primary quantities in the microdynamical approach to the theory of crystal plasticity, but it may well be involved in any rate-dependent aspects of deformation.

Since it is the glide of dislocations that usually accounts for most of the strain, the glide velocity is the dynamical aspect most studied. However, cross-slip and climb velocities can also be important, especially as rate-determining factors when obstacles to glide in the primary slip plane have to be bypassed. We shall therefore consider briefly all three aspects of dislocation velocity.

6.4.1 Glide Velocity

In the absence of energy dissipation, an isolated dislocation would tend to accelerate indefinitely under applied stress were it not for a relativistic limitation to the velocity of sound. A direct extension of the elastic theory of dislocations predicts that, as the velocity of the dislocation increases, not only does a kinetic energy term appear in its total energy but the self energy of the dislocation also increases in a way that can be described roughly as relativistic (Hirth and Lothe 1982, Chap. 7). The self energy approaches infinity and the stress field undergoes a relativistic contraction in the direction of motion as the dislocation velocity approaches the velocity of sound (the velocity for transverse or longitudinal waves, v_s or v_p, respectively, applies depending on whether screw or edge dislocations are concerned). However, in practice, there are many dissipative processes that generally limit the velocity to the sub-relativistic domain (Hirth and Lothe 1982, Chaps. 7, 15, 16; Weertman and Weertman 1980).

Some of the dissipative processes are intrinsic effects, applying to an isolated dislocation in an otherwise perfect crystal. One such process is the scattering of phonons associated with the thermal vibrations in the crystal, which becomes important at high velocities, perhaps generally in excess of about $10^{-3}\,v_s$ or $1\ \mathrm{ms^{-1}}$, and which gives rise to a viscous drag on the dislocations (Granato 1984, and other references therein; Hirth and Lothe 1982, p. 211; Kocks et al. 1975, p. 85). However, when there is a significant Peierls potential, this gives rise to a more important viscous drag on dislocations, effective at much lower velocities and treatable in terms of kink mobility (Hirth and Lothe 1982, Chap. 15). Hirth and Lothe (p. 545) deduce a relationship which, in the approximation that the spacing of the Peierls valleys and the jump distance for kink migration are each equated to the Burgers vector b, can be written as

$$v \approx \frac{2\tau b^4 v}{kT}\exp\{-(E_{kn} + E_{km})/kT\} \tag{6.12}$$

where τ is the resolved shear stress, v the attempt frequency (of the order of the atomic vibration or Debye frequency, say, $\sim 10^{13}\ \mathrm{s^{-1}}$), and E_{kn}, E_{km} the activation energies for kink nucleation and migration, respectively. The term $\tau b^4 v/kT$ can be written as bv' where v' has the character of a vibration frequency of the dislocation line under stress τ (in effect, $\tau b^3 = hv'$ and $kT = hv$, where h is the Planck constant). In addition to the linear stress dependence introduced in this term, there is also a stress dependence in the activation energy. Thus, the kink nucleation energy

E_{kn}, although depending in form on the form assumed for the Peierls potential, can commonly be approximated as

$$E_{\mathrm{kn}} = E_k \left\{ 1 - \left(\frac{\tau}{\tau_p} \right)^{\frac{3}{4}} \right\}^{\frac{4}{3}} \tag{6.13a}$$

according to Kocks et al. (1975, p. 187), where E_k is the kink energy (Sect. 6.2.4) and τ_p the Peierls stress. A value for E_{km} seems to be more difficult to estimate (Hirth and Lothe 1982, p. 534); it is probably somewhat less than E_{kn} but may still be quite significant (Hirsch 1985). Further development of the theory needs to take into account the finite lengths of dislocation segments between nodes that limit the free run of kinks (Hirth and Lothe 1982, p. 545). For a discussion of the role of kinks in the dislocation velocity in germanium and silicon, see, for example, Louchet and George (1983) and Jones (1983).

In the majority of situations, the intrinsic behavior just considered will probably be masked by the influence of extrinsic factors affecting dislocation velocity. These factors may include any of the dislocation interactions listed in Sect. 6.3 which have a local pinning effect on the dislocation. The treatment by Hirth and Lothe (1982, Chap. 16) of the velocity of jogged dislocations provides an approach to the treatment of a wider range of local pinning effects.

More generally, if the glide velocity of a dislocation is determined by rate of thermally activated crossing of barriers of any sort, and if we express the velocity as

$$v = (\Delta A / l) v \tag{6.13b}$$

where A is the area swept out by a dislocation segment of length l in a single activation event, and v is the frequency of such events, given by (3–12 g), then we obtain

$$v = \frac{2 v_0 \Delta A}{l} \sinh \frac{\tau b \Delta A}{2kT} \exp \left\{ - \frac{\Delta E^* - \tau b \left(\Delta A^* - \frac{1}{2} \Delta A \right)}{kT} \right\} \tag{6.14}$$

where, in addition to ΔA and l just defined, v_0 is the attempt frequency (now of the order of the vibration frequency for a dislocation segment, say, 10^{10}–10^{11} s^{-1} Friedel 1964), ΔE^* the activation energy and ΔA^* the activation area.

In practice, it is usually difficult, a priori, to assign values to the parameters l, ΔA, ΔA^* and ΔA^* in (6.14) and so it is common to adopt a more empirical approach and attempt to fit experimental observations to expressions that are analogous to the approximate forms in Eqs. (3.12d) and (3.12e) in Sect. 3.2.4, namely:

$$v = A_0 \tau^m \exp \left(- \frac{Q}{RT} \right) \tag{6.15}$$

and

$$v = v_0 \exp \left(- \frac{Q - \tau b \Delta A^*}{RT} \right) \tag{6.16}$$

where A_0, v_0 are empirical constants, Q is an empirical activation energy (enthalpy), m and ΔA^* are the empirical parameters $(\partial \ln v/\partial \ln \tau)_T$ and $(RT/b)(\partial \ln v/\partial \tau)_T$, respectively, and R is the gas constant (note that ΔA^* and m are formally related by $m \equiv \tau b \Delta A^*/RT$). In Eqs. (6.15) and (6.16) the activation area and activation energy have been multiplied by the Avogadro constant, relative to the values in (6.14), and the entropic part of the energy has been subsumed under v_0, A_0.

Following Sect. 3.2.4, (6.16) should be the more appropriate form to use when $\tau b \Delta A \gg 2RT$. This situation tends to arise at relatively high stress and low temperature, when there are important viscous drag effects such as the Peierls resistance or the analogous low-temperature drag on screw dislocations in iron, thought to arise from the extension of core into a zonal structure in three intersecting planes (Hirsch 1960; Hirth and Lothe 1982, p. 369; Louchet 1979; Mitchell et al. 1963). Alternatively, if viscous drag effects are still controlling the dislocation velocity at relatively low stress and high temperature with $\tau b \Delta A \ll 2RT$, then the form in (6.15) should be the more appropriate and m could be expected to be close to unity.

Turning to the experimental situation, there have been various direct measurements of mean dislocation velocity in glide as a function of stress and temperature, using etching, X-ray topography or transmission electron microscopy for tracking the dislocations (for reviews, see Alexander and Haasen 1968; Gilman 1969; Haasen 1978, Chap. 11; Sprackling 1976, Chap. 9). The stress to which the velocity is related is commonly, and logically, taken to be the "effective stress" τ_e acting locally on the dislocation. The effective stress is estimated as the applied stress minus the internal stress τ_i arising from other dislocation, τ_i being taken to be equal to $\alpha G b \rho^{1/2}$ (when α is a numerical constant near to unity, G the shear modulus, b the Burgers vector, and ρ the dislocation density; a rationalization for this expression will be given in Sect. 6.6.2). When the experimental results are fitted to the form (6.15), the exponent m is found to vary widely, from values of approximately unity for silicon and germanium to values of 100 or more for copper. The value of m can also vary markedly with effective stress, so that several velocity/stress regimes can be distinguished as the stress is increased, commonly with $m \sim 1$ at very low or very high velocities and with higher values of m at intermediate velocities (Haasen 1978, p. 263).

Observed values of $m \approx 1$ are consistent with a predominance of viscous drag effects, and determinations of activation area may help to identify the controlling factor. Thus, in the case of germanium and silicon where a value of $m \approx 1$ is found at around 700–1,000 K and applied stresses of 10–100 MPa (still a relatively low-temperature regime for these materials), the activation area, calculated from $\Delta A^* = mkT/\tau b$, is of the order of a few times b^2 (Louchet and George 1983). This result is suggestive of the dislocation motion being controlled by a local effect in the core, such as kink nucleation or migration. In contrast, measurements on most metals and ionic crystals at relatively low temperatures, analyzed in a similar way, tend to give values of m much greater than unity and values of ΔA^* large compared with b^2, indicating that there is some factor other than a viscous drag involving a simple core interaction controlling the motion (for data, see Sprackling 1976, Chap. 9). Values of m substantially greater than unity suggest that the population of sites of effective thermal activation is also stress dependent and perhaps evolving with dislocation motion, an understanding of which depends on detailed microstructural study.

Transmission electron microscope observations in situ in metals (for example, Caillard and Martin 1983) reveal, in fact, behavior that is in marked contrast to the picture of a steady viscous drag. The observed motion tends to be unsteady on a relatively large scale in that periods of no motion alternate with periods during which relatively large areas are swept out in rapid motion; that is, the dislocation advances in spurts. This behavior can be explicitly taken into account by expressing the mean velocity v as:

$$v = \frac{\Delta s}{t_0 + t_g} \quad \text{or} \quad v = \frac{\Delta A}{l(t_0 + t_g)} \tag{6.17}$$

(Kocks et al. 1975, p. 93; Philibert 1979; Poirier 1985, p. 94),, where Δs is the distance or ΔA the area swept out by a dislocation segment of length l in a typical interval of time consisting of t_0 spent waiting at an obstacle and t_g spent in actual motion. In view of the possible variety of obstacles and drag processes and, hence, variation in the relative importance of the t_0 and t_g terms, a complex phenomenology can be expected, preventing simple interpretation of the empirical expressions in Eqs. (6.15) and (6.16) and making theoretical prediction difficult, although some attempt has been made by Morris and Martin (1984a, b).

6.4.2 Cross-Slip Velocity of Screw Dislocations

For undissociated dislocations, motion in the cross-slip plane will be controlled by the same factors as that in the primary slip plane, apart from the effects of change in the resolved shear stress and of any change in the Peierls stress or other crystallographically controlled factor in cases in which the cross-slip plane is not equivalent crystallographically to the primary plane. However, when dissociation of the screw dislocations occurs, the relative case of their movement into or within the cross-slip plane may be strongly affected. Effects of this kind have been especially studied in metals (see Vitek 1985, for a recent review) but they are probably widespread; for example, Poirier and Vergobbi (1978) suggest such an effect in olivine.

Thus, if the dissociation yields a partial dislocation having a Burgers vector nonparallel to the cross-slip plane, then separate movement of this partial dislocation in the cross-slip plane would be very difficult and result in forming a high-energy fault surface. In order for movement to occur in the cross-slip plane, the partial dislocations in such a case need to be recombined locally during the cross-slip motion through the formation of a constriction in the extended dislocation. Such a process can be thermally activated and models have been proposed by Schoeck and Seeger (1959), Seeger et al. (1959) and Wolf (1960), on the one hand, and by Friedel (1959, 1964, p. 164) and Escaig (1968a, b), on the other hand, for situations where the dislocation is less or more dissociated in the cross-slip plane than in the primary plane, respectively (see also Schoeck 1980; see also Vanderschaeve and Escaig 1979). The theoretical discussions have centered mainly on estimating the activation energy for the dislocation motion in the cross-slip plane, predicting that in some cases the activation energy will depend on the stress and in other cases not.

In the case of only slightly dissociated dislocations, the energy barrier associated with the constriction mentioned above can be viewed formally as a sort of Peierls barrier and the motion of the constriction as analogous to the migration of a kink. This view is particularly appropriate in cases where the dissociation is difficult to resolve but the core can be regarded as being extended somewhat outside the cross-slip plane, as in a zonal dislocation (cf. b.c.c. metals: Hirth and Lothe 1982, p. 369; cf. b.c.c. metals: Louchet 1979; Vitek 1985).

6.4.3 Climb Velocity of Edge Dislocations

We consider only the case of a pure edge dislocation under the influence of a normal stress σ applied parallel to its Burgers vector (for the case of a mixed dislocation and a general stress, see Hirth and Lothe 1982, p. 562). The first estimate of the climb velocity v can be made using the formalism of Sect. 3.2.4 if we regard the mobile entity as a dislocation segment of length equal to the structure repeat distance b_1 parallel to the dislocation and we write $v = b_2 v$ where b_2 is the structure repeat distance in the direction of climb and v the jump frequency. From relation (6.6b) the climb force on such a segment is $\sigma b b_1$, which can be written in molar terms as σV_m where V_m is the molar volume. In the case of relatively low stresses such that $\sigma V_m \ll RT$, we can then apply (3.12d) to obtain

$$v = b_2 v = \frac{b_2 V_m \sigma v_0}{RT} \exp\left(-\frac{\Delta G^*}{RT}\right) \tag{6.18}$$

where ΔG^* is the molar activation energy. If the thermally activated event is essentially one of volume self-diffusion, we can treat $v_0 \exp(-\Delta G^*/RT)$ as the diffusional jump rate Γ which, from (3.32), is equal to $D/\alpha f b_3^2$ where D is the diffusion coefficient, αf a numerical factor of order unity, and b_3 the jump distance. Thus, in the approximation that αf is put equal to unity and b_2, b_3 equal to the Burgers vector b, (6.18) becomes

$$v \approx \frac{D V_m \sigma}{bRT} \tag{6.19}$$

An alternative approach, analogous to that used in Chap. 5 (especially Sect. 5.7), is to suppose that, during the climb, material is transported to or from a reservoir consisting of the surface of a cylinder of radius R, R being the average distance to neighboring dislocations or other structural discontinuities or surfaces that can act as sinks or sources. The steady-state solution of the transport Eqs. (3.18) and (3.22) in such a case is

$$J/l = \frac{2\pi\lambda\Delta\mu}{\ln(R/r)} \tag{6.20}$$

(Carslaw and Jaeger 1959, p. 189), where J is the flux from a length l of dislocation when the driving potential difference between the cylinder surface and the dislocation is taken to be the difference in chemical potential $\Delta\mu$ (assuming σ and T as independent variables), r the effective radius of the dislocation core, and λ the quantity M/V_m where M is the mobility of the diffusing material and V its molar volume (see Sect. 3.4). The flux J is related to the climb velocity v through $J = vbl/V_m$, which substituted in (6.20) leads to

$$v = \frac{2\pi}{\ln(R/r)}\frac{M\Delta\mu}{b} \tag{6.21}$$

or

$$v = \frac{M\Delta\mu}{b} \tag{6.22}$$

The mobility M in (6.22) is related to the diffusion coefficient D by the Einstein relation (3.27), $M = D/RT$. The value of $\Delta\mu$ is obtained from the change in the work term in the chemical potential as $\Delta\mu = V_m\sigma$ if there is no potential barrier at the source or sink. Substituting these quantities in (6.22) leads again to

$$v \approx \frac{DV_m\sigma}{bRT} \tag{6.23a}$$

or, in the approximation that $V_m \approx Lb^3$, where L is the Avogadro number, to

$$v \approx \frac{Db^2\sigma}{kT} \tag{6.23b}$$

The two results in Eqs. (6.19) and (6.23a) are, of course, equivalent apart from minor differences in the approximations due to slight differences implicit in the posing of the two climb models.

In applying (6.23a) to elements, V_m and D are the molar volume and self-diffusion coefficient for the atomic species. In the case of compounds, V_m is the molar volume of the molecular species constituting the crystal, and D is the effective self-diffusion coefficient for this species, which can be obtained from the self-diffusion coefficients of the constituent atomic species.

Although direct experimental testing of (6.23a) would be difficult, it is widely accepted for application to problems such as climb creep and the growth of prismatic loops when it is thought that pipe diffusion is negligible.

6.5 Dislocation Populations and Their Evolution

In a real crystal undergoing plastic deformation, the dislocations tend to be neither uniformly distributed nor equally mobile, and the density of the dislocation assemblage and its configuration tend to evolve during straining. These structural and behavioral factors have to be taken into account in any microdynamical theory of flow. Microstructural observation therefore plays a vital part in guiding the proper development of an adequate theory of macroscopic flow. Observations are required both at the optical microscope scale, to reveal deformation banding, subgrain formation, etc., at this scale, and at the submicroscopic scale, to reveal the actual dislocation configurations, the formation of pile-ups, dipoles, loops, tangles, cellular arrangements, etc., often described as "substructure".

Even a crude description of the dislocation population must, in general, take account of two aspects, which may be designated as the mean density and the "cellularity" (Kocks 1985a). These two aspects will be dealt with in the first two subsections to follow, and the more profound microstructural reorganizations involved in recovery and recrystallization will be touched upon in the third subsection.

6.5.1 The Configuration of the Dislocation Assemblage

The configuration of individual dislocation segments can reflect the nature of the barriers to dislocation motion and of the interactions between dislocations. This configuration can be revealed by transmission electron microscopy (TEM). When the crystal is oriented in the electron beam so that a Bragg diffraction condition is near to being satisfied, the slight variations in orientation due to the variation in distortion in the long-range elastic stress field give rise to local increase or decrease in the intensity of diffraction near the dislocation core compared with elsewhere, and hence to the dislocation being imaged by the diffraction contrast effect Figs. 6.10, 6.11, 6.12, 6.13, 6.14, 6.15, 6.16).

The observation of long straight dislocation lines of simple crystallographic orientation points to a strong interaction between the dislocation and the structure itself. This interaction may consist of a large amplitude of the Peierls potential in the case of a nonextended dislocation, or it may involve an effective deepening of the Peierls valley by extension of the dislocation core into planes other than the slip plane, as in the case of climb dissociation (Sect. 6.2.5). Examples of straight dislocations suggesting structural control are shown in Figs. 6.10a and 6.11a,

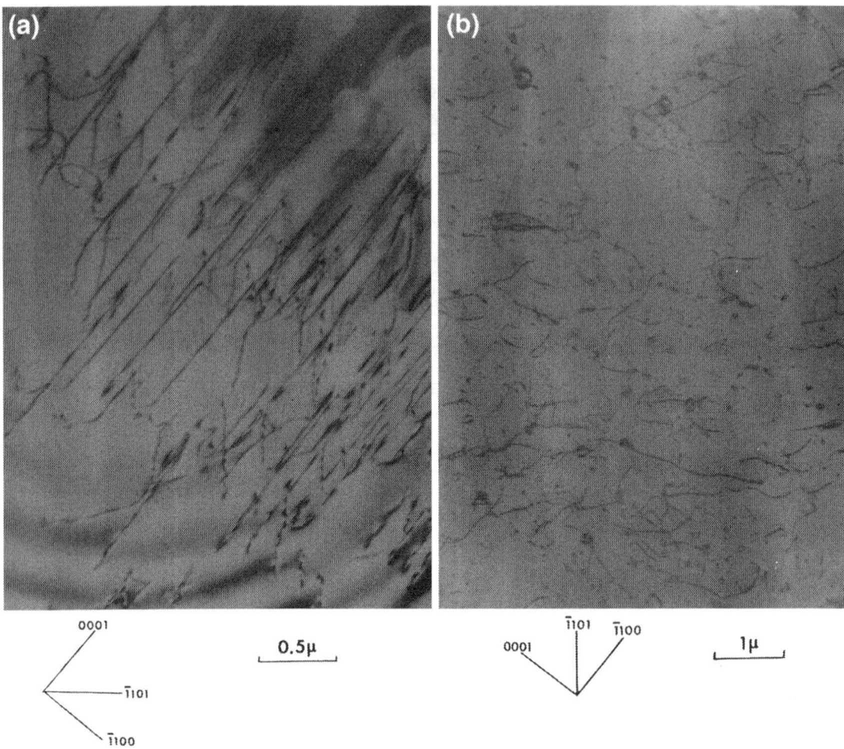

Fig. 6.10 a Straight dislocations in quartz crystal deformed at 575 °C. **b** Wavy dislocations in quartz crystal deformed at 800 °C (copies of Figs. 5.10 and 5.20, respectively, from Morrison-Smith 1973)

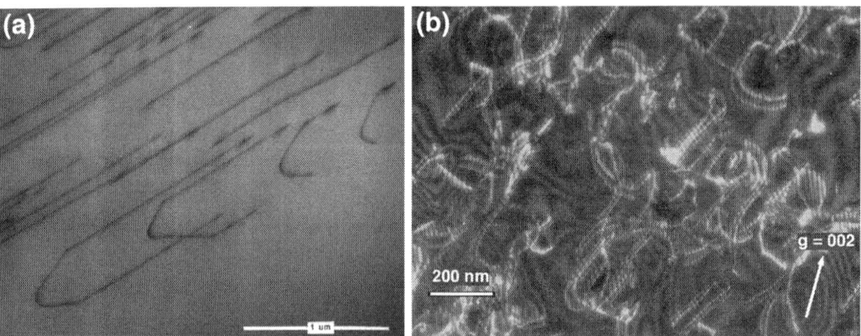

Fig. 6.11 a Straight screw dislocations in olivine (supplied by Dr. J. D. Fitz Gerald). **b** Dissociated dislocations in dolomite plagioclase (from Stünitz et al. 2003)

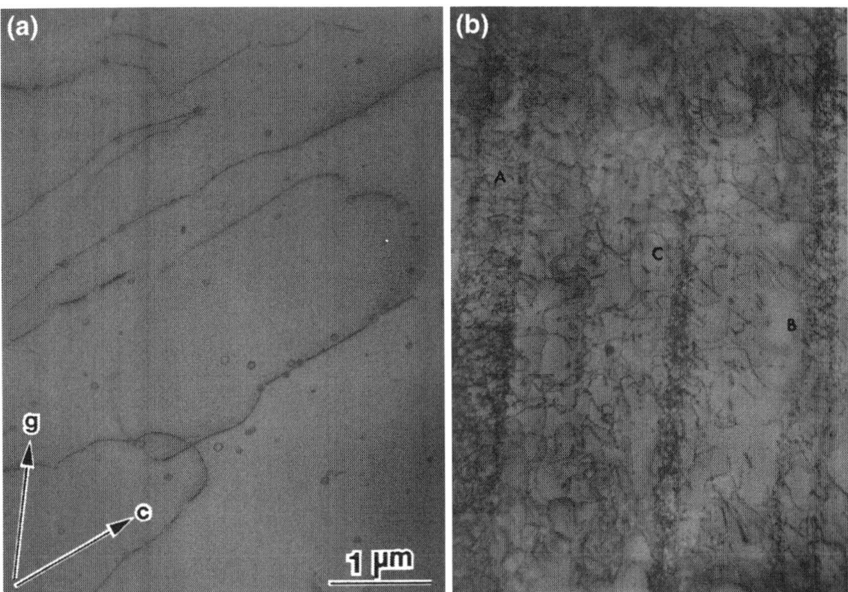

Fig. 6.12 a Effect of an obstacle on the motion of dislocations (a copy of Fig. 9 from McLaren et al. 1989). b Dislocation cell structure in quartz (copy of Fig. 7.4 from Morrison-Smith 1973)

Fig. 6.13 Dislocation structure in a low-angle tilt boundary (supplied by Dr. J. D. Fitz Gerald)

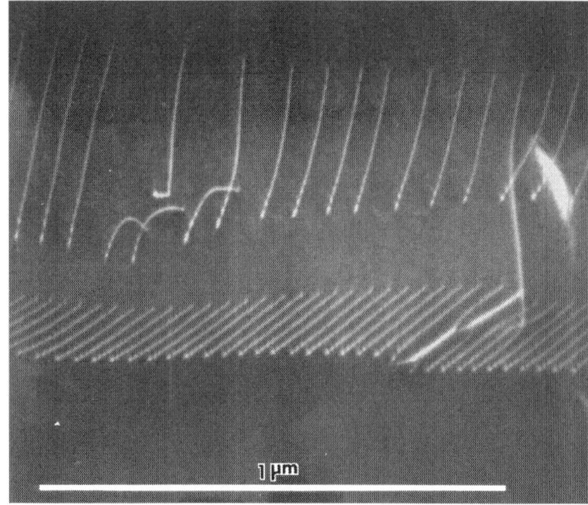

while the curved characters of the dislocations illustrated in Fig. 6.10b suggests that structural control is unimportant in this case. Dissociation of dislocations can also be revealed in TEM images (Fig. 6.11b).

The effect of localized obstacles on the motion of dislocations can be seen in TEM images such as Fig. 6.12a. However, special techniques may have to be used

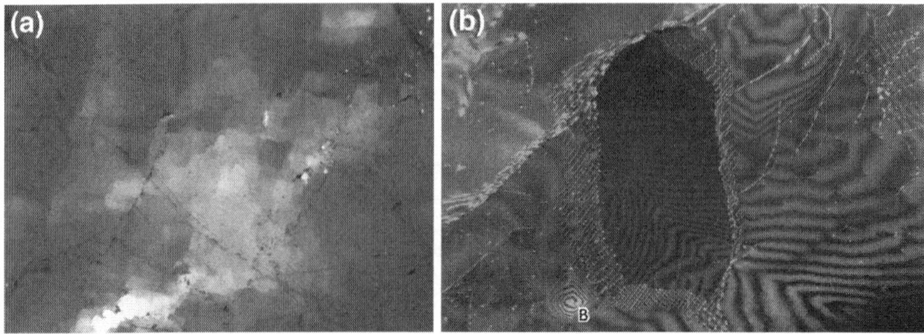

Fig. 6.14 a Subgrains revealed in optical microscopy in plane-polarized light in naturally deformed olivine (scale: the long edge is 1 mm) (supplied by Dr. J. D. Fitz Gerald). b Dislocation structure marking the boundaries of a subgrain in experimentally deformed quartz (copy of Fig. 3c in Fitz Gerald et al. 1991)

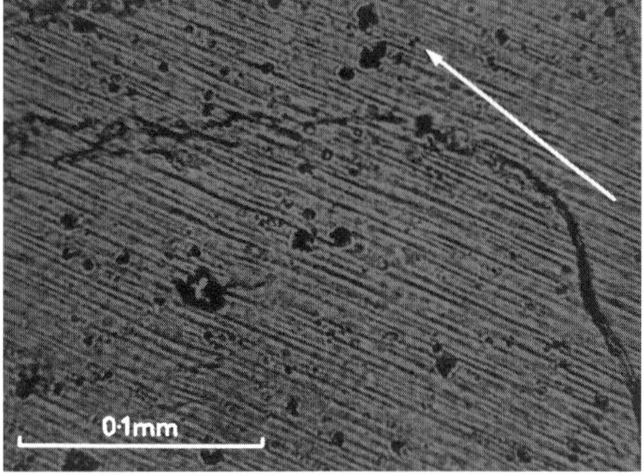

Fig. 6.15 Deformation lamellae in synthetic quartz revealed in plane-polarized light with x10 objective; the trace of $(10\bar{1}0)$ is marked (copy of Fig. 4a from McLaren et al. 1970)

in order to obtain an image of the dislocation configuration as it actually exists while the crystal is under internal stress, as will be discussed further later in this subsection.

Passing from the individual dislocation line to the spatial distribution of the dislocation assemblage, the primary observation is that there is commonly a marked heterogeneity in this distribution, which may take different forms. On the one hand, there may be a variation in dislocation density without obvious variation in the organization of the dislocations relative to each other, giving rise to a *cell structure* such as illustrated in Fig. 6.12b, in which the walls defining the cells consist of tangles of dislocations. On the other hand, the dislocations may be organized into walls which consist of orderly arrays containing one or more sets of parallel dislocation lines of given sign, forming what are called *subgrain boundaries*, because of the small change in orientation across them (Fig. 6.13). Subgrain

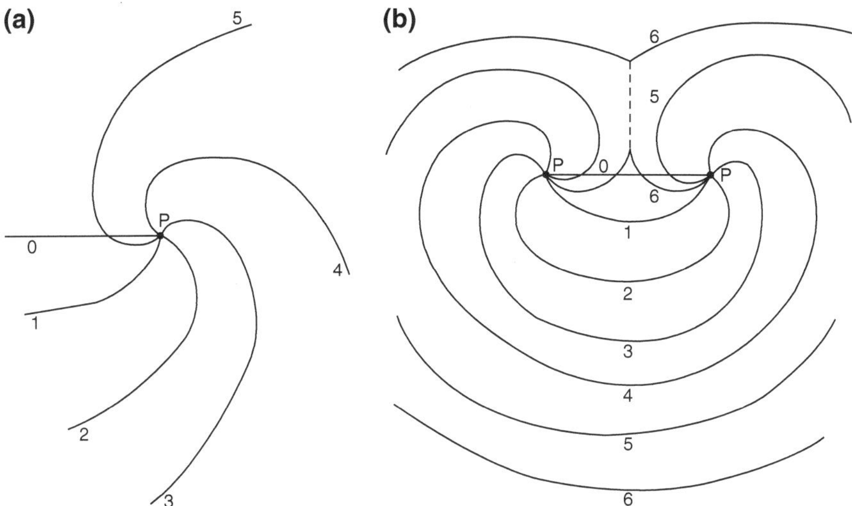

Fig. 6.16 a Dislocation source initiated as a dislocation 0 pinned at point P; the subsequent positions 1, 2, 3,…of the dislocation line represent increasing length of the dislocation, augmenting the dislocation density of the crystal. b A "two-armed" source, known as a Frank-Read source

boundaries may form for reasons of energy minimization, their formation being a process that occurs more readily at higher temperatures and probably usually involves some climb of the dislocations. However, although an energy minimization argument for cell formation can be given (Weertman and Weertman 1983a, p. 1292), cell structure may also form for kinetic rather than energetic reasons (Kocks 1985a). Geometric constraint may also be an influential factor in dislocation distribution, especially in requiring an excess of dislocations of given sign in order to accommodate imposed bending; such dislocations are sometimes called geometrically necessary dislocations.

When there is a well-defined substructure, the dislocations in the cell walls or subgrain boundaries are sometimes referred to as "bound" dislocations and those lying within the cells or subgrains, often forming themselves a three-dimensional network, as "free" dislocations. However, it cannot be concluded without further evidence that these two sets of dislocation are immobile and mobile, respectively. In fact, the reverse may be the case, as shown for example, by Martin and coworkers in careful TEM studies on various metals undergoing creep, using techniques both of pinning the dislocations before releasing the applied stress and of direct observation during deformation in the electron microscope (Caillard and Martin 1982a, b, 1983; Clément et al. 1984; Kubin and Martin 1980; Morris and Martin 1984a, b). They showed that the mobile dislocations in these materials traverse the subgrains very rapidly after breaking away from subboundaries in which they reside for most of the time ($t_0 \gg t_g$ in expressions (6.17)). The breakaway process probably involves thermally activated cross-slip or climb, or in alloys, escape from solute pinning. Also the mobile dislocations may eventually traverse many subgrains, the spacing of the dislocation sources being greater than the subgrain size, and the subboundaries themselves may have some mobility, contributing a minor component to the strain. Martin and coworkers also show that there is an internal stress field extending some way into the subgrain interior from the subboundary dislocations, which are therefore not fully mutually compensating in respect of their long-range elastic stresses (not in "equilibrium" in the sense of Frank 1955). In minerals, insofar as mutual interaction of dislocations on multiple slip systems is important, there may well be similar behavior.

The parameter most frequently used to characterize the cellularity of the dislocation substructure is the cell or subgrain diameter. However, depending on ideas of what most influences the flow strength, other parameters have also been used, especially the dislocation link length within sub boundaries or sub boundary mesh size (Ardell and Przystupa 1984; Lin et al. 1985; Morris and Martin 1984a; Öström and Ahlblom 1980; Öström and Lagnerborg 1980).

Coarser-scale structure reflecting heterogeneity in deformation behavior is seen at the microscopic or grain scale. Microscopically visible slip bands (Sect. 6.1; Fig. 6.2), themselves representing a heterogeneity in dislocation activity, are often distributed heterogeneously, especially in grains in polycrystals, giving rise to various sorts of *deformation bands*. These bands may be defined by differences in the amount of activity of a primary slip plane, by the localized activity of a secondary slip system, by the alternation of active slip systems, and so on, and they form regions of corresponding heterogeneity in distribution of dislocations. At higher temperatures and lower strain rates, subgrain formation may also become evident, the optical observations detecting coarser-scale subgrains than the submicroscopic.

Optical or scanning electron microscope (SEM) observations on slip band structures generally require careful polishing of the specimen prior to deformation. The distribution of the dislocations themselves, such as those defining subgrain boundaries, can also be revealed by suitable etching after the deformation provided the dislocation density is relatively low; for techniques see Wegner and Christie (1985a, b). The optical transparence of minerals permits further techniques of observation in transmission optical microscopy, not available for metals, which exploit birefringence and stress-optical effects. For example, small orientation changes revealed by observation of thin sections between crossed polarisers permit subgrain boundaries to be located (Fig. 6.14), and other optical features such as "lamellae" (Fig. 6.15) reveal the presence of localized concentrations of dislocations or other local heterogeneities (Christie and Ardell 1974; McLaren et al. 1970). At low dislocation densities, the dislocation configuration itself can also be revealed by decoration techniques (Kohlstedt et al. 1976). For further information on observation techniques, see Nicolas and Poirier (1976), and Hobbs et al. (1976).

6.5.2 *Dislocation Density and Multiplication*

During plastic deformation, very marked increases can occur in the dislocation density ρ, specified as the average total length of dislocation line per unit volume. For example, commencing with a dislocation density of around 10^{10} m^{-2} or so, such as might typically be found in annealed or as-grown metallic and ionic crystals of ordinary quality, the density can increase by several orders of magnitude with 10 % strain. Such marked growth in density is generally thought to arise from the presence of specific, persistent dislocation *"sources"* that can produce a succession of new dislocation segments which augment the total population.

Dislocation growth or multiplication results from the differential movement of parts of an existing dislocation line. A persistent source is formed when the parent dislocation line is pinned at a certain point by an obstacle or by the deviation of the line into another plane in which it is immobile. The adjacent freely moving segment of the dislocation line will tend to sweep circularly around the pinning point, forming an ever-lengthening spiral of dislocation line which is added to with each additional revolution at the source (Fig. 6.16a). Various versions of this basic multiplication process have been proposed, including climb sources.

Thus, if the mobile segment is bounded by a second pinning point at which spiraling in the opposite sense occurs, the two spirals annihilate where they meet and so outwardly expanding loops can be formed (Fig. 6.16b). Such a "two-armed" source is known as a *Frank–Read source* and is often invoked for dislocation multiplication. However, the presence of a second suitable pinning point within a field of view that is not obscured by additional substructure may well be generally rather fortuitous, as is indicated by the observation of Caillard and Martin (1983) that in aluminum "one-armed" sources can be identified but not Frank–Read sources. Other types of dislocation source are discussed by Bilby (1955).

A spontaneous generation of a dislocation loop in its slip plane with the aid of thermal fluctuations is highly improbable (Cottrell 1953, p. 53). However, in the presence of high internal stresses in the neighborhood of precipitates, it is evidently possible to generate a prismatic dislocation loop in climb in the absence of pre-existing dislocation, as shown by loops generated at bubbles in quartz (McLaren et al. 1983).

If the number of dislocation sources were assumed to be proportional to the length of dislocation line already present, then, in the absence of annihilation processes, we would expect the dislocation density to increase at a rate proportional to the existing density and to the rate of straining, that is, $d\rho \propto \rho d\gamma$, leading to the relation

$$\rho = \rho_0 \exp(\gamma/\gamma_e) \tag{6.23c}$$

between the dislocation density ρ and the shear strain γ, where ρ_0 is the dislocation density at $\gamma = 0$ and γ_e is the strain needed for an e-fold increase in ρ (cf. Haasen 1978, p. 269). If, alternatively, the density of sources were proportional to the density of network links, that is, to $\rho^{3/2}$ (see Ardell and Przystupa 1984) so that $d\rho \propto \rho^{3/2}d\gamma$, then the exponential relationship would be replaced by

$$\rho = \rho_0/\left\{1 - (\sqrt{e} - 1)(\gamma/\gamma_e)/\sqrt{e}\right\}^2 \qquad (6.23d)$$

which increases catastrophically as $\gamma \to \sqrt{e}\gamma_e/(\sqrt{e} - 1)$. In either case, the tendency for exponential or catastrophic growth in dislocation density with strain will be moderated by two factors: (1) the circumstance that the dislocations may not be all equally mobile and (2) the occurrence of a certain rate of elimination of dislocations by escape at or incorporation into boundaries or by recovery or recrystallization. As a consequence the dislocation density will tend to level off eventually but a quantitative description of its behavior is very difficult to obtain.

6.5.3 Recovery and Recrystallization

(see also Sects. 3.3.2 and 3.3.3)

When the temperature is high enough for dynamic recovery processes to be thermally activated, the build-up in dislocation density due to multiplication during straining is counteracted by the mutual annihilation of dislocations, possibly at a rate that is proportional to ρ^2 (Ardell and Przystupa 1984; Johnson and Gilman 1959). There is then a tendency for a steady dislocation density to be approached. The occurrence of recovery also involves the reorganization of the remaining dislocations into lower energy configurations, especially into well-organized subgrain boundaries (it may be noted, in passing, that the analogous process in static recovery, observed as polygonization at the microscopic scale by Cahn (1949, 1951), is sometimes regarded as the first experimental evidence for the existence of dislocations). The presence of well-organized walls of dislocations, such as shown in Fig. 6.13, is therefore widely taken as evidence that recovery has occurred.

As in the case of static recovery, the quantitative treatment of dynamic recovery requires a measure for the recovery. Since the changes in structure are difficult to document fully even if relevant parameters such as mean dislocation density, subgrain size, and subboundary mesh size can be identified, the integrated effect of the structural changes as expressed in change in the flow stress is commonly used in specifying the recovery. However, it should be recognized that the use of such a measure is an empirical expedient which does little to elucidate the processes at the dislocation scale. Following Mitra and McLean (1966, see Poirier 1985, p. 105), the rate of dynamic recovery r can then be determined from stress drop tests, at least in situations where the flow stress is governed primarily by mutual dislocation interaction (athermal regime; Sect. 6.6.1), as

$$r = \lim_{t \to 0} \frac{\Delta\tau}{\Delta t} \qquad (6.24)$$

where $\Delta\tau$ is a small drop in applied stress from τ to $\tau - \Delta\tau$ and Δt is the time interval before straining is observed to recommence after the drop. This quantity tends to be markedly higher than similar measures of static recovery, indicating that the stress itself is assisting the recovery processes (for example, through the action of the climb force, Sect. 6.2.1). There is thus the possibility that the temperature dependence of r, expressed as an activation energy, may include a stress term as well as the expected activation energy for self-diffusion.

Dynamic recrystallization (Sect. 3.3.3) is a further process that may modify the build-up and structure of the population at elevated temperature. In the case of rotation recrystallization, its occurrence may not have an obvious mechanical effect because the new grains are formed in an apparently continuous transition from the earlier formed subgrain boundaries. For example, in the experimental deformation of marble at 1,000–1,300 K and constant strain rate, no abrupt changes in flow stress are observed to coincide with the appearance of new grains evidently formed by rotation recrystallization (Griggs et al. 1960; Schmid et al. 1980).

In contrast, the onset of migration recrystallization, often observed in hot working of metals, can have a marked mechanical effect, usually a softening, sometimes cyclic (McQueen 1977; Mecking and Gottstein 1978; Roberts 1984; Sellars 1978). For this type of recrystallization to occur, a sufficient build-up in stored energy associated with the dislocation substructure is necessary and high angle boundaries must be sufficiently mobile. Thus, factors that inhibit recovery (such as low stacking fault energy in f.c.c. metals) are favorable to its occurrence.

Where both types of recrystallization are observed in the same material, two distinct domains of behavior are found, with rotation recrystallization tending to occur at lower temperatures and strain rates (or stresses) than migration recrystallization, presumably due to inhibition of grain boundary mobility by impurities in this domain (Guillopé and Poirier 1979; Tungatt and Humphries 1981, 1984); Fig. 6.17. A succinct review of dynamic recrystallization is given by Poirier (1985, pp. 179–190).

6.6 Dislocation Theories of Flow in Single Crystals

6.6.1 Introduction: Athermal and Thermal Models

In preceding sections, we have discussed the factors governing both the motion of the individual dislocation and the collective properties of the population of dislocations in a deforming crystal. We now combine these considerations in discussing how the macroscopic flow stress is determined during a deformation that occurs primarily by dislocation glide. This procedure leads to the establishment of theoretical flow laws for crystal plasticity.

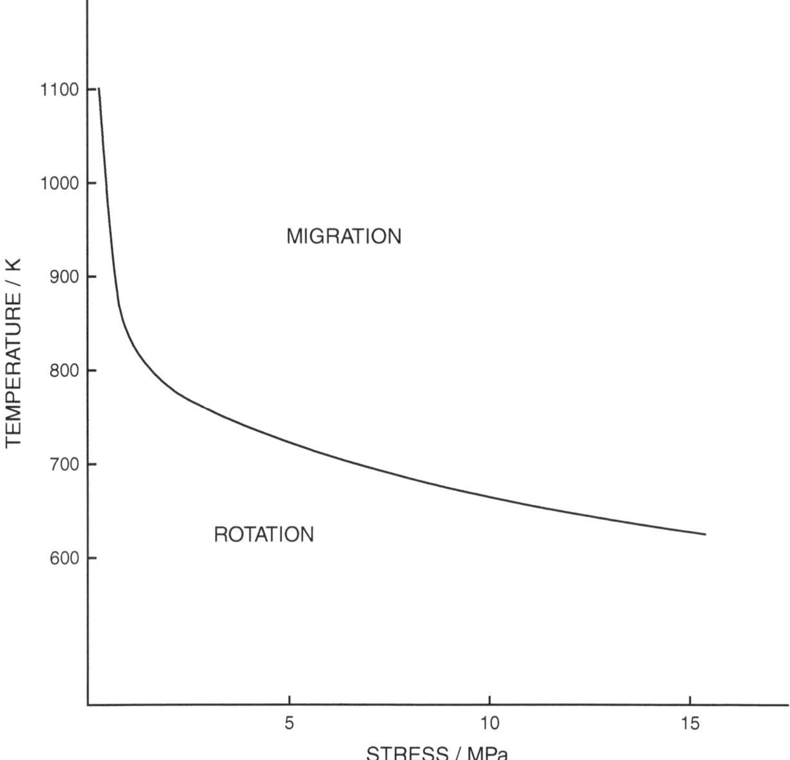

Fig. 6.17 Boundary between rotation and migration recrystallization in NaCl deformed in creep (after Fig. 6.13 from Poirier 1985)

The macroscopic flow behavior is governed by two complementary considerations:

1. The requirement of a certain instantaneous flow stress to bring about the plastic straining.
2. The modification of the structure as a result of the straining, leading in turn to a modification of the flow stress. (The term structure is to be understood as covering all aspects of the actual configuration of the dislocation assemblage and the distribution of other entities such as point defects, solute atoms and dispersed phases). The progress of the deformation is then determined by the simultaneous interplay of these two factors and can be simulated by an iteration if suitable quantitative expression can be given to them.

Because of the many factors that may influence the resistance to motion of individual dislocations and the many aspects of structure that may influence the integration of these resistances to determine the macroscopic flow stress, there exists in the literature a rather bewildering array of models for explaining the plastic behavior of crystals, covering various types of materials and different temperature and strain-rate regimes. Only a bare outline will be attempted of these models here, not only for reasons of space but also because, until there is an adequate observational basis for the structural and mechanistic assumptions required for particular theories, there is little practical point in their elaborate development beyond what is useful for guiding further experimental work. This observational basis for theoretical development is still very inadequate, on the whole, for most minerals.

In attempting to classify the theoretical models in a broad view, two types of distinction in behavior can be made. The first concerns temperature sensitivity, distinguishing between:

1. athermal regimes—weak temperature dependence
2. thermal regimes—strong temperature dependence.

The second type of distinction concerns the predominant sort of dislocation interaction determining the flow stress, distinguishing between:

1. models in which the mutual interaction between dislocations plays the main controlling part
2. models in which the interaction of the dislocation with other structural entities is dominant.

The second distinction can also be expressed as one between population control and glide control in the sense that, in the former case, the control depends in an essential way on there being a multiplicity of dislocations while, in the latter case, factors of a different kind are exerting control on the glide behavior of individual dislocations.

Before considering particular groups of models, it is interesting to look, in a general way, at the implications of the existence of the athermal and thermal regimes of behavior. As pointed out in Chap. 4, it is often observed that there are both low-temperature and high-temperature regimes, separated by an athermal regime, although among particular materials the actual temperature ranges of the regimes may vary widely relative to some characteristic temperature such as the melting point. In the light of the present chapters on deformation mechanisms, this tripartite tendency (Fig. 6.18) is seen to apply particularly where dislocation mechanisms are involved. In the thermal regimes, the temperature sensitivity of the flow stress can here be presumed to reflect the existence of thermally surmountable barriers controlling dislocation movement, of which there is potentially a wide range, as we have seen in earlier sections. However, the existence of two separate thermal regimes would seem to imply that there are two distinct classes of such barriers, reflected in two ranges of values of the activation energy E, as follows:

1. Values of E small relative to the binding or cohesive energy per atom, say, less than 1 eV (100 kJ mol^{-1}). Such values may be typical of the Peierls energy, the kink nucleation energy or the solute interaction energy in many materials, especially metals or simple ionic crystals. Since the rate at which barriers are

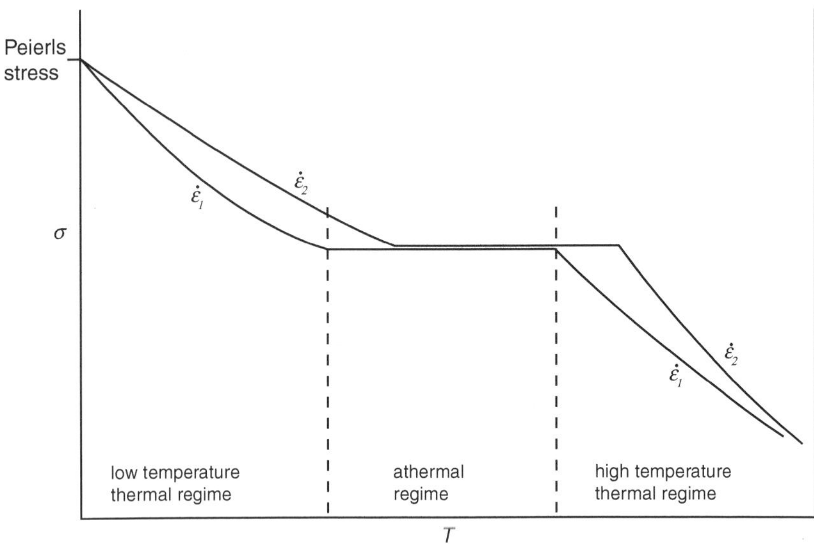

Fig. 6.18 Schematic representation of temperature and strain-rate dependence of flow stress in dislocation flow. $\dot{\varepsilon}_1$ and $\dot{\varepsilon}_2$ are strain rates ($\dot{\varepsilon}_2 > \dot{\varepsilon}_1$)

surmounted by thermal activation becomes significant when kT reaches a certain fraction of E, processes involving these relatively small values of E will tend to be important only at relatively low temperatures.

2. Values of E comparable with the binding energy per atom, generally greater than 1 eV. Such activation energies are typical of the formation or migration energies of point defects such as vacancies or Schottky or Frenkel defects that may be involved in, for example, diffusion. The rates of processes involving such energies will only become important at relatively high temperatures.

A distinction between low-temperature and high-temperature thermal regimes might thus be rationalized in terms of two such ranges of E values for the controlling deformation mechanisms. Insofar as there is actually a clear separation of the two ranges of E values, there is then also the possibility of having an intermediate, athermal regime in which the barriers involved in the low-temperature thermal regime are no longer effective but in which the temperature is not yet high enough for processes characteristic of the high-temperature thermal regime to be activated at a significant rate, thus explaining the existence of the three regimes distinguished in Sect. 4.1 and Fig. 6.18. However, the rationalization in terms of two ranges of activation energies may be an oversimplification since other factors, such as whether the activation area ΔA^* in (6.16) corresponds to the involvement of few or many atoms, or whether the t_g or t_0 term in (6.17) predominates in determining the dislocation velocity, may also be relevant in characterizing the different regimes.

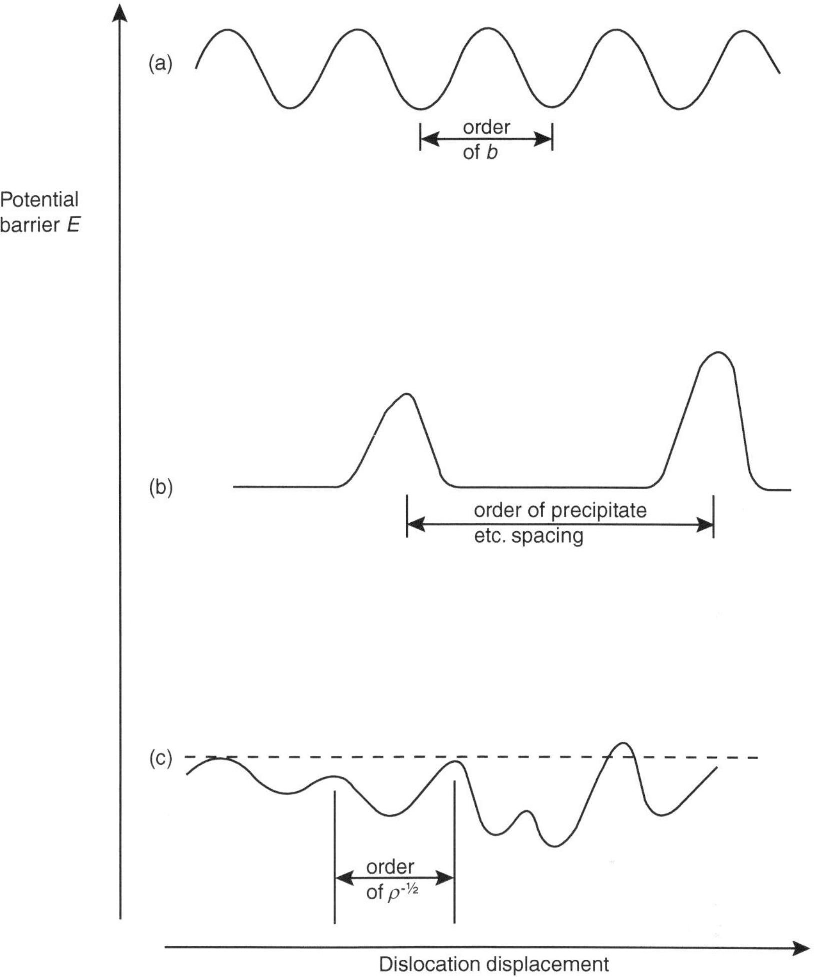

Fig. 6.19 Types of potential barriers to dislocation motion. **a** Barriers related to the Peierls potential. **b** Discrete obstacles to dislocation motion such as dispersed phases. **c** Barriers deriving from interaction with other dislocations

In the following subdivision of models into athermal and thermal groups, it is convenient not to distinguish initially between low-temperature and high-temperature regimes, allowing the relevance to one or other regime to emerge as the analysis proceeds. However, it is useful to make a preliminary distinction between three general types of barriers to dislocation motion (Fig. 6.19), the nature and distribution of which are basic to the character of the deformation behavior, as follows:

(a) The barriers are spaced on a scale of the order of the Burgers vector b and arise from the Peierls potential, possibly with modifications due to the presence of point defects such as solute atoms. Insofar as it can be separately distinguished, the contribution of these barriers to the flow stress will be designated τ_p (it should be made clear that τ_p here is not being used in strictly the same sense as in Sect. 6.2.3 where the definition of the Peierls stress applied primarily at absolute zero temperature).

(b) The barriers consist of obstacles such as dispersed phases and the misfit elastic fields around solute atoms, the dimensions and spacings of these barriers being large compared with b. Their distinguishable contribution to the flow stress will be designated τ_f, signifying a frictional or viscous drag.

(c) The barriers are the interactions with other dislocations, which are often represented by an internal stress field τ_i made up of the long-range stress fields of the dislocations but which, as will be discussed in Sect. 6.6.3, can be interpreted in other ways also. The contribution of these barriers to the flow stress will be here designated as τ_d.

The magnitudes of the activation energies ΔE^* and activation areas ΔA^* for particular cases can thus vary widely, leading to corresponding variations in the characteristics of the mean dislocation velocity (Sect. 6.4) and eventually, through the Orowan relation (6.4a), being reflected in the macroscopic flow behavior. However, it must be borne in mind that the stress dependence of the flow rate may reflect a role of stress in the mean dislocation density as well as one in the dislocation velocity.

Insofar as separate components of the flow stress can be related to the particular classes of barriers to dislocation motion listed above, the macroscopic flow stress τ can be viewed as a summation of these components,

$$\tau = \tau_p + \tau_f + \tau_d$$

However, such linear additivity may not always apply, as in the case of solid solutions for which the strain-hardening effects that would normally be treated as increases in the component τ_d are observed to depend on the presence of the solute, the effect of which would otherwise be treated as a contribution to the component τ_f (Kocks 1984, 1985b).

We first consider athermal models. Since the flow stress component τ_p is normally rather sensitive to temperature, it is not of primary concern in the athermal regime, where attention focusses on the components τ_f and τ_d, treated in the next two subsections.

6.6.2 Athermal Models Based on Discrete Obstacles

If the moving dislocations have to cut through or circumvent fixed structural obstacles without much aid from thermal fluctuations, the applied stress itself must supply the force that is required locally for overcoming or circumventing the obstacles. In this case, the obstacles will give rise to an athermal, frictional resistance to the dislocation motion (the term "frictional" is used in analogy with ordinary sliding friction in the sense of being relatively rate insensitive, but it is not fully analogous because of the absence of the strong normal-stress dependence characteristic of ordinary friction). It is the objective of the models treated in this subsection to calculate the stress component τ_f that must be applied to the crystal in order to bring about a macroscopic plastic deformation by dislocation motion in the presence of the more or less athermal frictional resistance from the discrete obstacles.

In practical terms, two categories of athermal obstacle models can be usefully distinguished, namely, those of solute hardening and of particle hardening (or strengthening). Both are based on heterogeneities in the crystal but at different scales, as distinguished in Sects. 6.3.1 and 6.3.2. For general reviews, see Brown and Ham (1971), Kocks et al. (1975), Gerold (1979), Martin (1980), Haasen (1978, Chap. 14, 1983), Strudel (1983), Ardell (1985), Humphreys (1985), Kocks (1985b) and Nabarro (1985). Haasen (1983) also considers the effects of long-range ordering in solid solutions.

Solute hardening. Solid solution effects can be athermal when the temperature is moderately low and the solute atoms are effectively immobile or only slightly mobile. The various ways in which solute atoms can impede the motion of dislocations and thus increase the flow stress are listed in Sect. 6.3.1. We shall be concerned here mainly with the elastic interactions since the effect of core interactions will tend to be more strongly thermally activated (Sect. 6.6.5).

It was noted in Sect. 6.3.1 that prior segregation of solute at pre-existing dislocations will tend to lock them so that a higher applied stress may be required to initiate dislocation movement than to sustain it later, giving rise to an initial yield drop in the stress–strain curve. We are here considering the effect of the solute on the dislocations when they are in motion, an effect that dominates the flow stress at small strains in many alloys of low initial dislocation density and low Peierls stress. Although thermal activation may be important at very low temperatures, as shown by a marked decrease in flow stress with rise in temperature, there tends to be a temperature range, around room temperature for common metals or ionic compounds, in which the flow stress at small strains is almost independent of temperature, a range known as the plateau hardening regime. Because the plateau hardening effect is relatively insensitive to strain rate, it can be referred to as a frictional effect rather than as a viscous drag. It is to the plateau hardening regime, therefore, that the main athermal solute hardening theories are directed.

The basic theoretical problem is to calculate, given a distribution of obstacles, the number of them per unit length that are interacting effectively with a given dislocation segment at any instant and to determine the applied stress needed to overcome the combined effect of the interactions. Two limiting cases are commonly considered (Haasen 1978, p. 327, 1983; Hirth and Lothe 1982, p. 681). The first, Friedel–Fleischer theory, envisages the dislocating line being forced by the applied stress against point obstacles of given strength, while the second, Mott–Labusch theory, envisages the line lying in a minimum energy configuration determined by the potential fields around diffuse obstacles.

The *Friedel–Fleischer theory* applies for relatively strong obstacles in dilute concentration and predicts a solute hardening component of the flow stress of

$$\tau_f = \frac{x^{\frac{1}{2}} F^{\frac{3}{2}}}{b^2 (2T)^{\frac{1}{2}}} \approx \frac{1}{b^3} \left(\frac{xF^3}{G} \right)^{\frac{1}{2}} \tag{6.25}$$

where x is the mole fraction of solute, F the maximum interaction force per solute atom or molecule (for estimates, see Kocks 1985b), T the line tension or energy per unit length of the dislocations ($2T \approx Gb^2$ from Sect. 6.2.2), G the shear modulus and b the Burgess vector. The *Mott–Labusch theory* applies to weak, diffuse obstacles in higher concentration and predicts a solute hardening component of the flow stress of

$$\tau_f = \frac{x^{\frac{2}{3}} F^{\frac{4}{3}} w^{\frac{1}{3}}}{2\alpha^{\frac{1}{3}} b^2 (Tb)^{\frac{1}{3}}} \approx \frac{1}{b^3} \left(\frac{x^2 F^4 w}{4\alpha G} \right)^{\frac{1}{3}} \tag{6.26}$$

where w is the "obstacle width" or interaction range for the solute and α is a numerical factor of order unity deriving from the form of the interaction potential. Both square-root and two-thirds power (or stronger) dependence on solute concentration have been observed for the flow stress at small strains in metallic solid solution alloys; for details and references, see Haasen (1983) who also reviews the effect of solutes on the form of the stress–strain curve for various classes of materials.

In spite of the agreement with observed concentration dependence, just noted, reviews have emphasized that other observations indicate that, in interaction with dislocations, the solute atoms can seldom be regarded as discrete entities or point obstacles as in the Friedel–Fleischer theory, except possibly at very low concentrations (Kocks 1985b; Nabarro 1985). Rather, the effect of the solute atoms is generally more smeared out, providing a series of extended potential troughs in the path of the dislocation. This effect is recognized in some degree in the Mott–Labusch approach but is emphasized more strongly in terms of a "trough model" by Kocks (1985b) who also recognizes a limited amount of solute mobility already within the plateau hardening regime.

Particle hardening. The effect of precipitates or second phase particles depends on the size and physical characteristics of the particles as well as on their spacing. According to the nature of the interaction with the dislocations (Sect. 6.3.2), there are two types of situation, distinguished by whether the dislocation cuts through the particle (by common convention, covered by the term precipitation hardening) or bypasses it (dispersion hardening):

1. In the *particle-cutting case*, the theory of the flow stress involves the same sort of statistical considerations as in the case of solute hardening effects, the difference lying in the obstacle now being larger and possibly stronger. So long as the spacing of the particles is sufficiently large relative to their effective

dimensions, the Friedel–Fleischer relation in (6.25) can be applied; rewriting this relation in terms of the spacing l of the obstacles in the slip plane, using $l = bx^{-\frac{1}{2}}$, we obtain an estimate of the flow stress component due to precipitation hardening as

$$\tau_f = \frac{F^{\frac{3}{2}}}{bl(2T)^{\frac{1}{2}}} \approx \frac{F^{\frac{3}{2}}}{b^2 l G^{\frac{1}{2}}} \tag{6.27}$$

[An analogous expression can be obtained from (6.26) for the case of diffuse precipitates interacting over a range comparable to their spacing but it is probably of more limited applicability; see discussion by Ardell (1985)]. The evaluation of F depends on the nature of the dislocation-particle interaction and may involve coherency, surface energy, ordering, stacking fault, and elastic modulus effects (Ardell 1985; Gerold 1979; Martin 1980, pp. 53–60). If F is taken as being proportional to d^n, where d is the width or diameter of a particle, and we introduce the volume fraction x_v of particles through $x_v \approx d^2/l^2$, then (6.27) leads to

$$\tau_f \propto x_v^{\frac{3n}{4}} l^{\frac{3n}{2}-1} \tag{6.28}$$

indicating that, for a given volume fraction of particles, the flow strength will increase as the precipitate coarsens, provided $n > 2/3$.

2. In the *bypassing case*, the theoretical model envisages that the dislocation line bows around the particles, leaving residual loops ("geometrically necessary dislocations") as it proceeds further. This process requires a flow stress component corresponding to the Orowan bow-out stress

$$\tau_f = \frac{\alpha G b}{l} \tag{6.29}$$

(Embury 1985; Haasen 1978, p. 250; Martin 1980, p. 62) where G, b, l are as defined for Eqs. (6.25) and (6.26) and α is a numerical factor of the order of unity. Such a relation is found to fit experimental observations on initial yield stress quite well in metals with varying concentrations of dispersed hard particles, especially if l is taken as the distance between grain surfaces rather than between their centers (Martin 1980, p. 63).

The opposite dependences of (6.28) and (6.29) on the spacing l (provided $n > 2/3$) indicate that during particle coarsening at constant volume fraction, known as "Ostwald ripening", (Haasen 1978, p. 207) a critical size and spacing of particles will tend to be reached beyond which bypassing becomes easier than cutting, this size corresponding to a maximum in the precipitation hardening of the material. The decrease in strength upon "over-aging" of age-hardening alloys is commonly attributed to such an effect, although this explanation is questioned by Ardell (1985).

With increasing plastic strain there are several possible effects consequent upon the accumulation of residual dislocation loops around bypassed particles that lead to more complex strain-hardening behavior in the bypassing case than in the particle-cutting case (normal strain hardening accompanying dislocation multiplication will be discussed in the next section). The residual loops themselves become involved in mutual dislocation interactions, or the accumulating internal stress around the particle may lead to prismatic punching (possibly initiated by the reorientation of the Orowan loop itself through cross-slip or climb) which also increases the dislocation density, or secondary slip systems may be activated in the vicinity of the particles. These various processes, reviewed by Martin (1980, p. 72) and Strudel (1983), are additional to the normal dislocation multiplication effects, as observed in pure metals and which are effective in a similar way in precipitation hardened materials. Thus, the strain hardening of particle-bearing materials through the effects of internal stress, over and above the precipitation hardening effect, tends to be a very complex phenomenon, leading to predicted shapes of stress–strain curve varying from linear to quadratic—see Strudel (1983) who discusses at some length the concept of internal stress as applied in the theory of the flow stress for particle-strengthened materials.

Order hardening. Where short-range ordering exists, there is a local resistance to the motion of a dislocation having a Burgers vector of the disordered structure because of the work done locally in destroying the ordering across the slip plane. An additional local stress of γ/b is then needed, where γ is the antiphase boundary energy per unit area, and so the crystal shows an effect analogous to a precipitation hardening in which the force F for cutting through the barriers derives from this local stress (Friedel 1964, p. 383).

When long-range ordering exists (Sect. 1.2.3), the independent movement of ordinary dislocations having a Burgers vector of the disordered structure will tend to be difficult and super dislocations (Sect. 6.2.5) will move much more readily, unless a high Peierls stress intervenes. The properties associated with the motion of the super dislocations will be analogous in many respects to those of extended dislocations except that the joining strip of antiphase boundary can more readily exist in arbitrary planes compared with the stacking fault strip in extended dislocations. Although the super dislocations may be able, in principle, to move locally without resistance from order destruction, there will be some hardening associated with the domain structure of the ordering, corresponding to an additional stress component of the order of γ/a, or somewhat less when the finite thickness of domain walls is taken into account, where a is the domain size and γ is of the order of $S\Delta E/b^2$, S being the Bragg–Williams order parameter defined by (1.1) and ΔE the reduction in energy per unit cell due to the ordering (Friedel 1964, p. 384; Haasen 1983, p. 1392; Hirth and Lothe 1982, p. 685). Also various yield point and aging phenomena can appear, for example, due to segregation effects, and the hardening may tend to peak before complete ordering is reached (Haasen 1983).

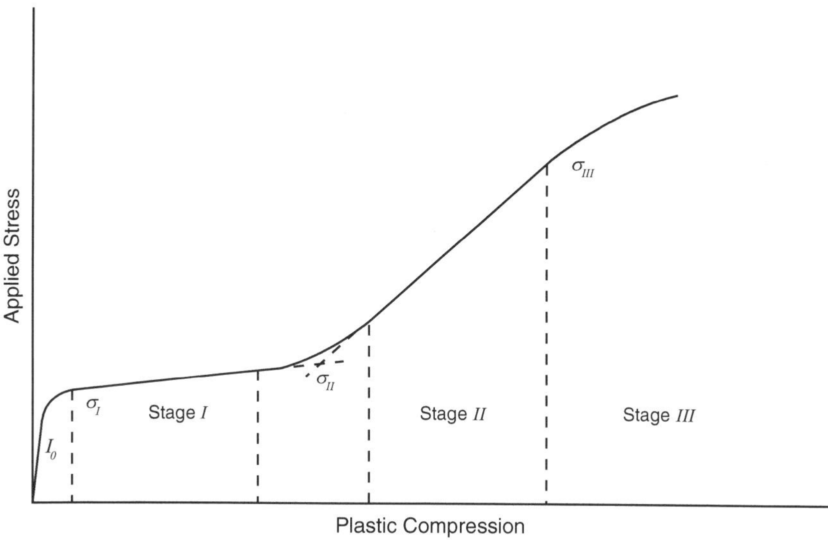

Fig. 6.20 Depiction of the three stages of strain hardening in NaCl (after Fig. 11.7 from Sprackling 1976)

6.6.3 Athermal Models Based on Mutual Dislocation Interaction

These models have occupied a dominating place in metallurgical writings, in spite of involving great theoretical difficulties. They have been developed especially in connection with the strain hardening of fairly pure metals. However, a wider relevance may be expected wherever dislocations can move freely and multiple slip systems operate, facilitating dislocation interaction and multiplication and the development of a three-dimensional network of dislocations.

Mutual interaction models have been developed most fully in attempting to explain the stress–strain curves of the pure f.c.c. metals (for reviews, see Basinski and Basinski 1979; Mecking 1981a; Nabarro et al. 1964; Weertman and Weertman 1983a). However, the same considerations have been found to be applicable to f.c.c. alloys, b.c.c. and close-packed hexagonal metals, ionic crystals of NaCl structure, and diamond-structure covalent crystals (see also Alexander and Haasen 1968; Haasen 1983; Sprackling 1976, Chap. 13). Three stages of strain hardening are widely recognized. These are illustrated in Fig. 6.20 and are characterized as follows:

Stage I: Microscopical observations indicate that in stage I the dislocation motion is mainly confined to a single slip system (the system for which the resolved shear stress is a maximum) and that the dislocations travel large distances, comparable to specimen or crystal dimensions. The extent of this stage, known as the "easy glide" region, tends to be greater at lower temperatures.

Relatively few dislocation sources are active and the dislocation density does not increase markedly with strain, a trend with which the low strain-hardening rate of the order of 10^{-4} G is correlated.

Stage II: This stage begins with the onset of more obvious activity of secondary slip systems intersecting the primary system. The primary system continues to contribute most to the strain but with much shorter distances of travel of the individual dislocations, as deduced from the lengths of slip traces at the surface, and with a much increased rate of strain hardening, now of the order of G/300 to G/200. The dislocation density increases markedly during this stage with the build-up of the three-dimensional dislocation network. As a result of dislocation reactions the network may contain many sessile segments. The flow stress in stage II shows very little dependence on temperature but its extent depends on the stacking fault energy (see stage III).

Stage III: As higher stresses are reached, the linear hardening regime of stage II gives way to a further stage of more or less parabolic shape in which the strain hardening rate steadily decreases toward values of the order of 10^{-3} G or less. In this third stage, the build-up of the dislocation network begins to be moderated by dynamical recovery processes that relieve dislocation pile-ups or that lead to the annihilation of dislocations or the sharper development of cell-like structures (Sect. 6.5.1). The most important step in such recovery processes in metals at low to moderate temperatures is widely thought to be cross-slip, which is evidenced in the appearance of wavy slip traces. Since dissociated dislocations must recombine to undergo cross-slip, a dependence on stacking fault energy can be introduced into the stress–strain curve at this stage, as well as some temperature dependence; in particular, the onset of stage III is earlier, and hence the length of stage II shorter, for higher stacking fault energies. At higher temperatures, dislocation climb becomes a potentially important dynamical recovery mechanism, introducing a stronger temperature and time dependence (Sect. 6.6.6).

In attempting to explain this athermal or only mildly temperature dependent behavior in terms of mutual dislocation interaction two views have been taken of the predominating characteristic of the interaction (Sect. 6.3.4). In the first, emphasis is put on the role of the long-range elastic stress field associated with the dislocation network. In the second, the short-range forest-cutting effect is emphasized.

According to models based on the effect of the long-range stress field, the flow stress τ is that which is required to counteract the internal stress τ_i due to the dislocation network. Assuming that the interactions between parallel dislocations are the most important and that the stress fields of all except the nearest dislocation cancel to zero at a given point, and noting that the relevant shear stress at a point due to a dislocation at a distance r is of the order of $Gb/2\pi r$ (Sect. 6.2.1), then, putting $r = \rho^{-\frac{1}{2}}$, we obtain the Taylor estimate of the internal stress,

$$\tau_i = \alpha Gb\rho^{\frac{1}{2}} \tag{6.30}$$

(Taylor 1934; Weertman and Weertman 1983a, p. 1281), where G is the shear modulus, b the Burgers vector, ρ the dislocation density and α a numerical constant near to unity (theoretical estimates of α vary from 1/2 to 1/3, while measurements on copper suggest $\alpha = 1/3$, as quoted by Haasen (19781, p. 25); see also Basinski and Basinski (1979), The flow stress is then obtained by putting $\tau = \tau_i$.

In models based on the forest-cutting effect, an account is taken of the work required for a moving dislocation to cut through the dislocations that cross the slip plane, involving mainly short-range interactions (Sect. 6.3.4). Since the stress needed will be proportional to the number of such intersections per unit distance traveled, the flow stress can again be expected to be proportional to $\rho^{\frac{1}{2}}$ (at least insofar as dislocation networks of different densities are self-similar or ρ is now taken to be the density of the forest dislocations). The proportionality factor will presumably depend on the nature of the impediment to dislocation movement arising from the intersection, for which there are a number of models (see summary by Weertman and Weertman 1983a, Sect. 5.5). Also, if there is significant thermal activation of the intersection process, some temperature dependence may be introduced in addition to that deriving from the shear modulus G.

An alternative to the forest-cutting view is to place emphasis on the "mesh length" or "link length" of the three-dimensional network of dislocations that builds up during activity of the various slip systems (see, for example, Burton 1982b; Kuhlmann-Wilsdorf 1985). Again, the average mesh length is inversely proportional to $\rho^{\frac{1}{2}}$ and so the formal theoretical implications tend to be the same.

Thus, regardless of whether long-range or short-range effects are involved, a flow stress determined by mutual dislocation interactions can be expected to have the form

$$\tau = \alpha G b \rho^{\frac{1}{2}} \tag{6.31}$$

where ρ is a dislocation density, which may refer to the total dislocation population or to some significant part of it, and α is a numerical factor of the order of unity. The particularities of specific models enter through the way in which the quantity ρ is expressed in terms of resolved shear strain γ in order to obtain a stress–strain relation $\tau(\gamma)$. The central aim of the models is to predict the strain-hardening rate

$$\theta = \frac{d\tau}{d\gamma} = \frac{\alpha G}{2}\left(\frac{b}{\rho^{\frac{1}{2}}}\frac{d\rho}{d\gamma}\right) = \frac{\alpha R G}{2} = \beta G \tag{6.32}$$

where $R = \left(b/\rho^{1/2}\right)(d\rho/d\gamma)$ and $\beta = \alpha R/2$ is a constant insofar as R is a constant. θ thus reflects the value of $d\rho/d\gamma$ at a given value of $\rho(\gamma)$ and hence involves the problems of dislocation multiplication (Sect. 6.5.2). Given (6.32), the stress–strain relation can then, in principle, be obtained by integration.

As pointed out by Hirth and Lothe (1982, Chap. 22) there has been little progress in developing fundamental theories of the flow stress and strain hardening on the basis of the properties of mutual dislocation interaction without introducing ad hoc assumptions about the characteristics of the dislocation population; that is, the form of development of the actual dislocation substructure cannot yet be predicted from first principles. All models of the flow stress and the strain hardening so far proposed have therefore been in some degree phenomenological in that they are based on assumptions about the basic parameters characterizing the dislocation population and are therefore in that degree descriptive rather than predictive. Although there is a large literature on such models, we can here only indicate their broadest aspects.

The dimensionless quantity R in (6.32) is essentially an expression of the ratio of dislocation storage to dislocation travel. This ratio is widely represented in the various models as a function of the ratio of two parameters observable or definable in terms of the microstructure. Thus in the often-quoted long-range model of Seeger for stage II hardening, R is expressed, apart from a numerical factor of order unity, as $(nb/L\Delta\gamma)^{1/2}$ where n is the number of dislocations released by a given source, as determined from the slip step height nb, at the surface, L is the mean free path of dislocations, as measured by the length of the slip traces on the surface, and $\Delta\gamma$ is the resolved shear strain in excess of a reference strain of the order of the strain at the onset of stage II; see brief treatments and references in Haasen (1978, p. 274) and in Weertman and Weertman (1983a, p. 1285). Weertman and Weertman (1983a, p. 1284) further describe a model for stage I hardening in which R is written as d/L where d is the active slip plane spacing and L is as before; however, they also quote the view of Hirsch that no really satisfactory model has yet been advanced for stage I hardening.

In a survey of the effects of mutual dislocation interaction in relation to the flow stress and the strain hardening, Kocks (1985a) has emphasized that the deformation should be viewed in terms of the two-dimensional motion of dislocations in the slip plane as a percolation process, rather than one-dimensionally as in the previous paragraph. As relevant microstructural parameters in what is essentially a forest-cutting model he has used the quantities l, the average spacing of the obstacles to dislocation motion (of order $\rho^{1/2}$), and λ, the average spacing of "hard spots" in the slip plane which are not penetrated during the "percolation" of the dislocation line across the slip plane as it surmounts the penetrable obstacles ("soft spots"); λ is in some way related to the scale of the cell structure (Sect. 6.5.1) that becomes obvious in stage III. Kocks then derives an expression for the strain hardening which is equivalent to that given by putting R equal to $(l/\lambda)^2$ in (6.32). The observed hardening rate in stage II is then obtained if $\lambda \approx 10\,l$.

Although it is generally agreed that, in stage III, dynamical recovery effects are causing the strain hardening to fall below the maximum rate reached in stage II (sometimes called the athermal hardening rate), the phenomenology of stage III is not fully understood and satisfactory models are still lacking. As already indicated, it is widely considered that cross-slip is an important factor in the dynamical recovery, but this view has been disputed (for example, Kuhlmann–Wilsdorf 1985) and there may be other factors involved. Thus, Kocks (1985a) has pointed to a possible role of the breaking of attractive junctions involved in the forest effects, a process that would also introduce a sensitivity to stacking fault energy and a greater temperature sensitivity, such as is observed. The formation of a more clearly defined cell structure in the dislocation population is also a feature of stage III and at least two structural parameters will therefore still be required, but they may have to be treated as being independent rather than as a single parameter through their ratio, as has been done for stage II (see, for example, Mecking 1981a).

At very large strains ($\gg 1$), the strain-hardening rate becomes small. In some cases, it appears to approach zero, corresponding to a "saturation" state, but, in other cases, no saturation limit is apparent up to strains of at least 10 (Hecker and Stout 1984). It has been suggested that the presence of solutes is important in this connection since they may retard dynamical recovery and so prevent the attainment of the balance between hardening and dynamical recovery that is implied in a saturation limit (Hecker and Stout 1984; Kocks 1984). See further discussion on large strains in Sect. 6.8.5.

6.6.4 Creep in the Athermal Regime

Under this somewhat self-contradictory heading we consider the time-dependent or thermally activated contributions to the deformation which have been ignored in the previous two sections as being of secondary importance relative to the more or less instantaneous, athermal deformation that occurs under an applied stress greater than the yield stress. We now consider the occurrence of further straining at a finite but decaying rate that may be observed if the stress is maintained at the same level and measurements are made with sufficient sensitivity. This transient creep in the athermal regime usually obeys a logarithmic law (Sect. 4.3.2).

The athermal part of the deformation can be imagined to have terminated with the dislocations being forced against barriers (internal stress and/or other obstacles) that are too high to be overcome by the applied stress. However, some of these barriers will be only slightly too high and, with the passage of time, may subsequently be surmounted with the aid of thermal fluctuations, leading to a contribution to deformation that will appear as transient creep. The decrease in the creep rate with time can be attributed either to a progressive elimination of the situations where a dislocation is resting against a relatively low barrier (exhaustion effect) or to a progressive raising of the barriers as a result of the strain hardening accompanying the additional strain (hardening effect). The exhaustion hypothesis was early explored by (Mott and Nabarro 1948) but the hardening hypothesis has become more favored (Friedel 1964, p. 305; Mott 1953; Weertman and Weertman 1983a).

Theories of logarithmic creep are commonly based on an expression for the strain rate $\dot{\gamma}$ such as may be obtained from Orowan's equation $\dot{\gamma} = \rho b v$, using (6.13b), putting $v = (\Delta A / l)\nu$, substituting (3.12f) for ν for the case $\tau b \Delta A \gg kT$ applicable for relatively high stress and low temperature, and finally putting $l = \rho^{-\frac{1}{2}}$. This approach leads to

$$\dot{\gamma} = \rho^{\frac{3}{2}}b\Delta A v_0 \exp[-(\Delta E^* - \tau b\Delta A^*)/kT] \tag{6.33}$$

where ρ is the dislocation density and the other symbols are as used in Sect. 6.4.1 and Fig. 6.19. The quantity ΔA^* is viewed as increasing progressively according to the exhaustion or hardening effect just discussed. In the hardening case we can write

$$\Delta E^* = \Delta \tau_d b\Delta A^* + U \tag{6.34}$$

where $\Delta \tau_d b\Delta A^*$ represents the work needed to overcome the local fluctuation in mutual dislocation interaction that is holding up the dislocation, and U is any additional barrier (such as a Peierls or other short-range barrier) that must also be overcome when moving the dislocation (note that the activation area ΔA^* in (6.34) has been taken as being identical with that in (6.33), but it is possible that in some cases a distinction will need to be maintained). On the view that the mutual dislocation interactions can be described in terms of an internal stress (Sect. 6.6.3), $\Delta \tau_d$ is the local fluctuation in internal stress τ_i to be overcome by thermal activation. We now suppose that the mutual dislocation interaction effect increases linearly with strain, putting $\Delta \tau_d = \theta \gamma$, where θ is the strain-hardening coefficient and γ is the creep strain. In this case, the relationship (6.33) can be integrated at constant stress to give the creep law

$$\gamma_{\text{tot}} = \gamma_{\text{inst}} + \gamma_0 \ln(1 + vt) \tag{6.35a}$$

(Friedel 1964, p. 306; Haasen 1978, p. 277) {Weertman 1983 #7385, whose expression for v lacks an explicit stress dependence}, where γ_{tot} and γ_{inst} are the total and instantaneous (elastic plus plastic) strains, respectively, and γ_0, v are given by

$$\gamma_0 = \frac{kT}{\theta b\Delta A^*} \tag{6.35b}$$

$$v = \frac{\rho^{\frac{3}{2}}b\Delta A v_0}{\gamma_0} \exp\left[-\frac{U - (\tau - \gamma_{\text{inst}}\theta)b\Delta A^*}{kT}\right] \tag{6.35c}$$

The relation in (6.35a) leads to a creep rate

$$\dot{\gamma} = \frac{\gamma_0 v}{1 + vt} \tag{6.35d}$$

Therefore, the initial creep rate is predicted to be $\gamma_0 v$ and the creep rate becomes γ_0/t for $vt \gg 1$.

The model just described is still a phenomenological one in that the parameters $\theta, \Delta A^*$ and ΔA have not been given precise physical meaning in microstructural terms, but it serves to illustrate the essential nature of logarithmic creep, which, in this model, derives from the linear strain dependence of the activation energy ΔE^* through (6.34).

Empirically, the determination of an apparent activation energy Q directly by temperature stepping procedures should indicate a continual increase in Q during the progress of logarithmic creep. However, because of the limited range over which creep rates can be measured, the observed values of Q at a given stress will fall in a limited range at any given temperature, even though a wide range of activation energies may exist; also the values typical of the observable range may be expected to increase with temperature, as can be seen by writing (6.33) as $\Delta E^* - \tau b \Delta A^* = kT \ln(\rho^{\frac{3}{2}} b \Delta A v_0 / \dot{\gamma})$ and noting that the In term will be effectively nearly constant(Weertman and Weertman 1983a).

Before concluding this section it is appropriate to emphasize the distinction between logarithmic creep and anelastic creep. Whereas logarithmic creep is observed in situations where the yield stress has been exceeded and some athermal plastic deformation brought about, anelastic creep is observed under small stresses, below the macroscopic yield stress, and is not limited to the athermal regime. Anelastic deformation is defined as reversible, time-dependent strain that is linearly related to the stress in the sense that the Boltzmann superposition principle is obeyed in summing the contributions to the current strain rate due to all the past stress history (Jackson 1986; Nowick and Berry 1972; Zener 1948). Models for anelastic dislocation creep involve the movement of dislocation segments within finite limits set by insurmountable barriers (see discussion of amplitude-independent internal friction from dislocation motion in Sect. 2 of Fantozzi et al. 1982).

6.6.5 Thermal Models Based on Viscous Drag

This first category of models for temperature-sensitive dislocation flow is concerned with situations in which the flow rate is determined primarily by the rate at which dislocations move against viscous drag forces that act more or less uniformly on them. Thermal recovery processes (Sect. 6.5.3) play a relatively minor role and so the term *glide-controlled creep* is sometimes applied (Poirier 1985, p. 101). It is a category that is especially relevant in the low-temperature thermal regime (Sect. 6.6.1), in cases in which the dislocation velocity is strongly influenced by Peierls or similar atom-scale barriers that are thermally surmountable in this temperature range and which tend to be largely of an intrinsic nature. A cross-slip controlled deformation might be included in here. However, the category also has application in the high-temperature thermal regime where effects such as solute drag are important; these tend to be extrinsic effects. Thus, it is a category that is potentially of importance in minerals in either thermal regime.

The two prototype models of the category are the microdynamical models of (Weertman 1957) and of (Haasen 1964). Both models are based on the Orowan equation for the strain rate,

$$\dot{\gamma} = \rho b v \tag{6.36}$$

Appropriate expressions for the dislocation velocity v and the mobile dislocation density ρ are derived and inserted in (6.36). In this way, the strain rate is obtained directly and the stress–strain curve is obtained by integrating (6.36), in both cases taking into account the dependences of ρ and v on the strain γ as well as on stress τ, temperature T, etc. Expressions for ρ and v have been discussed in general in Sects. 6.5.3 and 6.4.1, respectively.

The model of (Haasen 1964) (see also Alexander and Haasen 1968), developed for application to silicon and germanium and similar materials, takes into account strain hardening by assuming that the stress τ acting at a dislocation, designated the "effective stress" τ_{eff}, is equal to the applied stress τ_{appl} minus an internal stress or "back stress", imagined to arise from the long-range stress fields of the other dislocations or otherwise to express the effect of their interaction. Then, from (6.30) or (6.31),

$$\tau = \tau_{\text{eff}} = \tau_{\text{appl}} - \alpha G b \rho^{\frac{1}{2}} \tag{6.37}$$

Haasen also uses the velocity relation in (6.16) and a development of (6.23a) for the dislocation density in which $\gamma_e^{-1} \propto v\tau_{\text{eff}}$. The stress–strain curve or creep curve is then obtained by numerical iteration using these three relationships together with (6.36).

One of the interesting features of Haasen's model is that it reproduces the striking yield drop that is often observed in stress–strain tests in silicon, germanium and other nonmetallic materials having a low initial dislocation density. This effect can be attributed to a softening arising from the circumstance that, as the dislocation density increases with straining, the dislocation velocity required for achieving the specified strain rate decreases and so the required stress falls; on continuing to larger strains strain hardening eventually takes effect and the required stress rises again. In a creep test the corresponding phenomenon is the generation of a sigmoidal creep curve. This type of behavior has been observed in quartz (Fig. 6.21) and the application of the Haasen model to quartz has been explored by Hobbs et al. (1972) and Griggs (1974). Such a yield drop effect is to be distinguished from that arising from the Cottrell solute locking effect (Sect. 6.3.1 and Fig. 6.9), tending to give a less sharp stress drop than in the latter case.

The model of Weertman (1957) is developed in a somewhat similar way. An expression for the dislocation velocity analogous to Eqs. (6.12) with (6.13a) and $E_{\text{km}} = 0$ is used and the dislocation density is treated as depending on a given density of sources and on the existence of pile-ups, for which certain parameters are assumed (see also Poirier 1985, p. 119). The microdynamical models can be applied in cases of viscous drag effects such as solute drag or jog drag by using an appropriate expression for the dislocation velocity derived from (3.15). Thus, if the mobility M of a dislocation segment that is dragging a solute atom is taken as equal to the mobility of the solute atom, that is, D/kT, where D is the diffusion coefficient of the solute in the crystal, and if the force acting on the dislocation segment is, from (6.6a), $\tau_{\text{eff}}bl$, where τ_{eff} is the effective stress (6.37) and l is the length of dislocation line per solute atom being dragged, then we have

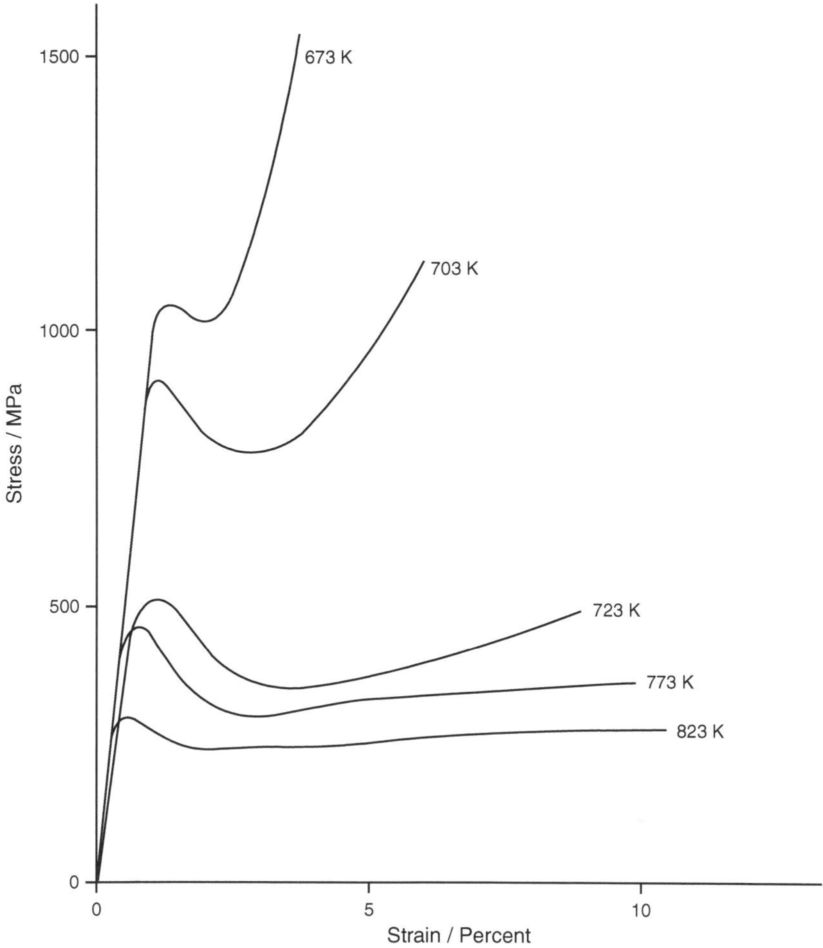

Fig. 6.21 Compression stress-strain curves for synthetic quartz crystal S1 under 300 MPa confining pressure at various temperatures, showing yield drop phenomenon (After Fig. 4 in Hobbs et al. 1972, 352 pp)

$$v = FM = \frac{Dbl\tau_{\mathrm{eff}}}{kT} \tag{6.38}$$

(Brion et al. 1971; Burton 1982a; Takeuchi and Argon 1976, 1979; Weertman 1977). See (Poirier 1985, p. 123) for a discussion on the appropriate choice of diffusion coefficient D in applications to concentrated solid solutions. It should be noted here, however, that there is a view (Kocks 1984) that solute drag is of negligible importance at high temperatures and that the role of solute hardening is effective through its influence on mutual dislocation interactions (Sect. 6.6.6).

The microdynamical models of the type considered in this subsection are limited to transient or primary creep so long as dislocation multiplication is envisaged to continue indefinitely without counterbalancing annihilation or immobilization. However, if a counterbalance is introduced, for example, in the form of a recovery process, the models have wider application, especially at high temperatures, and steady state creep can be simulated. Thus, Weertman (1975) introduced recovery effects in the form of the climb of dislocations out of the pile-ups, and Alexander and Haasen (1968) introduced an empirical, stress-dependent recovery term in the dislocation density. The resulting steady-state creep rate tends to be more highly stress dependent than a linear or Newtonian creep since, even though the velocity term in the Orowan equation may be linearly dependent on stress, additional stress dependence can be introduced in the mobile dislocation density term which incorporates both the multiplication and the recovery effects (Weertman 1975). However, recovery processes involving climb no longer represent aspects of viscous drag control of the dislocation dynamics and the models incorporating recovery thus take on a hybrid character, sharing their recovery aspects with the models of the next section. In practice, there may often be a transition from viscous drag or glide control to recovery or climb control with the approach to steady state.

6.6.6 *Thermal Models Based on Mutual Dislocation Interaction*

This category of models includes, and is to a large extent comprise of, models that are commonly referred to as models of *recovery-controlled creep*. These models have been discussed mainly in connection with secondary creep or steady state behavior. However, we shall attempt to introduce them in a slightly more general context.

Insofar as the mutual dislocation interaction that supports the applied stress and controls the movement of the gliding dislocations is athermal (Sect. 6.6.3), the rate control in time-dependent deformation arises mainly through the thermally activated modification (recovery) of the density and configuration of the interacting dislocation network. In this case, the role of the thermal activation is more indirect than in the models of the previous category (Sect. 6.6.5), in which the primary rate-controlling factor was the viscous drag in the thermally activated movement of the gliding dislocations; that is, in recovery-controlled creep, the thermal activation promotes the lowering of the barriers to dislocation motion rather than the surmounting of given barriers.

The notion of dynamical recovery has already been invoked in discussing stage III of the stress–strain curve in the athermal regime (Sect. 6.6.3). In that regime, the recovery is envisaged as being primarily a function of strain although it may also involve a minor degree of temperature sensitivity from thermal activation. If a saturation hardening is attained, it is only after very large strains ($\gg$1). However, on entering the thermal regime at higher temperatures, the dynamical recovery processes become more strongly temperature dependent, possibly with a transition in mechanism, such as from cross-slip control to climb control, and the recovery is now envisaged as being primarily a function of time. A steady state condition is then more readily achieved, often after quite small strains (~ 0.01–0.1). We therefore begin by discussing in a general and largely phenomenological way the deformation kinetics when time-dependent recovery is important, taking the case of creep, for simplicity, and generalizing the treatment of Sect. 6.6.4.

As in Sect. 6.6.4, we assume that the applied stress is supported primarily by the mutual dislocation interactions, as represented by the flow stress component τ_d, and that τ_d tends to increase due to strain hardening during creep. We now assume, in addition, that τ_d tends to decrease due to recovery as time progresses, and that the resultant effect can be written as

$$\Delta\tau_d = \theta\gamma - rt$$

and hence (6.34) becomes

$$\Delta E^* = b\Delta A^*(\theta\gamma - rt) + U \tag{6.39}$$

where γ is the plastic strain, t the elapsed time, and the parameters θ, r, respectively, the strain-hardening rate and the recovery rate. Both θ and r can, in principle, be determined empirically, θ by carrying out a stress–strain test at a relatively high strain rate at the conclusion of the creep test and r as set out in Sect. 6.5.3 (Poirier 1985, p. 105). Using the above expression for $\Delta\tau_d$ in (6.34) and following the same procedure as in deriving (6.35a) the following relation for the strain is obtained:

$$\gamma_{\text{tot}} = \gamma_{\text{inst}} + \gamma_0 \ln\left[1 + v\tau_r\left(e^{\frac{t}{\tau_r}} - 1\right)\right] \tag{6.40a}$$

where γ_{tot} and γ_{inst} are the total and instantaneous (elastic plus plastic) strains, respectively, and γ_0, τ and v are given by

$$\gamma_0 = \frac{kT}{\theta b\Delta A^*} \tag{6.40b}$$

$$\tau_r = \frac{kT}{rb\Delta A^*} \tag{6.40c}$$

$$v = \frac{\rho^{\frac{3}{2}}b\Delta A v_0}{\gamma_0}\left[1 - \exp\left(-\frac{\tau b\Delta A}{kT}\right)\exp\left[-\frac{U - b\Delta A^*(\tau - \theta\gamma_{\text{inst}})}{kT}\right]\right] \tag{6.40d}$$

For the case $\tau b\Delta A \gg kT$, (6.40d) reduces to (6.35c) and for the case $\tau b\Delta A \ll kT$ (and therefore also $\tau b\Delta A^* \ll kT$), (6.40d) becomes

$$v = \frac{\rho^{\frac{3}{2}}\tau b^2(\Delta A)^2 v_0}{\gamma_0}\exp\left(-\frac{U + \theta\gamma_{\text{inst}}b\Delta A^*}{kT}\right) \tag{6.40e}$$

In these expressions the symbols are as used in Sect. 6.4.1 and Fig. 6.19. The relation (6.40a) leads to a creep rate

$$\dot{\gamma} = \frac{v\gamma_0 e^{\frac{t}{\tau_r}}}{1 + v\tau_r\left(e^{\frac{t}{\tau_r}} - 1\right)} \tag{6.41}$$

The initial creep rate ($t = 0$) is therefore again $\gamma_0 v$ and the creep rate approaches $\gamma_0/\tau_r = r/\theta$ as $t \to \infty$. The latter result,

$$\dot{\gamma}_s = \frac{r}{\theta} \tag{6.42}$$

for the steady-state strain rate $\dot{\gamma}_s$ in recovery-controlled creep is the well-known *Bailey-Orowan equation* (Bailey 1926; Orowan 1946). It is usually derived more directly by equating $\theta\Delta\gamma$ and $r\Delta t$ in the steady state. It should be reiterated here that the Bailey–Orowan equation for the steady-state creep rate rests on the view that the strain hardening or dislocation multiplication aspect is dependent only on the strain and the recovery or dislocation elimination aspect dependent only on the elapsed time through thermal activation; this extreme view has been criticized as overlooking the possibility of strain-driven recovery or strain softening effects, the incorporation of which would require a different approach (see summary of these views in Roberts (1984)).

The quantity τ_r can be viewed as a sort of relaxation time for primary creep. When $t \gg \tau_r$, (6.40a) approximates the form

$$\gamma_{\text{tot}} = \gamma_{\text{inst}} + \gamma_0 \ln v\tau_r + \frac{r}{\theta}t \tag{6.43}$$

where the term $\gamma_0 \ln v\tau_r$ represents an additional contribution to the strain during primary creep above what would have been contributed in steady-state creep alone. In principle, the evaluation of this term, of the initial creep rate $\gamma_0 v$, and of the steady state creep rate γ_0/τ_r enables the three parameters γ_0, v and τ_r in Eqs. (6.40a) and (6.41) to be determined experimentally. However, it must be borne in mind that the underlying theoretical model of a deformation controlled primarily by mutual dislocation interaction is one in which the concepts of strain hardening, recovery, and the thermally activated surmounting of the dislocation interaction barriers have been incorporated in a more or less phenomenological way. Attempts to interpret these concepts more mechanistically have been largely confined to the case of steady-state creep, (6.42), to which the remaining discussion here will be restricted, concentrating on the quantity r while assuming the form of (6.32) for θ, that is, $\theta = \beta G$ with β constant.

On the mutual dislocation interaction model the flow stress τ is taken to be equal to $\tau_d = \alpha Gb\rho^{\frac{1}{2}}$ and so r is given by

$$r = \left(\frac{\partial\tau}{\partial t}\right)_\gamma = \frac{1}{2}\frac{\tau}{\rho}\left(\frac{\partial\rho}{\partial t}\right)_\gamma \tag{6.44}$$

If we now assume a first order kinetics of recovery in which $(\partial\rho/\partial t)_\gamma$ is proportional to the dislocation density ρ itself and to the dislocation climb velocity v_c, and inversely proportional to the climb distance needed to bring about annihilation, say half the dislocation spacing, $\frac{1}{2}\rho^{-\frac{1}{2}}$, then

$$\left(\frac{\partial\rho}{\partial t}\right)_\gamma = \frac{\alpha'\rho v_c}{\rho^{-\frac{1}{2}}} = \alpha' v_c \rho^{\frac{3}{2}}$$

where α' is a numerical constant. Inserting this expression in (6.44) and recalling $\tau = \alpha G b \rho^{\frac{1}{2}}$ leads to

$$\dot{\gamma}_s = \frac{r}{\theta} = \frac{\alpha'}{2\alpha\beta}\frac{\tau^2 v_c}{G^2 b}$$

In the case of diffusion-controlled climb, we can, from (6.23a), put $v_c \approx 2Db^2\tau/kT$ (assuming $\sigma = 2\tau$) to obtain finally

$$\dot{\gamma}_s = \beta'\frac{GbD}{kT}\left(\frac{\tau}{G}\right)^3 \tag{6.45}$$

where $\beta' \approx \alpha'/\alpha\beta$ is a numerical constant.

The form of the creep relation (6.45) suggests that in recovery-controlled creep the steady-state strain rate can be expected to depend fairly strongly on the stress. The cube stress exponent can be arrived at in a number of ways and has been described as the "natural" exponent for a high-temperature creep law (Weertman 1975). While various materials do conform approximately to (6.45) in their stress dependence, many others show a stronger stress dependence in high-temperature creep, leading Dorn to propose the semiempirical law

$$\dot{\gamma}_s = A\frac{GbD}{kT}\left(\frac{\tau}{G}\right)^n \tag{6.46a}$$

where A and n are empirical constants, found to be related according to

$$A \approx (1025)^{n-2.7} \tag{6.46b}$$

(Brown and Ashby 1980; Poirier 1985, p. 85). Various theoretical strategems have been proposed for rationalizing values of $n > 3$, based on more sophisticated recovery models (for example Poirier 1985, p. 110; Weertman 1975). It may also be noted that $n = 3$ in (6.45) derives, in part, from an assumption of linear strain hardening and that the alternative of a parabolic athermal strain hardening would immediately lead to $n = 4$. However, it has been found to be difficult to rationalize values of n greater than 5 and doubt has been expressed that a power-law representation of the creep behavior is the most appropriate in such cases (Poirier 1985, p. 111)

In Eqs. (6.45) and (6.46a) the temperature dependence of $\dot{\gamma}_s$ derives mainly from the temperature dependence of the diffusion coefficient D, which, in the case of compounds, will be the effective diffusion coefficient for the molecular species (Sect. 3.5.3). The activation energy obtained from an Arrhenius plot for the steady-state creep rate would in this case be expected to coincide with that for self-diffusion. Such an equality has indeed been widely observed at high temperature in pure metals (Poirier 1985, p. 44; Weertman and Weertman 1983b) as well as in various other materials, but it is not always found. For example, in olivine the experimental activation energy for high-temperature creep is higher than that found for the diffusion of any of its atomic components (Jaoul et al. 1981).

Thus, insofar as high temperature, steady-state creep by dislocation glide is recovery controlled, which it probably almost always is in some sense, the rate-controlling process may often be the volume self-diffusion involved in climb, as assumed for (6.45), but it is evidently not always so. Several alternatives may be mentioned. Thus, at intermediate temperatures it is possible that pipe diffusion along dislocation cores may be important, leading to a lower activation energy. Alternatively, and again at intermediate temperatures, the recovery process may be cross-slip controlled, for which the activation energy would probably also be lower than that for self-diffusion; the possible importance of cross-slip control has been the subject of controversy (Poirier 1976, 1978, 1979; Sherby and Weertman 1979). The observed tendency for the experimental activation energy for steady-state creep to decrease as the temperature decreases could thus correspond to a transition to either of these two situations; in the case of pipe diffusion the stress exponent n might also be expected to increase by 2 {Evans 1979}. Even if the recovery is climb controlled, the climb process itself may not be entirely diffusion controlled, but may depend also on the rate at which the diffusing material can be attached or removed at the dislocation core. The attachment/removal can be expected to occur mainly at jogs in the dislocation and so the climb rate could be limited by the jog nucleation rate if the latter were relatively low. Since the climb rate will be proportional to the product of the probability of an atom arriving/departing by diffusion at the dislocation and the probability of a jog being present at which to attach/detach the atom, the activation energy for climb will be the sum of the activation energies for diffusion and for job nucleation. A negligible value of the activation energy for jog nucleation is thus a prerequisite for the activation energies of steady state climb-controlled creep and diffusion to be equal. However, even with zero jog nucleation energy, a further source of inequality of the activation energies could arise in the case of compounds through there being a chemical reaction type of barrier involved in the attachment/detachment of atoms at the dislocation core; this barrier would derive from there being transitory higher energy configurations at the jog as individual atomic components of the compound are attached/detached.

Further theoretical development thus requires more specific assumptions about mechanisms at the microstructural scale, where many possibilities arise and where further developments in the future could follow from new understanding of mutual dislocation interaction mechanisms (Sect. 6.6.3). The archetype of climb-controlled recovery models is that of Weertman (Weertman 1955, 1957, 1968; see also Weertman and Weertman 1983b) but many variants of such a model have been put forward (for example, Burton 1982a; Gittus 1975; Sherby and Weertman 1979; Spingarn et al. 1979). Other creep models could be imagined to arise from forest-cutting or cell-wall-trapping views of mutual dislocation interaction (for example, Caillard and Martin 1982b; for example, Caillard and Martin 1982a, 1983) where thermally activated intersection or reaction could play a rate-controlling role. The latter role would be (Morris and Martin 1984b) akin to a quasi-viscous drag control, as considered in Sect. 6.6.5. Composite models that incorporate an element of viscous drag control as well as mutual dislocation interaction have been proposed by Gibbs (1966), Ahlquist et al. (1970), Lagneborg (1972) and others (see Gittus 1975, Chap. 3). In general, more definitive theoretical developments require a more specific basis in microstructural observations and models, taking into account where appropriate the organization of dislocations into cell or subgrain structures and its stress dependence (although Weertman and Weertman (1983b) make the point that an $n = 3$ power law tends to be predicted regardless of model details if the dislocation structures at different stress are self-similar in the sense that micrographs are superposable with change of magnification only). In particular, postulates about internal stress fields lack real validity until supported by observations such as those of Morris and Martin (1984a, b). In conclusion, it must be emphasized again (cf. Poirier 1985, p. 114) that the determination of rheological parameters such as the stress exponent n and the experimental activation energy Q gives poor constraint on the microscopic mechanisms of deformation in the absence of microstructural observations.

6.6.7 Thermal Models for Precipitate and Particle Effects

The presence of fine precipitates or dispersed hard particles, especially with sub-micron dimensions, can profoundly increase the creep strength of materials, leading to important technological applications, as, for example, in the use of nickel-based "superalloys" in gas turbines, or the development of the oxide-strengthened metals such as sintered aluminum powder (SAP) and thoria-dispersed (TD) nickel. Such materials can show peculiarities in creep behavior such as difficulty in establishing a steady-state creep rate or a tendency for the appearance of a marked tertiary or accelerated creep stage (the latter effect commonly arises from instability of the material, for example, the re-solution of precipitates at elevated temperature). Especially notable is the common tendency for the experimentally determined stress exponent in power-law creep to be abnormally high (values of 7–40 or more are reported), and the experimental activation energy may also be abnormally high, substantially exceeding the activation energy for self-diffusion. Because of these peculiarities, models for these cases are here dealt with separately from the other models for behavior in the thermal regimes. Reviews of creep in fine precipitate or particle-bearing materials have been given by Martin (1980), Haasen (1983) and Strudel (1983).

As in the athermal case (Sect. 6.6.2) we need first to make a broad distinction between situations in which the precipitates or particles can be sheared through by the passage of dislocations at stresses below the Orowan bypass stress and situations in which the particles remain undeformed. For this purpose, we shall designate the particles as shearable particles and hard particles, respectively, using the term particle now to refer both to the precipitates and to particles introduced in other ways, as in fabrication by sintering.

Creep deformation in the presence of shearable particles involves a viscous drag on the motion of the dislocations when the deformation is viewed at a scale larger than the particle spacing; that is, the deformation is of the type discussed in Sect. 6.6.5. At stresses below the level for athermal cutting of the particles (Sect. 6.6.3), there can still be a certain rate of cutting as a result of thermal activation and hence a certain dislocation velocity and strain rate. Kocks et al. (1975, pp. 147–163, 196–225) have discussed in considerable detail the kinetics of such a cutting process and have proposed that, if the strain rate is represented in the form

$$\dot{\gamma} = \gamma_0 \nu_0 \exp\left[-\frac{E(\tau)}{kT}\right] \tag{6.47a}$$

where γ_0 is a term containing the mobile dislocation density and the area swept out following each successful activation event (cf. (6.33)) and ν_0 is the attempt frequency (Kocks et al. 1975, p.124 estimate ν_0 to be of the order of 10^{10}–10^{11} s^{-1} in this situation). Then, the activation energy E can be represented in the form

$$E = E_0 \left[1 - \left(\frac{\tau}{\tau_0}\right)^p\right]^q \tag{6.47b}$$

where τ is the applied stress, τ_0 the threshold stress for athermal flow, E_0 the (Gibbs) energy needed to cut through the particle in the absence of thermal activation, and p, q numerical factors ($0 \le p \le 1$, $1 \le q \le 2$; typically $p = 2/3$, $q = 3/2$). Such a particle-shearing model is probably most relevant at stresses not markedly below the threshold for athermal flow.

In practice, experimental studies in metal-based systems involving particles that are shearable at certain stress levels show that significant creep rates can also be observed at substantially lower stresses than those envisaged in the previous paragraph. In these cases, the particles are probably not being sheared and processes of particle circumvention involving diffusion are thought to be rate controlling, as in crystals with hard particles.

The creep of crystals containing hard particles can be approached in two ways. On the one hand, if the resistance to deformation is viewed as arising from the presence of an internal stress—a rather phenomenological approach—then the creep can be discussed in terms of recovery in which this internal stress is reduced by thermally activated processes, allowing further deformation in the course of time, as in Sect. 6.6.6. The theoretical development of this approach requires the identification and treatment of the rate-controlling recovery processes taking place in the complex configuration of dislocations that presumably exists in the neighborhood of the particles. On the other hard, a more direct and mechanistic approach in terms of the dislocation dynamics can be taken, the creep rate being related directly to the climb rate of the mobile dislocations as they circumvent the particles. However, again the concept of an internal stress tends to enter in determining the local driving stress for the dislocation motion. In either case, the concept of there being an internal stress can be used to rationalize the high values of the observed stress exponent n and activation energy Q. Thus, if, instead of writing the observed strain rate as

$$\dot{\gamma} = A\tau^n \exp\left(-\frac{Q}{RT}\right)$$ (6.48)

in terms of the applied stress τ, where A is a constant and

$$n = \left(\frac{\partial \ln \dot{\gamma}}{\partial \ln \tau}\right)_T, \quad Q - R\left(\frac{\partial \ln \dot{\gamma}}{\partial (1/T)_\tau}\right),$$ (6.49)

we relate the observed strain rate to the local processes by writing

$$\dot{\gamma} = A'\tau_e^{n_e} \exp\left(-\frac{Q_e}{RT}\right)$$ (6.50)

where $\tau_e = \tau - \tau_i$ (τ_i is the internal stress), A' is a constant and

$$n_e = \left(\frac{\partial \ln \dot{\gamma}}{\partial \ln \tau_e}\right)_T, \quad Q_e = -R\left(\frac{\partial \ln \dot{\gamma}}{\partial (1/T)}\right)_{\tau_e},$$ (6.51)

then it can be shown, following Saxl and Kroupa (1972), that

$$n = n_e \tau \frac{1 - \left(\frac{\partial \tau_i}{\partial \tau}\right)_T}{\tau - \tau_i}$$ (6.52)

$$Q = Q_e - n_e RT^2 \frac{\left(\frac{\partial \tau_i}{\partial T}\right)_\tau}{\tau - \tau_i}$$ (6.53)

It follows that the experimental quantities n and Q can be substantially larger than the quantities n_e and Q_e that are effective locally.

The two approaches outlined in the previous paragraph for the case of hard particles correspond more or less, on the scale of the crystal, to the recovery and viscous drag models of creep discussed in Sects. 6.6.5 and 6.6.6, respectively. However, the distinction may ultimately prove to be an artificial one because of the common dependence on dislocation climb, generally agreed to be the rate-controlling process (Strudel 1983). It seems that climb can be especially rapid in the neighborhood of hard particles, presumably due to short-circuit diffusion at the particle interface or some sort of coordinated pipe diffusion on the scale of the particles. It has even been suggested (Strudel 1983) that the climb activity may be so intense that it contributes significantly to the strain itself, as a diffusion creep mechanism (Chap. 5). However, in other cases, diffusion is thought to be inhibited in the neighborhood of particles (Martin 1980, pp. 167–180). In any case, the activation energy may be relatively high because of the additional need to eliminate attractive junctions (Guyot 1980).

6.7 Dynamics of Mechanical Twinning

Mechanical twinning is generally viewed as taking place by the orderly propagation of a partial dislocation, with some mechanism such as the pole mechanism of Cottrell and Bilby (1951) and Thompson and Millard (1952) for ensuring that the twinning dislocation progresses from plane to adjacent plane with each sweep. The motion of the twinning dislocation will be impeded by similar interactions to those involved for slip dislocations but, as (Friedel 1964, p. 176) has pointed out, these interactions can be much weaker than for slip dislocations because of the smaller Burgers vector of a partial dislocation, resulting in lower Peierls potential, lower energies of interaction with solutes and other obstacles, etc.; also the twinning dislocation is not subject to strain-hardening interactions with a growing dislocation density. Consequently, it is often observed that twinning can proceed rapidly at relatively low stresses, as, for example, in calcite at room temperature. For a general discussion of mechanical twinning in terms of dislocations, see Hirth and Lothe (1982, Chap. 23).

Although twinning can propagate easily, there can be considerable nucleation barriers against its initiation. Friedel (1964, p. 176) suggests that the initiation stress needed should be of the order of γ/b, where γ is the energy per unit area of the stacking fault generated by the twinning dislocation and b its Burgers vector. For metals, this quantity is often of the order of 1 % of the shear modulus. Hirth and Lothe (1982, pp. 757, 825) deduce similar or somewhat higher nucleation barriers from a consideration of the thermally activated nucleation of partial dislocations. Therefore, local stress concentrations can be expected to play an important part in the initiation of twinning. The existence of an initiation stress that is high relative to the propagation stress would provide an explanation of various phenomena that are common in twinning, such as sharp stress drops or serrated stress–strain curves and acoustic effects ("twinning cry", as observed in tin). Some particular dislocation processes that may be involved in the nucleation of twins in the common metals are discussed by Mahajan (1981) and references given to others.

Mechanical twinning is more widely observed at relatively low temperatures. This observation is consistent with twinning being essentially an athermal process in most respects, initiated more readily when the thermal activation of slip is at a minimum and applied stress at a maximum. However, the sensitivity of the nucleation to structural imperfections probably rules out there being a well-defined critical shear stress for twinning. As a result of the generally athermal character, there has been little discussion of the role of twinning in deformation in thermal regimes. However, under the assumption that the nucleation will be rate-controlling, Frost and Ashby (1982, p. 10) have proposed a rate equation for twinning, analogous to (6.47a), which can be written in the form

$$\dot{\gamma} = \gamma_0 v_0 \exp\left[-E_N\left(1 - \frac{\tau}{\tau_0}\right)/RT\right] \tag{6.54}$$

where γ_0 is a term containing the density of nucleation sites and the strain resulting from each nucleation event, v_0 an attempt frequency, E_N the (Gibbs) activation energy for nucleation, τ_0 the stress needed to initiate the twinning athermally, and τ the applied stress; $\dot{\gamma}$ is the strain rate, T the temperature, and R the gas constant. Such an expression can only be expected to be relevant where thermally activated nucleation of twinning is significant, for example, possibly when the Burgers vector of the twinning dislocation is very small. An extensive discussion of thermal and athermal twin nucleation is given by Christian (1965, p. 777 et seq.).

Shear transformations, such as orthoenstatite–clinoenstatite, can be expected to involve similar factors to those involved in mechanical twinning. However, in addition, the dynamics of shear transformation will be influenced by the change in Gibbs energy accompanying the phase change. The influence of the latter will presumably be equivalent to the superposition of an internal stress assisting the process of transformation (or opposing it if the thermodynamic conditions still fall in the stability field of the first phase), but the magnitude of this effect will also depend on the value of the shear stress itself if the equilibrium boundary is affected by nonhydrostatic stress (Coe 1970).

Because of the finite plastic strain in a twin or transformation lamella, there may be elastic accommodation strains required to maintain continuity with the surrounding untwined or untransformed material, depending on the geometry involved. These requirements are especially obvious in the grains in a polycrystalline aggregate, the general consideration of which follows in the next section (Sect. 6.8). One of the consequences of the compatibility requirements is seen in the commonly lenticular shape of mechanical twins in polycrystals, the aspect ratio of the lenticles being higher the larger the twinning strain. A case of interlamellar stresses arising from shear transformation in feldspar is discussed by Yund and Tullis (1983).

6.8 Crystal Plasticity in Polycrystalline Aggregates

6.8.1 Introduction: The Compatibility Problem

Having considered at some length the various aspects of the crystal plasticity of individual crystals, we now discuss the deformation of polycrystalline aggregates in which the constituent grains are deforming by crystal plasticity processes. The deformation properties of the aggregate cannot be derived by a simple averaging of the properties of the grains, treated as separate individual crystals with their orientations taken into account, because of two factors:

1. the requirement of compatibility of strain from grain to grain, so that the continuity of the polycrystalline aggregate is maintained during the deformation
2. the influence of the presence of the grain boundaries on the behavior within the grains.

These are two separate factors since the strain compatibility requirement depends primarily only on the relative orientations of the grains and not on their dimensions, whereas the grain boundary effects introduce a grain size dependence into the behavior of the polycrystalline aggregate. We shall first consider the strain compatibility aspect.

Although it is well known that the individual grains of a polycrystalline aggregate do not deform homogeneously even when the macroscopic deformation of the aggregate is statistically homogeneous (Barrett 1943, p. 325; Boas and Hargreaves 1948), it is useful initially to view the stress and strain in each grain as being effectively homogeneous to a first approximation. Taking this view, there are two classical models for relating single crystal to polycrystal behavior, which serve as limiting cases or bounds to the actual behavior:

1. The model of Sachs (1928), in which it is assumed that the stress in each grain is equal to the macroscopic stress (as if the grains were loaded in series).
2. The model of Taylor (1938), in which it is assumed that the strain in each grain is equal to the macroscopic strain (as if the grains were deformed in parallel).

Strictly, neither model is physically realistic since the Sachs model leads to violation of the equations of continuity at the grain scale due to misfit at grain boundaries, while the Taylor model leads to similar violation of the equations of equilibrium. A physically valid model has to incorporate some degree of heterogeneity in both stress and strain at the grain scale, and we now consider ways in which this aim has been approached.

The actual departures from a state of homogeneous stress can be represented in terms of an internal stress, by postulating that the actual local stress results from a superposition of a uniform applied stress and a locally varying internal stress. The internal stress may give rise to a residual stress when the macroscopic stress is removed and so be, in principle, measurable. The internal stress will, of course, be a function of the strain unless a steady state is established. Its mean value will be a measure of the amount by which the applied stress has to exceed the flow stress of an average individual grain in order to achieve macroscopic flow. For use of the notion of internal stress in the polycrystal context, see Leffers (1981) and Berveiller et al. (1981).

Heterogeneity of strain within the grains of a polycrystalline aggregate has often been modeled by postulating a core and mantle structure for the grains. In this view, the core of the grain is regarded as undergoing a more or less homogeneous strain, approximating the macroscopic strain, while the complicated intergranular adjustments are concentrated in a mantle region in the vicinity of the grain boundary. Such a model is sometimes treated as a two-phase view of the polycrystalline aggregate (Mecking 1981b). In a careful analysis in core and mantle terms, the strain in the core region must be acknowledged as itself departing from the macroscopic strain by a small amount that is related to the internal stress discussed above (see, further, Sect. 6.8.4). The mantle region is distinguished from the core by a more pronounced development of multiple slip and lattice rotation and, in metals, is reported to be of the order of tens of micrometers in thickness (Leffers 1981). The core and mantle distinction has been especially emphasized by some writers in relation to high temperature creep (for example, Gifkins 1974, 1978) but it is not clear how widely useful the concept is. Another approach to the description of the heterogeneity of deformation and its accommodation is in terms of geometrically necessary dislocations (Ashby 1970, 1971; Mecking 1981b).

6.8.2 Multiplicity of Micromechanisms: The von Mises Criterion

If two grains, initially in contact along a common boundary, are individually subjected to simple shearing on planes that are not parallel between one grain and the other, then after the deformation the grains will no longer fit together at a common boundary without further adjustments in shape or rotation. In order to maintain continuity at the grain boundary it is necessary, in general, that adjoining grains of different orientation undergo deformation by a combination of shears that gives the equivalent of more than one simple shear in each grain.

In attempting to answer the question of how many microshear mechanisms are required to operate within a grain in order to achieve strain compatibility in an aggregate, it is usual to start by enunciating the criterion of von Mises (1928). Viewing the micro mechanisms as interpenetrable or superposable simple shears, von Mises showed that, in order to achieve an arbitrary homogeneous strain, five independent shears are required (independent in the sense of making possible certain deformations that cannot be achieved with any combination of the other available shear mechanisms). Applying this criterion to deformation by multiple slip in an aggregate of randomly oriented grains in which the strain in each grain is the same as the macroscopic strain (homogeneous strain or Taylor model), it requires that, in general, five independent slip systems must operate in each grain. Methods for determining the number of independent systems in a given set of crystallographic slip systems, and some results, are set out by Groves and Kelly (1963), Kocks (1964) and Paterson (1969).

It is to be noted that, for symmetry reasons, the number of independent slip systems within a given set of crystallographically equivalent slip systems may be substantially less than the multiplicity of the set. For example, there are 12

$\{110\}\langle 1\bar{1}0 \rangle$ slip systems in holohedral cubic crystals but only two are independent; the activity of a third system does not make possible any deformation that cannot be achieved with the first two. The inadequacy of the $\{110\}\langle 1\bar{1}0 \rangle$ slip systems for producing a general deformation is illustrated by the impossibility of achieving change of length parallel to any $\langle 111 \rangle$ direction with such systems.

The von Mises criterion of five independent slip systems has been widely evoked as a necessary condition for the ductility of polycrystalline aggregates. Thus, in polycrystalline NaCl-structure materials, the transition from brittleness to ductility at elevated temperatures has been correlated with the onset of $\{100\}\langle 1\bar{1}0 \rangle$ slip in addition to the normal $\{110\}\langle 1\bar{1}0 \rangle$ slip, an addition which raises the total number of available independent slip systems from two to five (Copley and Pask 1965; Groves and Kelly 1963; Pratt 1967). The criterion is, of course, not in itself a sufficient condition for ductility since there must also be adequate mobility of the dislocations and interpenetrability of the slip activity.

The von Mises criterion refers to a deformation that is homogeneous at the microscopic or grain scale. When the heterogeneity of deformation at this scale is taken into account, the question arises as to how far the criterion can be relaxed as a result of the additional scope for maintaining the mutual fit between grains through compensating local deformations. Thus, if grains of a particular small range of orientations were incapable of undergoing the required strain for lack of suitable slip systems, the remaining majority of grains might be able to undergo additional compensating deformation so that the undeformable grains could be accommodated as hard inclusions. The possibility of such an effect has been demonstrated by Hutchinson (1977) in a self-consistent numerical treatment of a polycrystalline aggregate of hexagonal material having active basal and prismatic slip systems (which comprise four independent slip systems) but lacking the pyramidal slip systems that would permit normal strain parallel to the c axis (see Sect. 6.8.4). Thus, it can be concluded that, in practice, not more than four independent slip systems are necessary for the deformation of a polycrystalline aggregate solely by intragranular slip. Whether this requirement can be reduced to three is not clear but the calculation of Hutchinson (1976) for NaCl-structure materials (discussed further in Sect. 6.8.4) suggests that this may not be possible. Nevertheless, intuitively, it would seem that with three independent slip systems active, the amount of activity needed of any fourth, accommodating micro mechanisms would, at least, be relatively small.

In many nonmetallic materials, especially in rocks, the grains are often of relatively low crystallographic symmetry and the number of available independent slip systems is therefore often small. For example, olivine has only three independent slip systems and one of these is relatively strong, while feldspars probably have even fewer (quartz and calcite do not present the same problem provided that more than one of their known sets of slip systems is active). Nevertheless, these substances can be deformed in polycrystalline form, suggesting that other micro mechanisms are active. Some possibilities for the latter are:

1. At sufficiently high temperatures, climb of dislocations may be important. Its role in compensating for a lack of slip systems is discussed by Groves and Kelly (1969). They show that, acting alone, six independent climb systems are required for a general, nondilatational strain, and that six different Burgers vectors, not necessarily crystallographically nonequivalent, are required to provide this set of climb systems. They also point out that, for dislocations of a given Burgers vector, climb motion produces two independent strain components additional to those due to glide in a specific plane, although the new strain components are not necessarily independent of those contributed by other active glide systems, and so further examination is needed to establish to what extent the climb increases the number of independent strain components overall. If dislocations can glide on any plane as well as climb, then a general strain is possible if there are three noncoplanar Burgers vectors (as in olivine).

2. Also at sufficiently high temperatures, sliding on grain boundaries introduces another mode of relative displacement in the aggregate. The extent to which this sliding can be counted as contributing to the available independent modes of deformation is not easy to see and would seem to depend on the nature of the mechanisms accommodating the sliding (see Sects. 5.2.2 and 7.1.3 on the interdependence of the sliding and accommodation processes). In the case of accommodation by diffusion, there can be an independent contribution to the overall deformation if the accommodation involves components of strain that cannot be produced by available climb or glide systems. However, in the case of accommodation by plastic deformation within the grains, the requirements on availability of slip systems for the accommodation of the sliding will be identical to the requirements for compatible deformation of grains in the absence of sliding and so the occurrence of slip-accommodated sliding at grain boundaries does not introduce any additional independent modes of deformation. Thus, on this view, sliding at grain boundaries only introduces additional independent modes of deformation insofar as it involves atom transfer mechanisms for accommodation which in themselves are independent mechanisms. Grain boundary migration is not a mechanism of deformation and so does not enter into the present kinematical considerations although it may have important dynamical consequences through its effect on the stresses needed to operate the various mechanisms.

3. Dilatancy of the aggregate permits accommodation by the development or elimination of voids. Void formation may consist of cavitation, as occurs in creep of metals at high temperature at atmospheric pressure, of microcracking, as in the deformation of rocks under confining pressure at any temperature (Paterson 1969), or of variation in pore space in an already porous rock (cf. deformation of sand). Whenever dilatancy is involved, a significant pressure dependence of the flow stress can be expected, as was observed, for example, in magnesium oxide by Paterson and Weaver (1970).

4. Twinning mechanisms may contribute in appropriate cases, although the single-sense property of the twinning shear makes twinning a less effective accommodation mechanism than slip. Kinking can also contribute in a similar way to

twinning if sufficiently penetratively distributed (Paterson 1969) but often it is probably better viewed as a special type of heterogeneity of deformation associated with the slip system on which it is based, contributing somewhat to a reduction in the number of slip systems required.

5. Minor accommodation requirements may be met by elastic strains accompanying variations in stress within the grains.

In the discussion, so far it has been implicit that the crystallographic orientation distribution of the grains is more or less random and the shape roughly equiaxed. When strong preferred orientation is introduced, such as through large strain or recrystallization, the accommodation requirements are relaxed gradually as the single crystal limit is approached. Also accommodation requirements become somewhat less stringent as the grain shape becomes highly anisotropic, another factor that may be significant in highly deformed, foliated or lineated materials (Kocks and Canova 1981; Mecking 1981b).

6.8.3 Grain Boundary Effects and Grain Size Dependence

In a homogeneously deforming polycrystalline aggregate or in an aggregate in which the heterogeneity scales with the grain size, there would not necessarily be any grain size dependence of the flow stress. Even the presence of the geometrically necessary dislocations need not lead to a grain size dependence (Mecking 1981b). However, in practice, it is widely observed that the flow stress tends to be higher for small grain sizes in aggregates deforming by intragranular crystal plasticity, especially in the athermal regime. This effect indicates that the presence of the grain boundaries has itself a strengthening influence, apart from any effect associated with the occurrence of multiple slip.

The grain boundary strengthening effect is usually attributed in some way to the accumulation of dislocations in the neighborhood of the boundary, presumably due to there being an impediment to the accommodation of the dislocations in this region because of the influence of the contiguous grain or to there being interaction with other dislocations already accumulated there. The effect is commonly expressed through an additive term in the flow stress σ, as follows:

$$\sigma = \sigma_0 + kd^{-\nu}$$

where σ_0 is the contribution to the flow stress of the aggregate purely from the multiple slip within the grains, d the grain size, and k, ν constants. The value of ν is commonly put as $\frac{1}{2}$, following Hall (1951) and Petch (1953); see other references in Friedel (1964, p. 267). However, this value is often not well defined experimentally owing to the practical difficulty of keeping other factors constant while preparing specimens of different grain size, and a value of $\nu = 1$ is said to be equally valid in many cases (for example, Dollar and Gorczyca 1981; for example, Kocks 1970; Mecking 1981b).

A model involving the pile-up of dislocations at the grain boundaries is usually used to rationalize the value $v = \frac{1}{2}$ (Hall 1951; Nabarro 1950; Petch 1953), although a model involving emission of dislocations from the grain boundary has also been proposed (Li 1963). The value $v = 1$ has been rationalized using a core and mantle model (Kocks 1970; Mecking 1981b) and an intermediate case has been treated by Armstrong et al. (1962), while Ashby (1970, 1971) has given a treatment in terms of geometrically necessary dislocations. In the pile-up model for $v = \frac{1}{2}, k$ is predicted to have the value $\sigma_c l^{\frac{1}{2}}$ where σ_c is the stress required to initiate flow in a second grain, adjacent to the boundary against which the pile-up is occurring, and l is the mean dislocation spacing within the grains (Friedel 1964, p. 268; Haasen 1978, p. 282). It follows that the grain size sensitivity will be most pronounced at small grain size and low dislocation density. For common metals, values of k of 0.1–1 MPa m$^{-1/2}$ are found when $v = \frac{1}{2}$ is used (Armstrong et al. 1962). In the case of the core and mantle model for $v = 1, k$ is predicted to be $4t\Delta\sigma$ where t is the width of the mantle region and $\Delta\sigma$ is the increase in flow stress in the mantle above σ_0 (Kocks 1970; Mecking 1981b). In Ashby's model, with $v = \frac{1}{2}$, $k = \alpha G(b\gamma)^{\frac{1}{2}}$ where α is a constant, G the shear modulus, b the Burgers vector and γ the shear strain in a grain. See further discussion on the value of k by Hansen (1983, 1985).

In passing from the athermal deformation regime to the thermal regime at higher temperatures, the grain size sensitivity tends to be reduced, that is, the value of v smaller, because of thermal recovery effects in the grain boundary neighborhood (Mecking 1981b). Eventually, with further increase in temperature, grain boundary weakening associated with the change from crystal plasticity to atom transfer and grain boundary sliding mechanisms begins to enter, leading finally to the strong inverse grain size dependences discussed in Chaps. 5 and 7.

6.8.4 Relation of Single Crystal to Polycrystal Flow Stresses

The quantitative problem in the theory of the flow of polycrystalline materials by crystal plasticity is to relate the macroscopic flow stress σ of the aggregate to the flow stress of the single crystal, expressed as the resolved shear stress τ on an active slip system; that is, to evaluate the numerical factor M in the relation

$$\sigma = M\tau \tag{6.55}$$

In an athermal regime, τ is the resolved shear stress after a given strain and, in a steady-state thermal regime, τ is the resolved shear stress for flow at a specified strain rate. In order to obtain the macroscopic stress–strain ($\sigma - \varepsilon$) curve from the plot of resolved shear stress τ versus resolved shear strain γ for the single crystal, a similar numerical factor, which we take for present purposes to be again M, has to be used to relate γ to ε (cf. Eqs. (6.1) and (6.2)), so that finally the strain-hardening rate becomes

$$\frac{\mathrm{d}\sigma}{\mathrm{d}\varepsilon} = M^2 \frac{\mathrm{d}\tau}{\mathrm{d}\gamma} \tag{6.56}$$

For stricter definition of the factor M, see Kocks (1970). The polycrystal factor M is, in general, a tensor but we consider here only the case in which the stress σ is taken to have only a single nonzero component and in which M can be treated as a scalar. This case applies to uniaxial tension and compression tests, as well as to the usual axisymmetric triaxial tests of rock mechanics. In the latter case, if σ_1 and σ_3 are the greatest and least principal stresses, the stress difference $\sigma_1 - \sigma_3$ is to be taken as σ since the superposition of a hydrostatic pressure has no effect in the theory.

Under the athermal, perfect plasticity assumption that slip in a given system obeys Schmid's law (Sect. 6.1), that is, occurs at a critical resolved shear stress regardless of the shearing rate, the Taylor and Sachs models (Sect. 6.8.1) provide upper and lower bounds for M, namely, the Taylor factor M_T and the Sachs factor M_S, respectively. The Sachs factor M_S is calculated simply as the mean, over all grain orientations, of the reciprocal of the maximum Schmid factor for the individual orientations, since only one slip system is assumed to operate in each grain. In calculating the Taylor factor M_T, the minimum set of five slip systems that will operate to give the prescribed strain in a grain of a given orientation must first be selected. This is done through an optimization procedure, consisting either of minimizing the internal work $\tau_k \mathrm{d}\gamma_k$ summed over the available slip systems k (Taylor 1938 #2695) or of maximizing the external work $\sigma\mathrm{d}\varepsilon$ (Bishop 1953, 1954; Bishop and Hill 1951), these extremum conditions being implicit in the formulation of the perfectly plastic model (Chin and Mammel 1969; Kocks 1970; Lister et al. 1978). Then, from the stress state corresponding to the optimal set of slip planes, the component parallel to σ can be determined and averaged over all grain orientations to give the macroscopic stress σ, from which M_T can be determined. Strain hardening can be taken into account by changing the value of τ incrementally as the strain is incremented, still under the assumption of athermal behavior. Certain uniqueness problems can be overcome by introducing a degree of rate dependence in the single crystal stress–strain relationship (Asaro and Needleman 1985).

The calculation of the value of M has been explored most extensively for the f.c.c.cubic crystal structure, with a single set of 12 crystallographically equivalent slip systems $\{111\}\langle 1\bar{1}0 \rangle$, capable in themselves of satisfying the von Mises criterion. Proceeding as just described, the upper bound or Taylor factor M_T is found to be 3.06 for simple extension or compression (principal stresses σ, 0, 0) and 1.656 for shear (principal stresses σ, $-\sigma$, 0 and putting $\sigma = M\tau$) (Bishop and Hill 1951). The lower bound or Sachs factor M_S is 2.24 for simple extension or compression and 1.12 for shear (Cox and Sopwith 1937; Sachs 1928). The value $M_T = 3.06$ also applies for the b.c.c. structure with active slip systems $\{110\}\langle 1\bar{1}1 \rangle$, decreasing toward 2.75 as pencil glide comes into effect (Kocks 1970).

The resolved shear stress needed to operate a given slip system is, in practice, affected by slip occurring on intersecting slip systems, an effect known as latent hardening, which is not taken into account in the formulation of Schmid's law. Therefore, Kocks (1960, 1970) has proposed that an appropriate single crystal curve to choose as the basis for calculating the polycrystal curve on the Taylor model is that for a symmetrical or "polyslip" orientation in which several slip systems operate simultaneously. Alternatively, a constitutive relation for the single crystal that explicitly takes into account the latent hardening can be introduced (Asaro and Needleman 1985). In contrast, the Sachs model assumes single slip in the grains and so, logically, a single crystal curve for single slip should be used for it. If these respective procedures are followed, it is found that the measured polycrystal stress–strain curve for f.c.c. and b.c.c. materials tends to fall much nearer to the calculated curve for the Taylor model than that for the Sachs model (Kocks 1970). However, the Sachs model is patently defective in not recognizing that in practice the grains mostly deform by some degree of multiple slip. It has therefore been proposed that in applying the Sachs model, a single crystal curve for a polyslip orientation should also be used. Such a "modified Sachs model" predicts a polycrystal stress–strain curve much nearer to the measured curve (Leffers 1979, 1981).

In spite of this approximate agreement between observation and the predictions of the Taylor and modified Sachs models (see Asaro and Needleman 1985 for a review of recent developments of the Taylor model), the theoretical situation cannot be accepted as being very satisfactory on at least two grounds:

1. Both models are conceptually defective in that they entail physically unacceptable discontinuities in either force or displacement at the grain scale.
2. There remain severe difficulties in application to lower symmetry materials in which the von Mises criterion is not met even with the operation of several crystallographically nonequivalent slip systems, for example, in hexagonal close-packed materials with basal and prismatic slip feasible but pyramidal slip too difficult to operate, or in olivine with only three independent slip systems, all nonequivalent and one rather difficult to operate.

It is desirable, therefore, to develop some way of allowing for heterogeneity of deformation at the grain scale, while meeting local continuity and equilibrium requirements. Two approximate methods that aim to meet this need are the self-consistent and finite-element approaches.

In the self-consistent approach, the individual grain is viewed as an inclusion in a matrix, the properties of which are assumed to be identical with the macroscopic properties. The local departures of stress and strain in the individual grain from some initially assumed macroscopic values are calculated for all grain orientations and averaged to obtain a new set of macroscopic values, which can then be used to refine the calculated local departures in the grains, and so on iteratively. The applications to athermal plasticity by workers such as Hutchinson (1970) and Berveiller and Zaoui (1978, 1981) are developments of earlier applications to elastic–plastic problems by Kröner (1961), Budianski and Wu (1962) and Hill (1965). An important extension to creep problems was made by Hutchinson (1976, 1977), from whose papers most of the following comments are derived.

The first conclusion from self-consistent theory is that the polycrystal factor M in (6.55) for athermal plastic deformation of f.c.c. materials falls very near to the upper bound calculated on the Taylor model, thus justifying the use of the Taylor model where five independent slip systems are available and the flow stress is insensitive to strain rate. However, the self-consistent results tends to fall significantly below the athermal upper bound when strain-rate sensitive flow is concerned, and increasingly so as Newtonian flow (power-law stress exponent $n = 1$) is approached. In the Newtonian case, which is formally analogous to the elastic case, the self-consistent result is found to fall between the upper and lower bounds of Hashin and Shtrikman (1963), suggesting that the self-consistent theory gives a fairly accurate estimate of M.

As already pointed out (Sect. 6.8.2), the self-consistent calculations show that deformation in a polycrystal can be achieved with four independent slip systems when heterogeneity of deformation on the grain scale is allowed, as revealed for the hexagonal close-packed case for $\tau_c \to \infty$. However, the result for NaCl-structure materials with $n = 1$ leads to $\sigma \to \infty$, indicating that the three independent $\{100\}\langle 1\bar{1}1\rangle$ slip systems alone are insufficient for plasticity even on the self-consistent model. There is still some reservation in applying this conclusion to real materials since in the self-consistent model the grains are still treated as deforming homogeneously within themselves, even though differently from one to another, and so there may be a little more freedom of accommodation by heterogeneity of deformation in a real material. Peirce et al. (1982) have, for example, made a beginning in treating heterogeneity of deformation within grains but their approach has not yet been applied to the polycrystal problem.

For examples of the application of finite-element numerical modeling, see Abe and Nagaki (1981) and Peirce et al. (1982, 1983).

So far, we have been considering the relation of the single crystal to polycrystalline flow stresses at given strains without taking into account the effects of the evolution of the structure, except for passing reference to strain hardening. We next consider the effects of large strains and of the preferred orientations, both crystallographic and in grain shape that may exist initially or develop during the deformation.

6.8.5 *Preferred Orientations and Large Strains*

The local strain compatibility requirements in a polycrystal depend on the grain shape. The considerations set down in the previous subsections apply primarily to grain shapes that are more or less equant. When grains of very inequant shape are involved, fewer independent slip systems are required (four for rod-shaped and three for disk-shaped grains) because the compatibility requirements at the boundaries of small area (edge of disk or end of rod) affect only a small fraction of the volume of the grain and so can be neglected (Honneff and Mecking 1978;

Mecking 1980; Kocks and Canova 1981). This effect may be important when large strains (say, >0.3–0.5) lead to inequant grain shape or when grains tend to form initially in platy or needle-like form, as in the case of minerals such as mica and sillimanite. The effect has been explored in copper by Tomé et al. (1984), who show that it may lead to a geometric softening at large strains.

Another effect of plastic strain is to produce a crystallographic preferred orientation. When this effect is strongly developed, the properties of the aggregate approach in some degree those of a single crystal and again the requirement on the number of independent slip systems may be reduced to some extent. The pattern of preferred orientation can be calculated by direct extension of the calculation of the stress–strain curve (previous subsection), proceeding incrementally, with the weighting of the distribution of orientations being updated at each increment. Such calculations, using the Taylor model applied to cases of geological interest, were carried out by Lister and coworkers (Lister et al. 1978, Lister and Paterson 1979, Lister and Hobbs 1980); for recent papers on the measurement and analysis of crystallographic preferred orientation, see Wenk (1985). Asaro and Needleman (1985) set out a more sophisticated development of the Taylor model, based on single crystal constitutive relations formulated by Peirce et al. (1982, 1983) and aimed at treating large deformations with arbitrary histories and predicting preferred orientations. They demonstrate another form of geometrical softening of the aggregate, which is associated with the development of crystallographic preferred orientation and which is additional to the effect of Tomé et al. (1984), mentioned previously.

Structural inhomogeneities or instabilities have been observed to develop in polycrystals at large strains, generally in the form of "shear bands", that is, regions in which large shear strains have been concentrated (for example, Grewen et al. 1977; Hatherly 1978; Hatherly and Malin 1979; Gil Sevillano et al. 1980; Hecker and Stout 1984). The development of these features is usually related to a lack of the stabilizing effect of strain hardening, arising both from a saturation in local strain hardening within the grains and from geometric softening effects (Asaro and Needleman 1984, 1985; Mecking 1980, 1981a; Tomé et al. 1984).

Dynamic recrystallization (Sect. 6.5.3) is another important process in polycrystals deformed at higher temperatures. It has been studied especially in hot working of metals at strain rates greater than, say, 10^{-3} s^{-1}; see review by Roberts (1984) but it can also occur in creep. It has, of course, important consequences for preferred orientation development.

6.8.6 Polyphase Aggregates

In this subsection, we are concerned primarily with polyphase materials in which the grains of the individual phases are comparable in size and volume fraction and more or less equiaxed in shape. The influence of small volume fractions of finely dispersed particles or precipitates on the crystal plasticity of the enclosing matrix has already been considered in the context of single crystal behavior (Sects. 6.6.2

and 6.6.7). Linear structures, such as those in fiber-reinforced materials, and lamellar structures, such as are produced in eutectic and eutectoid systems, will not be considered specifically (see Kelly 1966; Kelly and Nicholson 1971; Le Hazif 1979; Piatti 1978).

We first mention some structural aspects of polyphase materials, restricting consideration mainly to two-phase materials for the sake of simplicity in bringing out the essential points, although analogous properties will generally apply also to materials of more than two phases. A preliminary point to note is that, in order for a more or less equiaxed grain shape of the phases to persist, the interfacial energies between the phases must not differ greatly from the grain boundary energies in the pure phases; otherwise, depending on the relative volumes of the phases, there will be a tendency to effects such as the spreading of one phase along three-grain edges or the grain boundaries of the phases (Smith 1964; Waff and Bulau 1979).

Given an equiaxed grain structure, one of the phases in a two-phase material can always be expected to be continuous. However, continuity in the second phase will only exist if its volume fraction is more than about one third. Of course, the second phase can be continuous at much smaller volume fractions in case of nonequiaxed grain structure or nonrandom distribution, as illustrated in the connectivity of pore space at low porosities. For the analysis of geometrical measurements of two-phase structures, see Underwood (1970), Exner (1983) and Exner and Hougardy (1988). A simple "law of mixtures"

$$X = X_A v_A + X_B v_B + \ldots \tag{6.57}$$

relating an aggregate property X to the phase properties $X_A \ldots$ and volume fractions $v_A \ldots$ which is applicable to properties such as density, cannot be applied to mechanical properties without further considerations. The mechanical properties depend on the distribution of the phases, which influences the boundary conditions applicable to the individual grains, and therefore a structure of the aggregate also has to be specified. However, limits on the mechanical behavior of the aggregate can be obtained by reference to two extreme cases of "parallel" and "series" arrangement of the phases (Fig. 6.22). These arrangements correspond, respectively, to the uniform strain case

$$\sigma = \sigma_A v_A + \sigma_B v_B + \ldots\ldots \tag{6.58}$$

and the uniform stress case

$$\varepsilon = \varepsilon_A v_A + \varepsilon_B v_B + \ldots\ldots \tag{6.59}$$

where σ, ε are the macroscopic stress and strain, respectively, for the aggregate and σ_A, ε_A, v_A, are the local stress, strain, and volume fraction for the phase $A \ldots$

Application of Eqs. (6.58) and (6.59) to linear elastic deformation leads to the Voigt and Reuss limiting cases, respectively, for the elastic constants of the aggregate:

$$E = E_A v_A + E_B v_B + \ldots\ldots \tag{6.60}$$

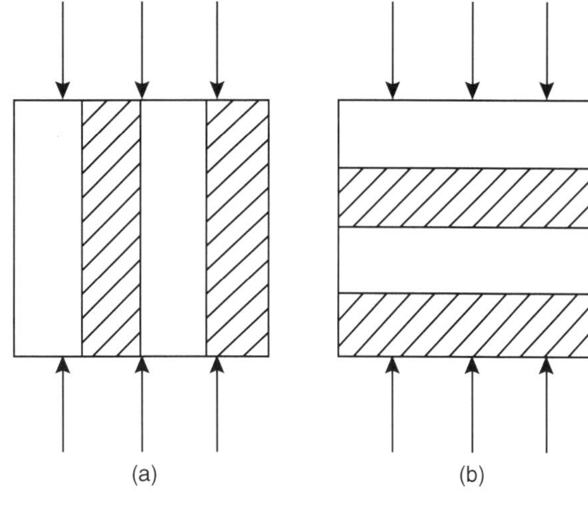

Fig. 6.22 a "parallel" and **b** "series" arrangement of phases in an aggregate, relative to a direction of loading indicted by the arrows

$$\frac{1}{E} = \frac{v_A}{E_A} + \frac{v_B}{E_B} + \ldots\ldots \tag{6.61}$$

where E, E_A, E_B, … are the elastic constants (that is, the stiffnesses, such as the Young moduli) of the aggregate and the phases A, B, …., respectively. Similar bounds apply to the viscosity in linear viscous deformation. In both cases, closer limits can be estimated by the method of Hashin and Shtrikman (1963) based on energy arguments rather than on a law of mixtures. It may be noted that in the case of a dilute suspension of undeformable particles in a viscous matrix of viscosity η_0, while the expression in (6.60) leads to infinite viscosity as an upper limit to the bulk viscosity, the expression in (6.61) leads to an approximate lower limit of $\eta_0(1 + v)$ where v is the volume fraction of the particles; this result may be compared with the Einstein (1906) value $\eta_0(1 + 2.5v)$, which, in turn, is lower than the values observed in concentrated suspensions (Jeffrey and Ascrivos 1976).

The expressions given in Eqs. (6.58) and (6.59) should also provide bounding estimates for the flow stress of a polycrystalline aggregate. In particular, (6.58) gives an upper limit corresponding, in principle, to the Taylor limit for poly-crystals (Sect. 6.8.4). Observations on an iron-silver alloy in the athermal regime indicate a variation of flow stress with composition that falls rather below this upper limit (Le Hazif 1980); curiously, these measurements also indicate only slight dependence of the flow stress on the volume fractions in the middle range where both phases are expected to be largely continuous. For a better estimate, self-consistent or finite-element calculations are required. The latter have been carried out in the case of the iron–silver alloy, leading to a satisfactory prediction of the observed behavior and showing that, although the actual strength falls significantly below the upper bound, it is still closer to that bound than to the lower bound (Durand and Thomas de Montpreville 1990; Le Hazif and Thomas de Montpreville 1981; Thomas de Montpreville 1983). The departure from the upper bound implies that there is significant heterogeneity of strain between the phases.

This heterogeneity has been observed directly in, for example, alpha–beta brass and nickel–silver alloys (Honeycombe and Boas 1948; Petrovic and Vasudevan 1978). Such a strain heterogeneity has also been observed in the thermal regime in oolitic limestone in which the finer-grained oolites are weaker than the coarser-grained matrix (Schmid and Paterson 1977) and finite-element calculations have been made for such cases (Ankem and Margolin 1982; Horowitz et al. 1981). A self-consistent scheme for treating multiphase materials has been given by Berveiller et al. (1981).

6.9 The Role of Pressure

6.9.1 General

Since plastic deformation is primarily a change of shape of a body without appreciable change of volume, only work terms involving shear stress are expected to be important in determining plastic behavior. Accordingly, plastic flow criteria for multiaxial stress states, such as the von Mises and Tresca criteria (Sect. 4.5), are expressed in terms of deviatoric stresses, with omission of any influence of the hydrostatic component of the stress. There is a similar omission in the Schmid law for the yield of single crystals (Sect. 6.1). Such a view is well supported experimentally in situations in which the hydrostatic component of the stress does not markedly exceed the deviatoric components. However, when the hydrostatic component of the stress is large or when second order effects are being considered, the influence of the hydrostatic component must also be taken into account and the associated changes in volume, or dilatancy, discussed. This topic is commonly referred to as the influence of the pressure on plastic flow and will be dealt with in its athermal and thermal aspects in Sects. 6.9.2 and 6.9.3.

In the ensuing discussion, the term "pressure" should, strictly, be taken as referring to the hydrostatic component of the stress, that is, the mean stress, since the effect of the pressure is associated, in a thermodynamic sense, with work done in volume changes. However, in practice, especially in the literature on experimental rock deformation and in geophysical applications, the term is commonly identified, somewhat loosely, with the confining pressure (Sect. 4.2) or the lithostatic pressure at a given depth in the Earth. The error incurred in this identification becomes less serious the larger the "pressure" is relative to the stress difference or the deviatoric component of the stress.

In the case of single crystals, the theoretical questions center on the influence of pressure on the various interactions that impede dislocation motion and on the rate of growth of dislocation populations. In the case of polycrystals, similar effects arise within grains and at grain boundaries. In addition, the initial structure of a polycrystal may be modified by the application of pressure, with further influence on the deformation behavior. This modification may involve effects associated with heterogeneity or anisotropy, such as the closure of cracks or pores, and the generation of internal stresses due to heterogeneity in elastic compressibility from grain to grain or to anisotropy of linear compressibility in adjoining grains of different orientation.

6.9.2 Pressure Effects in the Athermal Regime

In single crystals the influence of pressure on the flow stress in the athermal regime results primarily from changes in the elastic modulus. On the basis that the flow stress τ can be expressed as $\alpha G b \rho^{\frac{1}{2}}$ (Sect. 6.6.3) and that the term $b \rho^{\frac{1}{2}}$ will not vary with pressure at a given dislocation content since it is dimensionless, it follows that for a given dislocation density, the influence of pressure p on the flow stress will be given by

$$\frac{1}{\tau}\frac{\mathrm{d}\tau}{\mathrm{d}p} = \frac{1}{G}\frac{\mathrm{d}G}{\mathrm{d}p}$$

and hence

$$\tau = \tau_0 \exp\left(\int_0^p \frac{1}{G}\frac{\mathrm{d}G}{\mathrm{d}p}\,\mathrm{d}p\right)$$

where τ_0 is the flow stress at zero pressure (or other reference pressure). For relatively small variations in G with p, we thus have

$$\tau = \tau_0 \exp\left(\alpha_1 \frac{p}{G}\right) \approx \tau_0\left(1 + \alpha_1 \frac{p}{G}\right) \tag{6.62a}$$

where

$$\alpha_1 = \frac{\mathrm{d}G}{\mathrm{d}p} \tag{6.62b}$$

In these expressions, the elastic isotropy approximation is assumed in specifying the shear modulus G. The value of $\mathrm{d}G/\mathrm{d}p$ is between 1 and 2 for most materials, thus giving a 1–2 % increase in flow stress on raising the pressure by 500 MPa if G is 50 GPa. In the case of solute and particle hardened materials (Sect. 6.6.2), since the interaction force F in expressions such as Eqs. (6.25) and (6.26) can be expected to be proportional to G, the flow stress can again be expressed as the product of G and a dimensionless term, so that the pressure dependence of the flow stress is still represented by (6.62).

The pressure effect represented in (6.62) applies only for a given dislocation configuration. It should apply to the stress at a given strain as determined in an instantaneous pressure step test, but it does not necessarily apply to the comparison of complete stress–strain curves determined at different pressures since the pressure may also influence the evolution of the dislocation density ρ with straining, and thus the strain hardening. Since the introduction of a dislocation assemblage of density ρ leads to a small relative volume increase, of the order of ρb^2 (Sect. 6.2.2), there should be some increase in the strain hardening rate under pressure for a given rate of increase of ρ. This effect can be estimated by equating the extra work done by the stress, $\mathrm{d}\tau \cdot \mathrm{d}\gamma$, to the work done in relative volume increase, $\frac{\mathrm{d}V}{V} = b^2 \mathrm{d}\rho$ where $\mathrm{d}\rho$ is the increment in dislocation density and $\mathrm{d}\tau$ the extra increment in stress during the strain increment $\mathrm{d}\gamma$. It follows from the internal work expended per unit volume of the specimen,

$$\mathrm{d}W = \tau\,\mathrm{d}\gamma - p\,\mathrm{d}V = \tau\,\mathrm{d}\gamma - pb^2\mathrm{d}\rho$$

that the strain-hardening rate will be

$$\theta = \frac{\mathrm{d}\tau}{\mathrm{d}\gamma} = \frac{\mathrm{d}\tau_0}{\mathrm{d}\gamma} + pb^2\frac{\mathrm{d}^2\rho}{\mathrm{d}\gamma^2} \tag{6.63}$$

where $\tau_0 = \mathrm{d}W/\mathrm{d}\gamma$ is the stress in the absence of a superposed hydrostatic pressure (the pressure dependence of G is not being taken into account at this point). If, to gain some idea of the relative values of the terms in (6.63), we put $\tau_0 = \alpha G b \rho^{\frac{1}{2}}$ and $\rho = \rho_0 \exp(\gamma/\gamma_e)$ from Eqs. (6.31) and (6.23a), we obtain

$$\theta = \frac{\alpha\delta}{2}G\left(1 + \frac{2\delta}{\alpha}\frac{p}{G}\right) \tag{6.64}$$

where $\delta = b\rho^{\frac{1}{2}}/\gamma_e$. Putting $b \sim 10^{-13}\,\mathrm{m}$, $\rho \sim 10^{12}\,\mathrm{m}^{-2}$ and $\gamma_e \sim 10^{-2}$ suggests a value of the order of 10^{-2} for δ, which, with $\alpha \sim 0.3$, would give a strain-hardening rate consistent with that observed in stage II and early stage III of stress–strain curves at atmospheric pressure (Sect. 6.6.3). Thus, (6.64) indicates that the extra strain hardening associated with the dilatancy from dislocations is negligible in the normal experimental range of pressures. It can be concluded that, in the athermal field, the pressure effect in the stress–strain curve will derive almost entirely from the pressure effect on the shear modulus G in the absence of effects associated with dynamical recovery at larger strains.

With the onset of dynamical recovery, other pressure effects may arise. Thus, it has been observed that in stage III of the stress–strain curve the effect of pressure can be to reduce the flow stress, as, for example, in NaCl (Aladag et al. 1970). The reduction has been attributed to an increase in the stacking fault energy with increase in pressure, facilitating the constriction of dissociated dislocations as required for cross-slip, although this explanation has also to be questioned (Puls and So 1980). See Poirier (1985, pp. 155–156) for further discussion and references.

The effects described in the previous paragraphs should apply for both single crystals and polycrystals. However, additional pressure effects are possible in polycrystals through intergranular interaction effects associated with anisotropic or inhomogeneous properties of the grains and through volume changes associated with porosity. Thus, if the linear compressibility (the relative change in linear dimensions per unit pressure change) is anisotropic in the grains, then pressurization may give rise to intergranular stresses large enough to cause yielding in the grains, and thus increase the dislocation density before the specimen is subjected to the macroscopic stress in the stress–strain test. A similar effect can arise in the presence of a second phase of different elastic properties (for example, Bullen et al. 1964). If there is a change in pore volume during the deformation, there will be an associated increment in the flow stress of $(p/V)(dV/d\varepsilon)$ and in the strain-hardening rate of $(p/V)\left(\frac{d^2V}{d\varepsilon^2}\right)$, where V is the specimen volume and ε the axial strain in an axisymmetric triaxial test (Edmond and Paterson 1972).

6.9.3 Pressure Effects in the Thermal Regime

We consider first the case of *viscous drag control* in single crystals (Sect. 6.6.5), expressed in $\dot{\gamma} = \rho b v$, assuming the dislocation density to be sufficiently low that the effective stress acting on the dislocations can still be taken as being equal to the applied stress τ. In order to gain some idea of the magnitude of the pressure effect and of what are the important factors likely to affect it, we consider a specific model based on the expression (6.14) for v for the case of relatively low stress, which gives the approximation

$$v = \frac{v_0 b \Delta A^2 \tau}{lkT} \exp\left(-\frac{\Delta E_G}{kT}\right) \tag{6.65}$$

and leads, through $\dot{\gamma} = \rho b v$, to

$$\tau = \tau_{\text{eff}} = \frac{\dot{\gamma} lkT}{v_0 \rho b^2 \Delta A^2} \exp\left(\frac{\Delta E_G}{kT}\right) \tag{6.66}$$

where the symbols are as in Sect. 6.4.1 and ΔE_G is the activation energy for dislocation glide. If we assume that $v_0 \propto (Gb)^{\frac{1}{2}}$ and $l/\Delta A^2 \propto b^{-3}$, noting that ρb^2 is dimensionless and therefore independent of pressure for a given dislocation content, and we write $\Delta E_G = (\Delta E_G)_0 + p \Delta V_G$ where $(\Delta E_G)_0$ is the activation energy at zero pressure and ΔV_G is the activation volume for dislocation glide, then we obtain for a given dislocation configuration and strain rate,

$$\frac{1}{\tau}\frac{d\tau}{dp} = -\frac{1}{2G}\frac{dG}{dp} - \frac{7}{2b}\frac{db}{dp} + \frac{\Delta V_G}{kT} \tag{6.67}$$

Integrating, and putting $(1/b)(db/dp) = -1/3K$, where K is the bulk modulus, we obtain finally, in analogy to (6.63)

$$\tau = \tau_0 \exp\left(\alpha_1 \frac{p}{G}\right) \tag{6.68a}$$

where

$$\alpha_1 = -\frac{1}{2}\frac{dG}{dp} + \frac{7}{6}\frac{G}{K} + \frac{G\Delta V_G}{kT} \tag{6.68b}$$

The first two terms in α_1, representing the effects of pressure on the shear modulus and on the specimen dimensions, respectively, are each of the order of unity. They are the only terms taken into account in considerations such as those of Seeger and Haasen (1958) and Jung (1981). However, in the case of viscous drag control, it is possible that they may be less important than the third term, representing the effect of the dilatancy associated with the activation event; for example, if we put $\Delta V_G = 0.01$ nm^3, the order of an atomic volume, and $G = 50$ GPa, then $G\Delta V_G/kT = 30$ at T $= 1,200$ K

In this case (6.69) becomes

$$\tau \approx \tau_0 \exp\left(\frac{p\Delta V_G}{kT}\right) \tag{6.69}$$

When the dislocation density is high enough that the mutual dislocation interactions cannot be neglected, then the applied stress τ is given by

$$\tau = \tau_{\text{eff}} + \tau_i$$

where τ_{eff} is given by the expression (6.66) and $\tau_i = \alpha Gb\rho^{\frac{1}{2}}$. In this case

$$\frac{d\tau}{dp} = \frac{d\tau_{\text{eff}}}{dp} + \frac{d\tau_i}{dp} \tag{6.70}$$

which, with Eqs. (6.63) and (6.67), leads to the parameter α_1 in (6.68a) becoming

$$\alpha_1 = \frac{3\tau_i - \tau}{2\tau}\frac{dG}{dp} + \frac{\tau - \tau_i}{\tau}\left(\frac{7}{6}\frac{G}{K} + \frac{G\Delta V_G}{kT}\right) \tag{6.71}$$

The two limiting cases correspond to the simple cases already discussed:

1. $\tau_i = 0$: the pure viscous drag model with low dislocation density, discussed in the previous paragraph.
2. $\tau_i = \tau$: the pure athermal case, discussed in the previous subsection, for which $\alpha_1 = dG/dp$.

It is thus to be expected that the pressure sensitivity of the flow stress will be much greater when the flow is controlled mainly by viscous drag on the dislocation motion (Peierls force, etc.) rather than by mutual dislocation interaction, provided the activation volume ΔV_G in the former case is at least of the order of an atomic volume.

It is to be reiterated that the pressure sensitivities represented by Eqs. (6.69) and (6.71) apply only under the condition that the dislocation content of the crystal does not change, that is, that ρb^2 is constant, elastic volume changes being taken into account. Such effects are therefore properly measured in instantaneous pressure step tests, from which α_1 is obtained as

$$\alpha_1 = G\left(\frac{\Delta \ln \tau}{\Delta p}\right),$$

or ΔV_G as

$$\Delta V_G \approx kT \left(\frac{\Delta \ln \tau}{\Delta p}\right)_\rho$$

where bulk elastic effects are neglected. When stress–strain curves, each measured entirely at a given constant pressure, are to be compared, additional pressure effects may enter through the evolution of the ρb^2 term. Then, using (6.70) with Eqs. (6.63) and (6.67) it can be shown that a term

$$\frac{3\tau_i - 2\tau}{2\tau}\left(\frac{G}{\rho}\frac{d\rho}{dp} - \frac{2}{3}\frac{G}{K}\right) \tag{6.72}$$

has to be added to (6.71), where ρ is now the dislocation density developed during straining at a given pressure p up to the strain at which the flow stresses are compared. Taking (6.72) into account in the respective extreme cases of $\tau_i = 0$ and $\tau_i = \tau$ (or, equivalently, considering individually the expressions in (7.67) and $\tau_i = \alpha G b \rho^{\frac{1}{2}}$ ab initio) shows that if the rate of growth of dislocation density up to a given strain is reduced at higher pressure ($d\rho/dp < 0$ in (6.72)), then the effect of increasing the pressure will be to raise the stress–strain curve in the viscous drag case and to lower it in the athermal case and to introduce corresponding transients in a step test. In the case of polycrystals, the additional pressure effects mentioned in the previous subsection may also come into play.

We now turn to the case of thermal models based on *mutual dislocation interaction* (Sect. 6.6.6). We also here restrict consideration to steady-state flow and, specifically, to the particular simple recovery-controlled creep model expressed in the flow law in (6.45), which leads to

$$\dot{\gamma}_s \propto G^{-\frac{3}{2}}b^{\frac{7}{2}}\tau^3 \exp\left(-\frac{p\Delta V_D}{kT}\right) \tag{6.73}$$

when it is assumed that $D \propto (Gb)^{\frac{1}{2}}b^2 \exp(-p\Delta V_D/kT)$ from Eqs. (3.31) and (3.33), putting $v_0 \propto (Gb)^{\frac{1}{2}}$ from Sect. 3.2.4 and putting $\Delta E_D = (\Delta E_D)_0 + p\Delta V_D$ for the activation energy, where $(\Delta E_D)_0$ is the zero pressure activation energy and ΔV_D the activation volume for the diffusion that is required for the dislocation climb. In a creep test, the pressure effect can then be expressed through

$$\frac{1}{\dot{\gamma}_s}\left(\frac{d\dot{\gamma}_s}{dp}\right)_\tau = -\frac{3}{2G}\frac{dG}{dp} + \frac{7}{2b}\frac{db}{dp} - \frac{\Delta V_D}{kT} \tag{6.74}$$

Integrating and putting $(1/b)(\mathrm{d}b/\mathrm{d}p) = -1/3K$, where K is the bulk modulus, we obtain a steady-state strain rate of

$$\dot{\gamma}_s = \dot{\gamma}_{s0} \exp\left(-\alpha_1 \frac{p}{G}\right) \tag{6.75}$$

and

$$\alpha_1 = \frac{3}{2}\frac{\mathrm{d}G}{\mathrm{d}p} + \frac{7}{6}\frac{G}{K} + \frac{G\Delta V_D}{kT} \tag{6.76}$$

Alternatively, if the pressure effect is expressed as $(1/\tau)(\mathrm{d}\tau/\mathrm{d}p)_{\dot{\gamma}}$ for a constant strain-rate test, a similar argument leads to a steady-state flow stress of

$$\tau_s = \tau_{s0} \exp\left(\frac{\alpha_1}{3}\frac{p}{G}\right) \tag{6.77}$$

where τ_{s0} is the steady-state flow stress at zero pressure and α_1, is again given by (6.77).

When ΔV_D is of the order of an atomic volume or more, the third term in (6.76) tends to be large compared with the other two terms, leading to the commonly used expression

$$\dot{\gamma}_s = \dot{\gamma}_{s0} \exp\left(-\frac{p\Delta V}{kT}\right) \tag{6.78}$$

The quantity ΔV in this expression corresponds to an experimentally determined or apparent activation volume ΔV^* (Sect. 4.4.2), which can then be expected to be approximately, but not exactly, equal to the activation volume for diffusion ΔV_D.

When effects such as viscous drag and cross-slip of dislocations are introduced in more realistic thermal models of flow, the interpretation of the experimentally determined activation volume for steady-state flow may be more complex. Only when such effects are relatively minor compared with the diffusion-controlled recovery effects can the empirical activation volume be expected to approximate the diffusion activation volume ΔV_D.

Even if a steady state has been attained at a given pressure, there will, of course, tend to be transient effects when a step change is made in the pressure (Fig. 6.23). The relationships in Eqs. (6.75) and (6.77) can only be applied to the strain rate and the stress measured after the transient effects have given way to a new steady state. For the simple recovery-controlled model underlying (6.73), an instantaneous change in stress given by (6.63) could be expected after a step change in pressure in a constant strain-rate experiment, but this stress jump may be difficult to resolve if the recovery rate is high. When the steady-state flow stress includes a viscous drag component, a larger instantaneous effect is to be expected, corresponding to (6.71) replacing (6.63).

Relatively few measurements have been made of pressure effects in either creep or diffusion in minerals or related materials. See, for example, Ross et al. (1979),

Fig. 6.23 Depicting the effect of **a** a step down in pressure from p_2 to p_1 ($p_2 > p_1$) and **b** a step up in pressure from p_1 to p_2 and the associated transient behavior

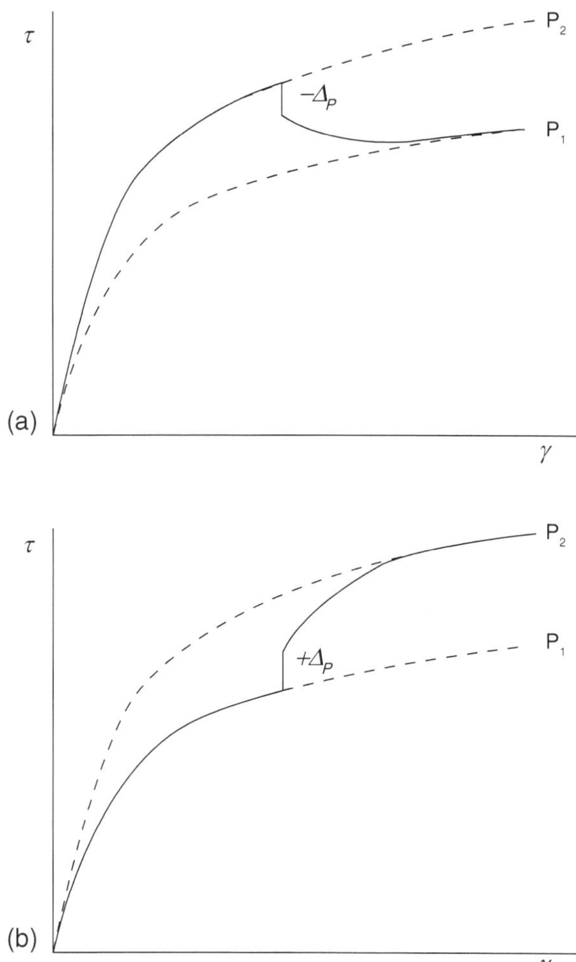

Kohlstedt et al. (1980), Heard and Kirby(1981), Green and Borch (1987), Li et al. (2004).

In view of this paucity of relevant measurements, some semiempirical procedures for estimating ΔV_D in the intrinsic diffusion region (pure crystals, high temperature) have been given by Sammis et al. (1981) and Poirier (1985, pp. 162–166) with a view to their use in estimating pressure effects in steady-state creep. It must also be borne in mind, for applications at very high pressures in the Earth, that activation volumes can be expected to decrease as the pressure increases in the range of nonlinear elasticity (Karato 1981a, b; Poirier 1985, pp. 166–167).

References

Abe T, Nagaki S (1981) Finite element modelling of Taylor slips in fcc polycrystals. In: Hansen N, Horsewell A, Leffers T, Lilholt H (eds) Deformation of polycrystals: mechanisms and microstructures, Roskilde, Denmark, Risø National Laboratory, pp 125–130

Ahlquist CN, Gasca-Neri R, Nix WD (1970) A phenomenological theory of steady state creep based on average internal and effective stress. Acta Metall 18:663–671

Aladag E, Davis LA, Gordon RB (1970) Cross-slip and plastic deformation of NaCl single and polycrystals at high pressure. Phil Mag 21:469–478

Alexander H, Haasen P (1968) Dislocations and plastic flow in the diamond structure. Solid State Phys 22:27–158

Amelinckx S (1979) Dislocations in particular structures. In: Nabarro FRN (ed) Dislocations in solids. North-Holland Publ. Co., Amsterdam, pp 67–460

Ankem S, Margolin H (1982) Finite element method (FEM) calculations of stress-strain behaviour of alpha-beta Ti-Mn alloys: Part I. Stress-straion relations. Part II. Stress and strain distributions. Metall Trans A 13A, 595–601:603–609

Ardell AJ (1985) Precipitation hardening. Metall Trans A 16A:2131–2165

Ardell AJ, Przystupa MA (1984) Dislocation link-length statistics and elevated temperature deformation of crystals. Mech Materials 3:319–332

Armstrong R, Codd I, Douthwaite RM, Petch NJ (1962) Plastic deformation of polycrystalline aggregates. Phil Mag 7:45–58

Asaro RJ, Needleman A (1984) Flow localization in strain hardening crystalline solids. Scripta Met 18:429–435

Asaro RJ, Needleman A (1985) Texture development and strain hardening in rate dependent polycrystals. Acta Metall 33:923–953

Ashby MF (1970) The deformation of plastically non-homogeneous materials. Phil Mag 21:399–424

Ashby MF (1971) The deformation of plastically non-homogeneous alloys. In: Kelly A and Nicholson RB (eds) Strengthening methods of crystals. Applied Science Publishers, London, pp 137–192

Bailey RW (1926) Note on the softening of strain hardening metals and its relation to creep. J Inst Metals 35:27–43

Barrett CS (1943) Structure of Metals. McGraw-Hill, New York, 567 pp

Basinski SJ, Basinski ZS (1979) Plastic deformation and work hardening. In: Nabarro FRN (ed) Dislocations in solids. North Holland Publ Co, Amsterdam, pp 261–362

Berveiller M, Zaoui A (1978) An extension of the self-consistent scheme to plastically-deforming polycrystals. J Mech Phys Solids 26:325–344

Berveiller M, Zaoui A (1981) A simplified self-consistent scheme for the plasticity of two-phase metals. Res Mechanica Lett 1:119–124

Berveiller M, Hiki A, Zaoui A (1981) Self-consistent schemes for the plasticity of polycrystalline and multiphase materials. In: Lilholt H (ed) Deformation of polycrystals: mechanisms and microstructures. Risø National Laboratory, Roskilde, Denmark, pp 145–156

Bilby BA (1955) Types of dislocation source. In: Report of the conference on defects in crystalline solids held at the H H Wills Physical Laboratory University of Bristol July 1944, London, The Physical Society, pp 124–133

Bishop JFW (1953) A theoretical examination of the plastic deformation of crystals by glide. Phil Mag 44:51–64

Bishop JFW (1954) A theory of the tensile and compressive textures of face-centred cubic metals. J Mech Phys Solids 3:130–142

Bishop JFW, Hill R (1951) A theory of the plastic distortion of a polycrystalline aggregate under combined stresses. Phil Mag 42:414–427

Blacic JD, Christie JM (1984) Plasticity and hydrolytic weakening of quartz single crystals. J Geophys Res 89:4223–4239

Boas W, Hargreaves ME (1948) On the inhomogeneity of plastic deformation in the crystals of an aggregate. Proc Roy Soc (Lon) A193:89–97

Borch RS, Green HW (1987) Dependence of creep in olivine on homologous temperature and its implications for flow in the mantle. Nature 330:345–348

Brion HG, Haasen P, Siethoff H (1971) The yield point of highly-doped germanium. Acta Metall 19:283–290

Brown AM, Ashby MF (1980) On the power-law equation. Scripta Met 14:1297–1302

Brown LM, Ham RK (1971) Dislocation-particle interactions. In: Nicholson RB (ed) Strengthening methods in crystals. Elsevier, Amsterdam, pp 12–135

Bucher M (1982) Core energy and Peierls stress of edge dislocations on (110) and (001) slip planes in NaCl. Phys Stat Sol B 114:383–392

Budianski B, Wu TT (1962) Theoretical prediction of plastic strains of polycrystals. In: Proceedings of 4th national congress of applied mechanics, ASME, pp 1175–1185

Bullen FP, Henderson F, Wain HL, Paterson MS (1964) The effect of hydrostatic pressure on brittleness in chromium. Phil Mag 9:803–815

Burton B (1982a) The influence of solute drag on dislocation creep. Phil Mag A46:607–616

Burton B (1982b) The dislocation network theory of creep. Phil Mag A45:657–675

Cahn RW (1949) Recrystallization of single crystals after plastic bending. J Inst Metals 76:121–143

Cahn RW (1951) Slip and polygonization in aluminium. J Inst Metals 79:129–158

Cahn RW (1953) Soviet work on mechanical twinning. Nuovo Cimento 4(10):350–386

Cahn RW (1954) Twinned crystals. Adv Phys 3:363–445

Caillard D, Martin JL (1982a) Microstructure of aluminium during creep at intermediate temperatures. I Dislocation networks after creep. Acta metall 30:437–445

Caillard D, Martin JL (1982b) Microstructure of aluminium during creep at intermediate temperatures. II In situ study of sub boundary properties. Acta metall 30:791–798

Caillard D, Martin JL (1983) Microstructure of aluminium during creep at intermediate temperatures. III The rate-controlling process. Acta metall 31:813–825

Carslaw HS, Jaeger JC (1959) Conduction of heat in solids, 2nd edn. Clarendon Press, Oxford, 510 pp

Carter CB (1984) What's new in dislocation association. In: Kubin L, Castaing J (eds) Dislocations 1984, Paris, Editions du CNRS, pp 227–251

Cherns D (1984) Climb of dissociated dislocations. In: Kubin L, Castaing J (eds) Dislocations 1984, Paris, Editions du CNRS, pp 215–226

Chin GY, Mammel WL (1969) Generalization and equivalence of the minimum work (Taylor) and maximum work (Bishop-Hill) principles for crystal plasticity. Trans Met Soc AIME 245:1211–1214

Christian JW (1965) The theory of transformations in metals and alloys. Pergamon, Oxford, 973 pp

Christie JM, Ardell AJ (1974) Substructures of deformation lamallae in quartz. Geology 2:405–408

Clément N, Caillard D, Martin JL (1984) Heterogeneous deformation of concentrated Ni-Cr FCC alloys: macroscopic and microscopic behaviour. Acta Metall 32:961–975

Coe RS (1970) The thermodynamic effect of shear stress on the ortho-clino inversion in enstatite and other coherent phase transitions characterized by a finite simple shear. Contr Mineral Petrol 26:247–264

Copley SM, Pask JA (1965) Deformation of polycrystalline MgO at elevated temperatures. J Am Ceram Soc 48:636–642

Cottrell AH (1953) Dislocations and plastic flow in crystals. Clarendon Press, Oxford, 223 pp

Cottrell AH, Bilby BA (1951) A mechanism for the growth of deformation twins in crystals. Phil Mag 42:573–581

Cox HL, Sopwith DG (1937) The effect of orientation on stresses in single crystals and of random orientation on strength of polycrystalline aggregates. Proc Phys Soc 49:134–151

Dana ES (1932) A textbook of mineralogy, 4th edn. Wiley, New York, 851 pp (revised and enlarged by Ford WF)

Di Persio J, Escaig B (1984) Dislocation cores in molecular crystals. In: Kubin L, Castaing J (eds) Dislocations 1984, Paris, Editions du CNRS, pp 267–282

Dollar M, Gorczyca S (1981) A new approach to flow stress—grain size relationship. In Lilholt H (ed) Deformation of polycrystals: mechanisms and microstructures, Roskilde, Denmark, Risø National Laboratory, pp 163–172

Durand L, Thomas de Montpreville C (1990) Etude du comportement mécanique des matériaux biphasés au moyen de la méthode des eléments finis. Res Mechanica 29:257–285

Edmond JM, Paterson MS (1972) Volume changes during the deformation of rocks at high pressures. Int J Rock Mech Min Sci 9:161–182

Einstein A (1906) Eine neue Bestimmung der Molekül-dimensionen. Ann Phys 19:289–306

Embury JD (1985) Plastic flow in dispersion hardened materials. Met Trans A 16a:2191–2200

Escaig B (1968a) Sur le glissement dévié des dislocations dans la structure cubique à faces centrées. J Physique 29:225–239

Escaig B (1968b) L'activation thermique des déviations sous faibles contraintes dans les strutures h.c. et c.c.. Phys Stat Sol 28:463–474

Exner HE (1983) Qualitative and quantitative surface microscopy. In: Cahn RW, Hassen P (ed) Physical metallurgy, 3rd edn. North Holland Physics Publ, Amsterdaam, pp 581–647

Exner H E, Hougardy HP (eds) (1988) Quantitative image analysis of microstructures. DGM Informations-Gesellschaft Verlag, Oberursel, 235 pp

Fantozzi G, Esnouf C, Benoit W, Ritchie IG (1982) Internal friction and microdeformation due to the intrinsic properties of dislocations: the Bordoni relaxation. Prog Mater Sci 27:311–451

Fitz Gerald JD, Boland JN, McClaren AC, Ord A, Hobbs BE (1991) Microstructures in water-weakened single crystals of quartz. J Geophys Res 96:2139–2155

Frank FC (1951) Capillary equilibria in dislocated crystals. Acta cryst 4:497–501

Frank FC (1955) Hexagonal networks of dislocations. In: Report of the conference on defects in crystalline solids held at H H Wills Physical Laboratory University of Bristol July 1954, London, Physical Society, pp 159–168

Friedel J (1959) Dislocation interactions and internal strains. In Rassweiler GN, Grube WL (ed) Internal stresses and fatigue in metals. Elsevier, Amsterdam, pp 220–262

Friedel J (1964) Dislocations. Pergamon Press, Oxford, 491 pp

Friedel J (1982) On the entropy of vibration of dislocations. Phil Mag A 45:271–285

Frost HJ, Ashby MF (1982) Deformation mechanism maps: the plasticity and creep of metals and ceramics. Pergamon, Oxford, 166 pp

Gerold V (1979) Precipitation hardening. In: Nabarro FRN (eds) Dislocations in solids. Vol 4: dislocations in metallurgy. North-Holland Publ Co, Amsterdam, pp 219–260

Gibbs GB (1966) Creep and stress relaxation studies with polycrystalline magnesium. Phil Mag 13:317–329

Gifkins RC (1974) Hot and strong. J Aus Inst Met 19:149–160

Gifkins RC (1978) Grain rearrangements during superplastic deformation. J Mater Sci 13:1926–1936

Gil Sevillano SJ, van Houtte P, Aernoudt E (1980) Large strain work hardening and textures. Prog Mater Sci 25:69–412

Gilman JJ (1969) Micromechanics of flow in solids. McGraw-Hill, New York, 294 pp

Gittus J (1975) Creep, viscoelasticity and creep fracture of solids. Applied Science Publishers, London, 725 pp

Granato AV (1984) Viscosity effects in plastic flow and internal friction. Scripta Met 18:663–668

Grewen J, Noda T, Sauer D (1977) Electron microscopic investigation of shear bands. Zeit f Metallkunde 68:260–265

Griggs DT (1974) A model of hydrolytic weakening in quartz. J Geoph Res 79:1655–1661

Griggs DT, Turner FJ, Heard HC (1960) Deformation of rocks at 500–800 °C. In: Griggs DT, Handin J (ed) Rock deformation. Geological Society of America, pp 39–104

Groh P, Kubin LP, Martin JL (eds) (1979) Dislocations et Déformation Plastique. Ecole d'été d'Yravals, 3–14 Sept 1979, Les Editions de Physique, 461 pp

Gross KA (1965) X-ray line broadening and stored energy in deformed and annealed calcite. Phil Mag 8(12):801–813

Groves GW, Kelly A (1963) Independent slip systems in crystals. Phil Mag 8:877–887

Groves GW, Kelly A (1969) Change of shape due to dislocation climb. Phil Mag 19:977–986

Guillopé M, Poirier J-P (1979) Dynamic recrystallization during creep of single-crystalline halite: an experimental study. J Geophys Res 84:5557–5567

Guyot P (1980) "Anomalies" dans le comportement à haute température des alliages à dispersion d'oxyde. Annales de Chimie Fr 5:74–76

Guyot P, Dorn JE (1967) A critical review of the Peierls mechanism. Canad J Phys 45:983–1016

Haasen P (1964) Versetzungen und Plastizität von Germanium und Silicium. In: Festkörperprobleme vol 3, Deusche Physikalische Gesellschaft, pp 167–208

Haasen P (1978) Physical Metallurgy (trans: Mordike J). Cambridge University Press, Cambridge, 381 pp

Haasen P (1983) Mechanical properties of solid solutions and intermetallic compounds. In: Cahn RW, Haasen P (ed) Physical metallurgy, 3rd edn. North-Holland Physics Publisher, Amsterdam, pp 1341–1409

Haasen P, Schröter W (1970) In: Fundamental aspects of dislocation theory. Nat Bureau of Standards, Washington, pp 1231–1258

Hall EO (1951) The deformation and ageing of mild steel: III Discussion of results. Proc Phys Soc B 64:747–753

Hall EO (1954) Twinning and diffusionless transformations in metals. Butterworths, London, 181 pp

Hansen N (1983) Flow stress and grain size dependence of non-ferrous metals and alloys. In: Baker TN (ed) Yield, flow and fracture of polycrystals. Applied Sci Publ, London, pp 311–350

Hansen N (1985) Polycrystalline strengthening. Met Trans A 16A:2167–2190

Hashin Z, Shtrikman S (1963) A variational approach to the theory of the elastic behaviour of multiphase materials. J Mech Phys Solids 11:127–140

Hatherly M (1978 The structure of highly deformed materials and the development of deformation textures. In: Proceedings of 5th international conference on textures of materials, Springer, Berlin, pp 81–91

Hatherly M, Malin AS (1979) Deformation of copper amd low stacking-fault energy copper-base alloys. Met Technol 6:308–309

Heard HC, Kirby SH (1981) Activation volume for steady state creep in polycrystalline CsCl: cesium chloride structure. In: Carter NL, Friedman M, Logan JM, Streams DW (ed) Mechanical behavior of crustal rocks. The Handin volume, Washington, DC, American Geophysical Union, pp 83–91

Hecker SS, Stout MG (1984) Strain hardening in heavily cold worked metals. In: Krauss G (ed) Proceedings of the seminar "deformation, processing, and structure" American Society of Metals, Metals Park, Ohio, pp 1–46

Heggie M, Jones R (1986) Models of hydrolytic weakening in quartz. Phil Mag A 53:L65–L70

Heidenreich RD, Shockley W (1948) Study of slip in aluminium crystals by electron microscope and electron diffraction methods. In: Report of a conference on strength of solids, University of Bristol, 7-9 July, 1947, London, The Physical Society, pp 57–75

Heinisch HL, Sines G, Goodman JW, Kirby SH (1975) Elastic stresses and self-energies of dislocations of arbitrary orientation in anisotropic media: olivine, orthopyroxene, calcite, and quartz. J Geoph Res 80:1885–1896

Hill R (1965) A self-consistent mechanics of composite materials. J Mech Phys Solids 13:213–222

Hirsch PB (1960) Proceedings of the 5th International Conference on Crystallography, Cambridge

Hirsch PB (1985) Dislocations in semiconductors. In: Loretto MH (ed), Dislocations and properties of real materials. The Institute of Metals, London, pp 333–348

Hirth JP (1985) A brief history of dislocation theory. Met Trans A 16A:2085–2090

Hirth JP, Lothe J (1982) Theory of dislocations, 2nd edn. Wiley, New York, 857 pp

Hobbs BE, McLaren AC, Paterson MS (1972) Plasticity of single crystals of synthetic quartz. In: Heard HC, Borg IG, Carter NL, Rayleigh CB (ed) Flow and fracture of rocks. American Geophysical Union, Washington, pp 29–53

Hobbs BE, Means WD, Williams P (1976) An outline of structural geology. Wiley, New York, 571 pp

Honeycombe RWK, Boas W (1948) Aust J Sci Res A1:70–84

Honneff H, Mecking H (1978) A method for the determination of the active slip systems and orientation changes during single crystal deformation. In Proceedings of 5th International Conference on TexMater, Springer, Berlin, pp 265–275

Horowitz FG, Tullis TE, Kronenberg A, Tullis J, Needleman A (1981) Finite element model of polyphase flow. EOS Trans AGU 62:396–397

Hull D (1975) Introduction to dislocations, 2nd edn. Pergamon Press, Oxford, 271 pp

Humphreys FJ (1985) Dislocation-particle interactions. In: (ed) Dislocations and properties of real materials, Institute of Metals, London, pp 175–204

Hutchinson JW (1970) Elastic-plastic behaviour of polycrystalline metals and composites. Proc Roy Soc Lond A 319:247–272

Hutchinson JW (1976) Bounds and self-consistent estimates for creep of polycrystalline materials. Proc Roy Soc Lond A 348:101–127

Hutchinson JW (1977) Creep and plasticity of hexagonal polycrystals as related to single crystal slip. Metall Trans A 8A:1465–1469

Jackson I (1986) The laboratory study of seismic wave attenuation. In: Hobbs BE, Heard HC (ed) Mineral and rock deformation: laboratory studies. American Geophysics Union, The Paterson Volume, Washington DC, pp 11–23

Jaeger JC, Cook NGW (1979) Fundamentals of rock mechanics, 3rd edn. Chapman and Hall, London, 593 pp

Jaoul O, Poumellec M, Froidevaux C, Havette A (1981) Silicon diffusion in forsterite: a new constraint for understanding mantle deformation. In: Anelasticity in the Earth, American Geophysics Union, Washington DC, pp 95–100

Jeffrey DJ, Ascrivos A (1976) The rheological properties of suspensions of rigid particles. J Amer Inst Chem Eng 22:417–432

Johnson WG, Gilman JJ (1959) Dislocation velocities, dislocations densities, and plastic flow in lithium fluoride crystals. J Appl Phys 30:129

Jones R (1983) Theories of dislocation mobility in semiconductors. J de Phys 44(C4):61–67

Jouffrey B (1979) Historique de la notion de dislocation. In: Kubin L, Martin JL (eds) Dislocations et Déformation Plastique. Ecole d'été d'Yravals 3-14 Sept 1979, Les Editions de Physique, pp 1–16

Jung J (1981) A note on the influence of hydrostatic pressure on dislocations. Phil Mag A 43:1057–1061

Karato S (1981a) Rheology of the lower mantle. Phys Earth Planet Int 24:1–14

Karato S (1981b) Pressure dependence of diffusion in ionic solids. Phys Earth Planet Int 25:38–51

Kelly A (1966) Strong solids. Clarendon Press, Oxford, 212 pp

Kelly A, Nicholson RB (eds) (1971) Strengthening methods in crystals. Elsevier, Amsterdam, 629 pp

Klassen-Neklyudova MV (1964) Mechanical twinning of crystals. Consultants Bureau, New York, 213 pp (translated from Russian by Bradley JES)

Kocks UF (1960) Polyslip in single crystals. Acta Metall 8:345–352

Kocks UF (1964) Independent slip systems in crystals. Phil Mag 10:187–193

Kocks UF (1970) The relation between polycrystal deformation and single crystal deformation. Metall Trans, pp 1121–1143

Kocks UF (1984) Solution hardening and strain hardening at elevated temperatures. In: Hansen N et al (eds) Deformation, Processing, and Structure, Metals Park, Ohio, American Soc for Metals, pp 89–107

Kocks UF (1985a) Dislocation interactions: flow stress and strain hardening. In: Dislocations and properties of real materials. The Institute of Metals, London, pp 125–143

Kocks UF (1985b) Kinetics of solution hardening. Met Trans A 16A:2109–2129

Kocks UF, Canova GR (1981) How many slip systems and which? In Hansen N et al (eds) Deformation of polycrystals: mechanisms and microstructures. Risø National Laboratory, Roskilde Denmark, pp 35–44

Kocks UF, Argon AS, Ashby MF (1975) Thermodynamics and kinetics of slip. Pergamon Press, London, 288 pp

Kohlstedt DL, Goetze C, Durham WB, Van der Sande J (1976) New technique for decorating dislocations in olivine. Science 191:1045–1046

Kohlstedt DL, Nichols HPK, Hornack P (1980) The effect of pressure on the rate of dislocation recovery in olivine. J Geoph Res 85:3122–3130

Kronberg ML (1961) Atom movements and dislocation structures in some common crystals. Acta Metall 9:970–972

Kröner E (1961) Zur plastichen verformung des vielkristalls. Acta Metall 9:155–161

Kubin L, Martin JL (1980) In situ deformation experiments in the HVEM. In: 5th international conference on strength of metals and alloys 1979, vol 3. Pergamon Press, Toronto, pp 1639–1660

Kuhlmann-Wilsdorf D (1985) Theory of work hardening 1934–1984. Metall Trans A 16A: 2091–2108

Labusch R, Schröter W (1975) Electrical and optical properties of dislocations in semiconductors. In: Huntley FA (ed) Proceedings of international conference on lattice defects in semiconductors, Freiberg, 1974, Institute of Physics, Bristol, pp 56–72

Lagneborg R (1972) A modified recovery creep model and its evaluation. Met Sci J 6:127–133

Le Hazif R (1979) Propriétés mécaniques des alliages biphasés. In: Dislocations et Deformation Plastique. Ecole d'été d'Yravals, 3-14 Sept 1979, Paris, Les Editions de Physique, pp 327–343

Le Hazif R (1980) Deformation plastique du systeme biphase fer-argent (limite plastique en fonction de la concentration). Scr Metall 14:987–988

Le Hazif R, Thomas de Montpreville C (1981) Relations entre les propriétés élasto-plastique des alliages biphasés isotropes à microstructure grossière et celles du leurs constituants. Res Mechanica Lett 1:61–65

Leffers T (1979) A modified Sachs approach to the plastic deformation of polycrystals as a realistic alternative to the Taylor Model. In: Haasen et al P (eds) Strength of metals and alloys, vol 2. Pergamon Press, Toronto, pp 769–774

Leffers T (1981) Microstructures and mechanisms of polycrystal deformation at low temperature. In: Hansen N, Horsewell A, Leffers T, Lilholt H (eds) Deformation of polycrystals: mechanisms and microstructures. Risø National Laboratory, Roskilde, Denmark, pp 55–71

Li JCM (1963) Petch relation and grain boundary sources. Trans AIME 227:239–247

Li L, Weidner D, Raterron P, Chen J, Vaughan M (2004) Stress measurements of deforming olivine at high pressure. Phys Earth Planet Int 143–144:357–367

Lin P, Przystupa MA, Ardell AJ (1985) Dislocation of network dynamics during creep deformation of monocrystalline sodium chloride. In: Strength of metals and alloys. Proceedings of 7th international conference on metals and alloys, vol 1. Pergamon Press, Oxford, pp 595–600

Lister GS (1979) Fabric transitions in plastically deformed quartzites: competition between basal, prism and rhomb systems. Bull Minéral 102:232–241

Lister GS, Hobbs BE (1980) The simulation of fabric development during plastic deformation and its application to quartzite: the influence of deformation history. J Struct Geol 2:355–370

Lister GS, Paterson MS (1979) The simulation of fabric development during plastic deformation and its application to quartzite: fabric transitions. J Struct Geol 1:99–115

Lister GS, Price GP (1978) Fabric development in a quartz-feldspar mylonite. Tectonophysics 49:37–78

Lister GS, Paterson MS, Hobbs BE (1978) The simulation of fabric development in plastic deformation and its application to quartzite. Model Tectonophys 44:107–158

Louchet F (1979) Plasticité des métaux de structure cubique centré. In: Kubin L, Martin JL (eds) Dislocations et Déformation Plastique. Ecole d'été d'Yravals, 3–14 Sept 1979, Les Editions de Physique, pp 149–160

Louchet F, George A (1983) Dislocation mobility measurements—an essential tool for understanding the atomic and electronic core structure of dislocations in semiconductors. J de Phys 44(C4):51–60

Lyall KD (1965) Plastic deformation of Galena. PhD thesis, The Australian National University, 224 pp

Mahajan S (1981) The nucleation and growth of deformation twins in metallic crystals. In: Ashby MF et al (eds) Dislocation modelling in physical systems. Pergamon Press, Oxford, pp 217–221

Martin JW (1980) Micro mechanisms in particle-hardened alloys. Cambridge University Press, Cambridge, 201 pp

Mataré HF (1971) Defect electronics in semiconductors. Wiley, New York

McLaren AC, Fitz Gerald JD, Gerretsen J (1989) Dislocation nucleation and multiplication in synthetic quartz: relevance to water weakening. Phys Chem Min 16:465–482

McLaren AC, Turner RG, Boland JN, Hobbs BE (1970) Dislocation structure of deformation lamellae in synthetic quartz: a study by electron and optical microscopy. Contr Mineral Petrol 29:104–115

McLaren AC, Cook RF, Hyde ST, Tobin RC (1983) The mechanisms of the formation and growth of water bubbles and associated dislocation loops in synthetic quartz. Phys Chem Miner 9:79–94

McQueen HJ (1977) The production and utility of recovered dislocation substructures. Metall Trans 8A:807–824

Means WD (1976) Stress and strain. Basic concepts of continuum mechanics for geologists. Springer, New York, 339 pp

Mecking H (1980) Deformation of polycrystals. In: McQueen HJ et al (eds) Strength of metals and alloys. Proceedings of ICSMAS 1979, vol 3. Pergamon Press, Toronto, pp 1573–1594

Mecking H (1981a) Strain hardening and dynamic recovery. In: Ashby MF (ed) Dislocation modelling of physical systems. Pergamon Press, Oxford, pp 197–211

Mecking H (1981b) Low temperature deformation of polycrystals. In: Hansen N et al (ed) Mechanisms and microstructures. Risø National Laboratory, Roskilde, Denmark, pp 73–86

Mecking H, Gottstein G (1978) Recovery and recrystallization during deformation. In: Haessner F (ed) Recrystallization of metallic materials, 2nd edn. Dr Riederer Verlag, Stuttgart, pp 195–222

Mitchell TE, Foxall RA, Hirsch PB (1963) Work-hardening in niobium single crystals. Phil Mag 8:1895–1920

Mitra SK, McLean D (1966) Work-hardening and recovery in creep. Proc Roy Soc (Lon) A295:288–299

Morris MA, Martin JL (1984a) Microstructural dependence of effective stresses and activation volumes in creep. Acta Metall 32:1609–1623

Morris MA, Martin JL (1984b) Evolution of internal stress and substructure during creep at intermediate temperatures. Acta Metall 32:549–561

Morrison-Smith DJ (1973) A mechanical and microstructural investigation of the deformation of synthetic quartz crystals. PhD thesis, The Australian National University, 222 pp

Mott NF (1953) A theory of work-hardening of metals. II: Flow without slip lines, recovery and creep. Phil Mag 44:742–765

Mott NF, Nabarro FRN (1948) Dislocation theory and transient creep. In: Report of a conference on strength of solids, held at H H Wills Physical Laboratory, Univ of Bristol, 7-9 July 1947, London, The Physical Society, pp 1-19

Nabarro FRN (1950) Influence of grain boundaries on the plastic deformation of metals. In: Harrison VGW (ed) Some Recent Developments in Rheology, London, United Trade Press (for British Rheologist's Club), pp 38–52

Nabarro FRN (1967) Theory of crystal dislocations. Clarendon Press, Oxford, 821 pp. (also Dover, New York, 1987)

Nabarro FRN (1984a) Dislocation cores in crystals with large unit cells. In: Kubin L, Castaing J (eds) Dislocations 1984. Core structure and physical properties, Paris, Editions du CNRS, pp 19–28

Nabarro FRN (1984b) Hollow dislocations in highly anisotropic crystals. S Afr J Phys 7:73–74

Nabarro FRN (1985) Soluton hardening. In: Dislocations and Properties of Real Materials, London, The Institute of Metals, pp 152–169

Nabarro FRN, Basinski ZS, Holt DB (1964) The plasticity of pure single crystals. Adv Phys 13:193–324

Nicolas A, Poirier JP (1976) Crystalline plasticity and solid state flow in metamorphic rocks. Wiley, London, 444 pp

Nowick AS, Berry BS (1972) Anelastic relaxation in crystalline solids. Academic, New York, 677 pp

Orowan E (1934) Zür Kristallplastizität. III. Über den Mechanismus des Gleitvorganges. Zeit Physik 89:634–659

Orowan E (1942) A type of plastic deformation new in metals. Nature 149:643–644

Orowan E (1946) The creep of metals. J West Scotland Iron Steel Inst 54:45–96

Öström P, Ahlblom B (1980) A dislocation link length model for strain-hardening in stage-II of polycrystalline metals of high stacking fault energy. Mate Sci Eng 43:115–124

Öström P, Lagnerborg R (1980) A dislocation link length model for creep. Res Mechanica 1:59–79

Pabst A (1955) Transformation of indices in twin gliding. Geol Soc Am Bull 66:897–912

Paterson MS (1969) The ductility of rocks. In: Argon S (ed) Physics of strength and plasticity. M.I.T. Press, Cambridge, pp 377–392

Paterson MS (1985) Dislocations and geological deformation. In: Dislocations and properties of real materials. The Institute of Metals, London, pp 359–377

Paterson MS, Weaver CW (1970) Deformation of polycrystalline MgO under pressure. J Am Ceram Soc 53:463–471

Paterson MS, Weiss LE (1966) Experimental deformation and folding in phyllite. Geol Soc Am Bull 77:343–374

Peirce D, Asaro RJ, Needleman A (1982) An analysis of nonuniform and localized deformation in ductile single crystals. Acta Metall 30:1087–1119

Peirce D, Asaro RJ, Needleman A (1983) Material rate dependence and localized deformation in crystalline solids. Acta Metall 31:1951–1976

Perez PJ, Tatibouët J, Vassoille R, Gobin PF (1975) Comportement dynamique des dislocations dans la glace. Phil Mag 31:985–999

Petch NJ (1953) The cleavage strength of crystals. J Iron Steel Inst 174:25–28

Petrovic JJ, Vasudevan AK (1978) Rolling deformation of two-ductile-phase Ag-Ni alloys. Mat Sci Eng 34:39–51

Philibert J (1979) Glissement des dislocations et frottement de réseau. In: Dislocations et Déformation Plastique. Ecole d'été d'Yravals 2-24 Sept 1979, Paris, Les Editions de Physique, pp 101–139

Piatti G Ed (1978) Advances in composite materials. Applied Science Publishers, London, p 405

Poirier JP (1976) On the symmetrical role of cross-slip of screw dislocations and climb of edge dislocations as recovery processes controlling high-temperature creep. Rev Phys Appl 11:731–738

Poirier JP (1978) Is power-law creep diffusion controlled? Acta Metall 26:629–637

Poirier JP (1979) Reply to "Diffusion-controlled dislocation creep: a defence". Acta Metall 27:401–403

Poirier J-P (1985) Creep of crystals. High-temperature deformation processes in metals. Ceramics and minerals. Cambridge University Press, New York, 260 pp

Poirier JP, Vergobbi B (1978) Splitting of dislocations in olivine, cross-slip controlled creep and mantle rheology. Phys Earth Planet Int 35:707–720

Polanyi M (1934) Ueber eine Art Gitterstörung, die einen Kristall plastisch machen könnte, Zeit. Physik 89:660–664

Pratt PL (1967) Strength and deformation of ionic materials. Geophys J R Astr Soc 14:5–11

Puls MP (1981) Atomic models of single dislocations. In: Dislocation modelling of physical systems. Proceedings of International Conference, Gainesville, Florida, USA June 22–27 1980. Pergamon Press, Oxford, pp 249–268

Roberts W (1984) Dynamic changes that occur during hot working and their influence regarding microstructural development and hot workability. In: Krauss G (ed) Deformation, processing, and structure. American Society for Metals, Metals Park, Ohio, pp 109–184

Ross JV, Ave Lallement HG, Carter NL (1979) Activation volume for creep in the upper mantle. Science 203:261–263

Sachs G (1928) Zur Ableitung einer Fliessbedingung. Z Verein deutsch Ing 72:734–736

Sammis CG, Smith JC, Schubert G (1981) A critical assessment of estimation methods for activation volume. J Geoph Res 86:10707–10718

Saxl I, Kroupa F (1972) Relations between the experimental parameters describing the steady state and transient creep. Phys Stat Sol (a) 11:167–173

Schmid E (1924) "Yield point" of crystals. Critical shear stress law. In: Proceedings of First International Congress for Applied Mechanics, Delft 1924, Technische Boekhandel en Drukkerij J Waltman Jr, p 342

Schmid E, Boas W (1936) Kristallplastizität mit besonderer Berücksichtigung der Metalle. Springer, Berlin, 373 pp

Schmid E, Boas W (1950) Plasticity of crystals (tr from German). F A Hughes and Co Ltd, London, 353 pp

Schmid SM, Paterson MS (1977) Strain analysis in an experimentally deformed oolitic limestone. In: Saxena S, Bhattacharji S (eds) Energetics of geological processes, Hans Ramberg Volume. Springer, Berlin, pp 67–93

Schmid SM, Paterson MS, Boland JN (1980) High-temperature flow and dynamic recrystallization in Carrara marble. Tectonophysics 65:245–280

Schoeck G (1980) Thermodynamics and thermal activation of dislocations. In: Nabarro FRN (ed) Dislocations in solids, vol 3. Moving dislocations. North-Holland Publ Co, Amsterdam, pp 63–163

Schoeck G, Seeger A (1959) The flow stress of iron and its dependence on impurities. Acta Metall 7:469–477

Seeger A (1984) Structure and diffusion of kinks in monatomic crystals. In: Dislocations 1984, Paris, Editions du CNRS, pp 141–177

Seeger A, Haasen P (1958) Density changes in crystals containing dislocations. Phil Mag 3:470–475

Seeger A, Berner R, Wolf H (1959) Die experimentelle Bestimmung von Stapelfehlenenergien kubischflachenzentrierter Metalle. Zeit Physik 155:247–262

Sellars CM (1978) Recrystallization of metals during hot deformation. Phil Trans Roy Soc Lon Ser A 288:147–158

Sherby OD, Weertman J (1979) Diffusion-controlled dislocation creep: a defense. Acta Metall 27:387–400

Smith CS (1964) Some elementary principles of polycrystalline microstructure. Met Rev 9:1–48

Spingarn JR, Barnett DM, Nix WD (1979) Theoretical descriptions of climb controlled steady state creep at high and intermediate temperatures. Acta Metall 27:1549–1561

Sprackling MT (1976) The plastic deformation of simple ionic crystals. Academic, London, 242 pp

Steeds JW (1973) Introduction to anisotropic elasticity theory of dislocations. Clarendon Press, Oxford, 274 pp

Strudel J-L (1983) Mechanical properties of multiphase alloys. In: Cahn RW, Haasen P (eds) Physical metallurgy, part II, 3rd edn. North-Holland, Amsterdam, pp 1411–1486

Stünitz H, Fitz Gerald JD, Tullis J (2003) Dislocation generation, slip systems, and dynamic recrystallization in experimentally deformed plagioclase single crystals. Tectonophysics 372:215–233

Takeuchi S, Argon AS (1976) Steady-state creep of alloys due to viscous motion of dislocations. Acta Metall 24:883–889

Takeuchi S, Argon AS (1979) Glide and climb resistance to the motion of an edge dislocation due to dragging a Cottrell atmosphere. Phil Mag A 40:65–76

Taylor GI (1934) The mechanism of plastic deformation of crystals. Part I. Theoretical, Proc Roy Soc (London), A 145, 362-387

Taylor GI (1938) Plastic strain in metals. J Inst Metals 62:307–324

Thomas de Montpreville C (1983) Modèles tridimensionels pour le calcul des comportments mécanique des alliages polyphasés. Res Mechanica 7:211–230

Thompson N, Millard DJ (1952) Phil Mag 43:422

Tomé C, Canova GR, Kocks UF, Christodoulu N, Jonas JJ (1984) The relation between macroscopic and microscopic strain hardening in f.c.c. polycrystals. Acta Metall 32:1637–1653

Tungatt PD, Humphries FJ (1981) An in situ optical investigation of the deformation behaviour of sodium nitrate—an analogue for calcite. Tectonophysics 78:661–675

Tungatt PD, Humphries FJ (1984) The plastic deformation and dynamic recrystalization of polycrystalline sodium nitrate. Acta Metall 32:1625–1635

Underwood EE (1970) Quantitative stereology. Addison-Wesley Publishing, 274 pp

Vanderschaeve G, Escaig B (1979) Glissement dévié des dislocations. In: Dislocations et Déformations Plastique. Ecole d'été d'Yravals, 3-14 Sept 1979, Paris, Les Editions de Physique, pp 141–148

Vitek V (1985) Effect of dislocation core structure on the plastic properties of metallic materials. In: Dislocations and properties of real materials. The Institute of Metals, London, pp 30–50

von Mises R (1928) Mechanik der plastischen Formänderung von Kristallen. Zeit für angewandte Math U Mech 8:161–185

Waff HS, Bulau JS (1979) Equilibrium fluid distribution in an ultramafic partial melt under hydrostatic stress conditions. J Geophys Res 84:6109–6114

Weertman J (1955) Theory of steady-state creep based on dislocation climb. J Appl Phys 26:1213–1217

Weertman J (1957) Steady-state creep of crystals. J Appl Phys 28:1185–1189

Weertman J (1968) Dislocation climb theory of steady-state creep. Trans Am Soc Metals 61:681–694

Weertman J (1975) High temperature creep produced by dislocation motion. In: Li JCM, Mukherjee AK (eds) Rate processes in plastic deformation of materials. Proceedings of John E. Dorn Symposium American Society on Metals, Oct. 1972, Metals Park, Ohio, pp 315–336

Weertman J (1977) Theory of internal stress for class I high temperature alloys. Acta Metall 25:1393–1401

Weertman J, Weertman JR (1964) Elementary dislocation theory. MacMillan, New York, 213 pp

Weertman J, Weertman JR (1980) Moving dislocations. In: Nabarro FRN (ed) Dislocations in solids. Amsterdam, North-Holland, pp 1–59

Weertman J, Weertman JR (1983a) Mechanical properties, mildly temperature-dependent. In: Physical metallurgy, 3rd edn. North Holland Physics Publishing, Amsterdam, pp 1259–1307

Weertman J, Weertman JR (1983b) Mechanical properties, strongly temperature-dependent. In: Physical metallurgy, 3rd edn. North Holland Physics Publishing, Amsterdam, pp 1309–1340

Weertman J, Weertman JR (1992) Elementary dislocation theory. Oxford University Press, Oxford, 213 pp

Wegner MW, Christie JM (1985a) General chemical etchants for microstructures and defects in silicates. Phys Chem Min 12:90–92

Wegner MW, Christie JM (1985b) Chemical etching of amphiboles and pyroxenes. Phys Chem Min 12:86–89

Wenk H-R (ed) (1985) Preferred orientation in deformed metals and rocks: an introduction to modern texture analysis. Academic, Orlando

Whitworth RW (1975) Charged dislocations in ionic crystals. Adv Phys 24:203–304

Wolf H (1960) Die Aktivierungsenergie für die Quergleitung aufspaltener Schraubenversetzungen. Zeit Naturforschung A 15:180–193

Yund RA, Tullis J (1983) Strained cell parameters for coherent lamellae in alkali feldspars and iron-free pyroxenes. N Jb Miner Mh Jg 1983:22–34

Zener C (1948) Elasticity and anelasticity of metals. University of Chicago Press, Chicago 170 pp

Chapter 7
Deformation Mechanisms: Granular Flow

7.1 Basic Concepts: Kinematics, Compatibility, Dilatancy

7.1.1 Introduction

In the previous two chapters we have considered deformation processes that involve the relative movement of individual atoms or molecules (Chap. 5) or of different parts of a given crystal or crystalline grain (Chap. 6). In this third chapter on deformation mechanisms we shall consider flow by the relative movement of more or less macroscopic entities, which may be whole grains or groups of grains such as particles of fractured rock. These entities can perhaps best be referred to generically as *granules* and their assemblage be said to constitute a *granular body*. The flow by relative movement of granules can then be called *granular flow*.

In considering the mechanisms of granular flow we shall take the somewhat unusual step of treating, under this same heading, both the nearly temperature independent phenomena of low temperature particulate and cataclastic flow and the strongly temperature sensitive phenomena of high temperature granular flow ("superplasticity") involving viscous or other accommodation processes. This approach derives from the view that the underlying kinematics of all these processes are similar, being based on the relative movement of granules as entities, even though the dynamics or rate control may depend on widely different processes, ranging from local friction or fracturing to atomic diffusion. The concern of this chapter will be with the various mechanisms of granular flow and no attempt will be made to expound the macroscopic flow theory, which is part of continuum mechanics.

The granules involved in the flow of the granular material are understood to be entities that persist identifiably for at least the duration of the deformation increment being considered. However, they may undergo some gain or loss of substance by diffusive exchange or fracturing, or they may themselves undergo deformation. They are described above as being macroscopic only in the sense of

M. S. Paterson, *Materials Science for Structural Geology*,
Springer Geochemistry/Mineralogy, DOI: 10.1007/978-94-007-5545-1_7,
© Springer Science+Business Media Dordrecht 2013

being large compared with atomic dimensions; their scale may still be such as would require microscopical observation.

For the sake of simplicity, discussion will, in general, be restricted to materials containing only one kind of solid granule. However, the granules will often not fully occupy the total volume of the material. In this case the granular material can generally be considered as consisting of two "phases" of extremely or even infinitely contrasting strength. The granules themselves will constitute the strong phase and the interstitial medium between them will constitute the weak phase, which may be just void space or a fluid or even a relatively weak solid. In order to define the granular material as a solid in the macroscopic sense, it is then necessary that there be three-dimensional connectivity through the contacts between the grains of the strong phase. Indeed, the distinction between a suspension and a granular solid may be expressed in terms of the existence of such a continuous "skeleton" of granular contacts through which shear stress is supported. However, the volume fraction of the granular phase may vary widely, from as low as 0.2 in unconsolidated sediments (Hamilton 1976) to unity in fully dense polycrystalline materials; that is, the porosity, defined as the volume fraction of interstices, may vary from around 0.8 to zero.

In the following subsections we shall attempt to set out a suitable kinematical framework in terms of which to discuss the mechanisms of granular flow and the factors that determine the dynamics of the flow. It will be useful to distinguish two aspects of the kinematics, as follows:

(1) the pattern of relative grain translations that eventually determine the macroscopic strains
(2) the questions concerned with the extension and boundaries of the grains, involving the overlaps or gaps that tend to develop and the rotations of the grains.

The main emphasis here is on the physics rather than the quantitative theory which is, in any case, still not well developed.

7.1.2 The Pattern of Relative Translations of Granules

In this first aspect of the kinematics we consider the relative movement of a granule viewed as a featureless entity of undefined volumetric extent. That is, in place of the real granule we consider only a *representative point* within it, and so, of the total geometry of the granular assemblage, only the arrangement of the representative points is at first taken into account. The only constraint that need be placed on the choice of the representative point is that it fall within a part of the original granule that persists as such during the deformation increment under consideration (it may be helpful at times to think of it as the centroid of the granule). The relative motions of the granules can then be described in terms of the displacements of the representative points relative to a set of external coordinates,

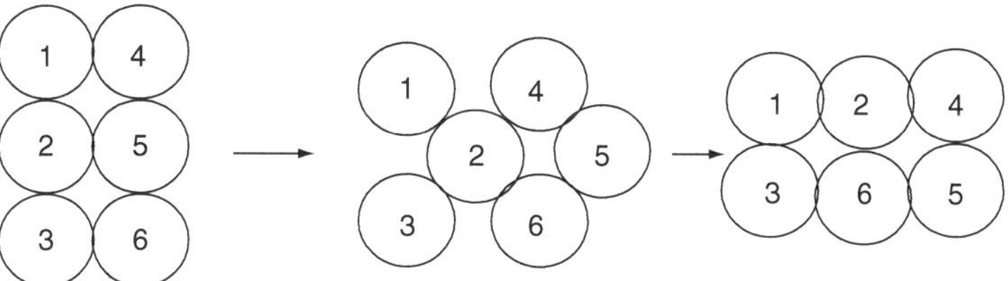

Fig. 7.1 Depicting how an array of spheres can undergo macroscopic shape change by exchange of neighbours

the origin of which may conveniently be chosen to coincide with the representative point of one of the granules. However, this description will differ from that for a continuous medium, familiar in the classical theories of elasticity and plasticity, because the constraints of continuity, whereby neighbouring points remain neighbouring, no longer apply. Indeed, it is an essential part of the mechanics of a granular material that granules can change neighbours in the course of the deformation.

The practical description of the relative translations of granules thus firstly involves the question of the scale of the local heterogeneity of movement leading to neighbour exchange. At the macroscopic scale the pattern of relative grain translations will be statistically homogeneous, more or less by definition, but this statistical aspect of the pattern contains little information about the local heterogeneities that essentially characterize the granular flow mechanism. The latter involves the individual relative translations of the granules. Thus a question that arises in relation to the description of the mechanism is how large a sample or *elementary representative group* of granules is required to define the essentials of the mechanism, the macroscopic strain being obtainable from the averaging of the deformations of these groups

In choosing an elementary representative group of granules it is necessary to take into account that the movement pattern may be continually changing. At any instant, relative movement of granules may be concentrated in certain locations, and these locations may change from one instant to another. Such a situation does not readily lend itself to very neat analysis. However, two aspects of the behaviour can be brought out by considering specific model situations, as follows:

(1) *Neighbour-exchange aspect.* The essential character of a neighbour-exchange process is illustrated in the notional, two-dimensional model of Fig. 7.1; here, grains 1 and 3 are initially separated by grain 2 but become immediate neighbours through the deformation, while grains 1 and 4 become separated. Since the relative translations to the new configuration are shown in the figure as occurring as directly as possible, it is evident that the amount of strain required for completion of a neighbour exchange is quite large, of the order of at least 0.3–0.4 natural strain (cf. Ashby and Verrall 1973). Thus deformation to fairly large strains is required in the observation and study of the neighbour-exchange process.

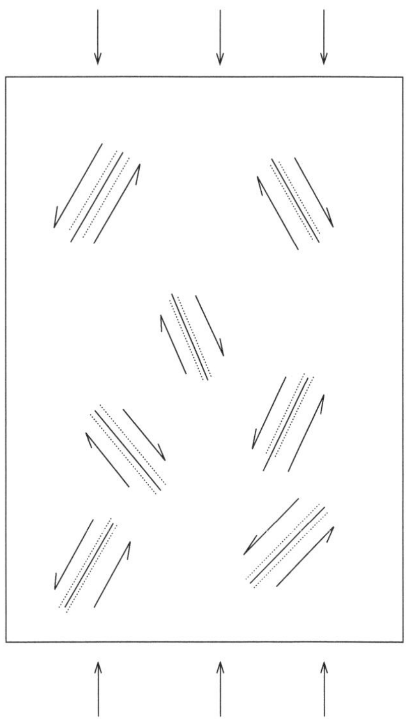

Fig. 7.2 Depicting how a granular body can be undergoing local transient shearing events, the pattern of which will be changing with time

(2) *Localized shearing aspect*. The relative granular displacement events such as idealized in Fig. 7.1 are unlikely to be distributed uniformly in space or time because of the irregularity and evolving character of the real granular structure. At any given instant, certain local structural configurations will be more favourable to relative granule movement than others, and, for reasons of movement compatibility, these relative movements will tend to occur as local shears extending through quite large groups of granules. That is, at any instant, the granular body will appear to be undergoing local shearing at isolated local sites, as depicted in Fig. 7.2 but the pattern or distribution of the local activity will tend to change with time. Such a continually changing kinematic picture has been emphasized by Rowe (1962) and (Horne 1965). Sometimes it may be useful to view the local shears as resulting from the passage of dislocations but, due to the absence of a long range periodic structure in the granular medium and to the continually changing pattern of local flow events, such dislocations will tend to be of an ephemeral nature.

So far, we have only considered the relative translations of the granules, or, specifically, of their representative points, insofar as these translations give rise to macroscopic change of shape, expressible in terms of deviatoric strain components. However, the relative translations may also contribute a macroscopic change of volume, or dilation, expressible in terms of a change in the mean spacing of the representative points. Further, even if the mean spacing were to remain constant (steady flow; no macroscopic dilation), there may be local dilations (positive or negative) associated with the local shearing events mentioned above, which would have dynamical implications, especially if a relatively incompressible interstitial phase were present.

For some attempts to give quantitative expression to the micro-geometrical or kinematic aspects of granular material deformation and to relate them to phenomenological theory, see Cambou (1982), Nemat-Nasser (1982, 1983), Oda (1982), Satake (1982), Cambou (1993).

7.1.3 Granule–Granule Relationships at their Boundaries

We now consider the actual geometry of the granules and their contacts rather than just the locations of the representative points. This geometry involves both the extension of the granules in space and their orientations (the latter can be defined, for a given granule, by the orientation relative to external coordinates of two material line segments intersecting in the representative point).

If two granules are initially in contact over a certain area of granule boundary, this contact will, in the absence of granule deformation, be preserved during an increment of aggregate deformation only if the relative translation vector for the granules is parallel to the contact area, and, in the case of non-spherical granules, if there is no relative rotation of the granules. Otherwise, the two granules will tend either to separate at the initially common boundary or to interfere there. In a real deformation in which three-dimensional connectivity is maintained, these two tendencies have to be limited in one or more of the following ways:

(1) the establishment of a self-consistent set of relative granule translations that are as far as possible both instantaneously parallel to the contact surfaces and yet such as will give rise to the required neighbour exchange.
(2) insofar as the relative granule translations are not parallel to the contact surfaces, the maintenance of sufficient granule contact for connectivity and control of dilatancy through the transfer of material by processes such as diffusion or fracturing and rearrangement of the fragments; in general cases, material may be introduced into or removed from the body itself in an exchange with the environment.
(3) alternatively to (2), the maintenance of connectivity and control of dilatancy through intragranular deformations; such deformations may become reversed in subsequent stages of the flow and, in any case, should sum algebraically to substantially less than the macroscopic strain (otherwise the deformation should be classified under Chap. 5 or 6)

The first of these three aspects of the deformation is of a primary nature, being an essential part of the granular flow process, while the second and third have a secondary or accommodation role, meeting the connectivity and dilatancy requirements (homologous with the compatibility constraints discussed in Sects. 5.2.2. and 6.8.1.

In contrast to the negligble or very small volume change during deformation of pore-free polycrystalline aggregates by crystal plasticity (Chap. 6), dilatancy is often an important aspect of deformation by granular flow mechanisms. Assuming that the solid volume of the granules themselves remains constant, the dilatancy derives from variation in the interstitial volume during the deformation. This variation can occur for two essentially different reasons, associated, respectively, with the contact stresses in the granules and with the pressure in the interstitial medium:

(1) as an accommodation mechanism associated with the relative translation of granules, as discussed above; thus, a looser packing of granules locally may be required to permit them to pass one another in the neighbour-changing process.
(2) as a direct response to a difference between the pressure in the interstitial medium (for example, pore fluid pressure) and the mean total macroscopic stress, assuming that the former pressure is independently controlled.

In the case in which the material of the granules can also be transported into or out of the specimen, there may be an additional dilatation or compaction associated with the net transfer.

7.1.4 Fabric and Memory

On general symmetry principles (Curie 1894; Paterson and Weiss 1961) it can be expected that in a granular body undergoing deformation the structural arrangement of the granules will be in some way anisotropic. The appropriate description of the structural arrangement, usually called a texture or fabric, will depend somewhat on the nature of the deformation mechanism, that is, on the dynamical factors governing the deformation. In the case of the low-temperature deformation of particulate media such as unconsolidated sediments, the fabric may be specified in terms of the contact normals between the particles; see, for example, Field (1963), Brewer (1964), Oda (1972a, c), Nemat-Nasser (1982) and Satake (1982). Thus, while a random dense packing of spheres may model some aspects of such a granular medium, the random character will lack the essential anisotropy of the medium actually undergoing deformation. In the case of polycrystalline material undergoing high-temperature granular ("superplastic") flow, the most important element of the fabric may be the configuration of the grain boundaries at which sliding is occurring.

Since the anisotropy of the fabric reflects the nature of the deformation process, the fabric serves as a "memory" of this process. However, as in the case of the crystallographic preferred orientation resulting from deformation by crystal plasticity, where later phases of a complex deformation history can overprint and obliterate the effects of the earlier phases, so in granular materials the fabric can be expected to constitute only a "fading memory". Nevertheless it may be a valuable source of information concerning the nature of the deformation processes.

7.2 Granular Flow Controlled by Frictional and Cataclastic Effects

7.2.1 Introduction

So far in this chapter we have considered purely kinematical or geometrical aspects of granular flow. We now introduce dynamical aspects, considering the factors that determine the resistance to flow under applied stress. This step immediately enables us to distinguish two broad categories of granular flow.

In the first category (microbrittle granular flow), considered in the immediately following sections under Sect. 7.2, the dynamics of the flow are controlled by frictional factors, including any local fracturing necessary to allow relative sliding of granules, that is, by local factors that are characteristic of the brittle field. In the second category (microplastic granular flow), considered in subsequent sections underSect. 7.3, the dynamics are controlled by local factors that are characteristic of the ductile field, involving either crystal plasticity within the granules (cf. Chap. 6) or atom transfer (cf. Chap. 5) or both. The distinction between these two categories is sufficiently profound that it is unconventional to consider them together in the same chapter. Yet they are logically related through the kinematical aspects considered under Sect. 7.1. In practice, their fields of application tend to be distinct because the first type of flow arises mainly at relatively low temperatures and the second at higher temperatures (although, in the case of solution transfer effects on the geological timescale, the temperature need not be very high).

In elaborating the first category further, it is useful to separate the case of pure particulate flow, involving only friction between intact granules, and the case of particulate flow with accompanying local fracture processes. We now consider these cases in turn, as well as touch on the question of what is meant by cataclastic flow and on the roles of temperature and pore fluid pressure.

7.2.2 Pure Particulate Flow

The simplest and, in many respects, archetypal model for granular flow is that of a cohesionless assemblage of rigid particles that remain intact during the flow. Such a model is often invoked in soil mechanics in connection with the flow of a cohesionless dry sand. It can also be taken as a reference case in developing more complex models in which additional physical processes are incorporated. We shall refer to it as the model of "pure particulate flow".

The basic physical premise of the pure particulate flow model is that the only source of energy dissipation is the friction between the granules/particles at their contacts during their relative movement. Such a model could therefore also be described as one of pure friction-controlled granular flow. The friction is taken in the simplest case to be described by the Amonton law according to which the tangential, frictional force component at the contact is proportional to the normal force component. In general, however, incomplete knowledge of the geometry of both the individual granules and their spatial arrangement makes it difficult to formulate a detailed micromechanical model for pure particulate flow. This problem has been discussed by Vaišnys and Pilbeam (1975) who identify several approaches which have been or, in principle, can be taken. If a sufficiently simple, regular array of granules is assumed, an analytical approach can be taken. Otherwise, alternative approaches are needed to cope with situations that may be either too complicated to analyse completely ("computation-limited") or lacking in a full specification of all details ("information-limited"). See comments in Myer et al. (1992) on a stochastic approach for clastic rocks.

The analysis of the behaviour of regular arrays of equal-sized rigid spheres brings to light certain elementary properties that may apply more generally (Deresiewicz 1958; Feda 1982, Sect. 4.3; Parkin 1965; Rennie 1959; Thornton 1979; Thornton and Barnes 1982; Thurston and Deresiewicz 1959; Trollope 1968). Starting with the contact forces between the spheres and calculating the corresponding macroscopic stresses, it can be shown that, for predominantly compressive stress states, deformation is initiated when the ratio of the extreme principal stresses, σ_1/σ_3, exceeds a critical value, which, in the case of a close-packed cubic array shortening in the (111) direction, is

$$\frac{\sigma_1}{\sigma_3} = 2\frac{1+\mu}{1-\mu} \tag{7.1}$$

(Feda 1982, p. 152). Recast in Coulomb form (Paterson and Wong 2005, Sect. 3.3), this expression leads to a coefficient of internal friction of the form

$$\tan\phi_i = \frac{1+3\mu}{2\sqrt{2(1-\mu^2)}} \tag{7.2}$$

It is to be noted that $\tan\phi_i$ has a finite value, $1/2\sqrt{2} \approx 0.35$, even when the contacts are frictionless ($\mu = 0$.) This result re-emphasizes the distinction between the nature of $\tan\phi_i$ and a physical coefficient of friction (Paterson and Wong 2005, p. 25). It expresses the fact that $\tan\phi_i$ covers the structural or interlocking aspect of the shearing strength as well as the truly frictional aspect. These two aspects of the strength can also be associated with non-dissipative and dissipative components of the deformation, respectively (Thornton and Barnes 1982). Thornton and Barnes have also analyzed the cases of lower-symmetry regular arrays, in which they show that the normality condition (Sect. 4.5) is not obeyed. Such results for regular arrays of spheres suggest similar properties for less regular arrays.

Since a regular array is too artificial a model to represent fully a real granular aggregate, it is important also to consider irregular packings and, eventually, mixed sizes and shapes of granules and the possibility of rolling as well as sliding at contacts. There are several approaches that may be taken. They include purely statistical, computer modelling and semi-empirical approaches.

In contrast to the analysis of a regular array, for which a specific description of the geometrical arrangement of the granules is assumed, in the statistical approach it is assumed that the granules are randomly arranged, although still being spheres of equal size in the simplest case. The geometry may be expressed in terms of the number of contacts on individual granules or of the variation in local porosity (Feda 1982, Sect. 4.3.2). In either case, it can be expected that at least two parameters will be needed for an adequate description of the array, for example, the mean porosity and the standard deviation of the local porosity, and so the way is opened for introducing two geometrical parameters into constitutive relations for granular aggregates. For further consideration, including the use of the thermo-dynamics of irreversible processes and observations on arrays of steel balls, see Feda (1982, Sect. 4.3.2), Mogami (1965, 1969) and Mogami and Imai (1967, 1969).

There is a total lack of specific information about the behaviour of individual granules in relation to their neighbours in a statistical model. In contrast, the situation at each granule in a finite array is available in computer modelling experiments (Cundall 1986, 1988a, 1988b, 1989; Cundall, et al. 1982; Cundall and Strack 1979a, 1979b, 1983; Hart et al. 1988; Thornton and Barnes 1986a). Since there appears to be a paucity of relevant physical observations on real granular materials at the microscopic scale, apart from a few studies on photoelastic models (Allersma 1982; Drescher 1976; Drescher and de Josselin de Jong 1972; Konishi, et al. 1982, and earlier references given by them; Oda and Konishi 1974a, 1974b), the computer modelling experiments provide a valuable source of ideas about the micromechanisms of granular flow, confirming and extending those from the photoelastic models. The granular material has most commonly been simulated by arrays of discs (for two dimensions) or spheres (for three dimensions), often of several sizes and arranged initially in more or less random ways. In the computation, specific assumptions are made about the nature of the interactions at the contacts (elastic and/or frictional).

The main conclusions about the mechanisms of pure particulate flow in a granular material from the model studies are the following:

(1) The forces are mainly transmitted through chains of aligned granules, inclined at relatively small angles to the maximum compressive principal stress, the other granules being more lightly loaded (Fig. 7.3).
(2) The shearing processes are concentrated amongst the lightly loaded granules, roughly defining domains within which there is less relative movement and more concentration of forces through chains.
(3) The granules rotate relatively to each other in a fairly coordinated fashion, thereby minimizing the total amount of frictional sliding and contributing in a

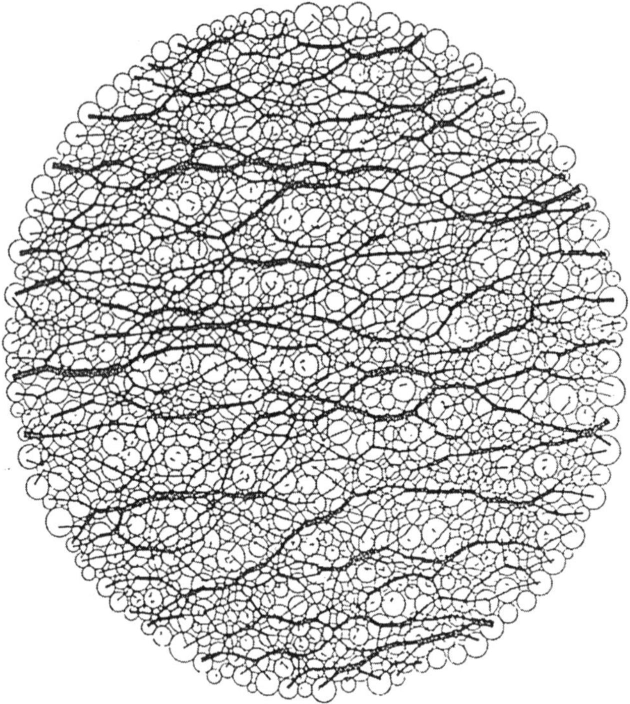

Fig. 7.3 A computer model of the distribution of forces between particles in a granular assemblage after 3 % shear strain. The lines represent the orientation of forces between particles and the thickness of the lines represent the magnitudes of the forces (copy of Fig. 3 in Thornton and Barnes 1986b)

major way to the deformation. Thus the granules along a chain tend to rotate alternately in opposite senses, leading to the eventual destruction of the chain and transfer of load. The rotations may be exaggerated in certain locations, giving a hinge-like effect with the appearance of a local strain discontinuity.

(4) During straining the number of contacts tends to decrease, especially in the direction of the least compressive principal stress, accompanied by the development of dilatation and anisotropy. The dilatation seems not to depend immediately on the intergranular friction, although this will have an eventual effect through its influence on the development of fabric.

(5) Upon unloading, residual stresses tend to remain, due to locked-in shear forces at contacts.

We now come to the microdynamical problem of integrating up from the behaviour at the grain scale to the macroscopic behaviour of the aggregate, that is, of deriving a constitutive relation or flow law from a given set of grain–grain forces. Such a step involves not only the mathematical difficulties arising from the complexity and variability of the pattern of contact forces and their proper description, but also the structural or geometrical problems of what this complex pattern actually consists of on a sufficiently large scale to correspond to an elementary representative group of granules that is typical of the macroscopic scale (Sect. 7.1.2). So far, the derivation of a flow law in physical terms from grain–grain forces, without artificial assumptions, has not been achieved. Instead, one can only point to a number of approaches that have been made by various workers which attempt to construct macroscopic theories that incorporate some elements of what is understood of the microscopic processes, with such additional assumptions as are thought to be rational or to be required in order to achieve a correspondence with macroscopic observations.

In considering the various theories it must be borne in mind that particular theories may be aimed only at describing particular aspects of the mechanical behaviour. The most important distinction is that between *steady flow*, in which the flow stress and the state of the aggregate are unchanged from one strain increment to the next, and *unsteady flow*, with simultaneously evolving structure and in which hardening or softening occurs, accompanied by dilatation or compaction. Because of the interdependence of flow stress and structure, any valid flow law can be expected to contain parameters (internal variables) that represent the structural factors, although some of these may not appear in the case of a steady flow. Some aspects of the problem of deriving a flow law for the aggregate from the intergranular contact forces as revealed in modelling experiments have been discussed by Cundall and Strack (1983). They suggest that the internal variables can be, at least in some degree, represented by "partitions" of the macroscopic stress. They show that, for example, the intergranular shear forces contribute only to the deviatoric components of the stress tensor, the isotropic part deriving solely from normal forces at contacts. Further, they show that the deviator can be partitioned into components that, respectively, represent the shear forces at actively sliding contacts (associated with the energy dissipation), the angular distribution of contacts (representing the fabric anisotropy), and the variation in magnitude of normal contact forces with angle (as a measure of the tendency of the "chains" to buckle). They also introduce the notion that the strength will depend on the structure in a way that can be measured by a "constraint ratio", defined as the ratio of the number of constraints (actual physical situations involving some relationship between a local force and a displacement resulting from it) to the number of degrees of freedom (independent relative translations or rotations of granules that are possible). However, further development is needed for a complete physical theory.

In the absence of a fully physically-based theory, there have been many theories that attempt to incorporate physical notions in some degree but supplement them with more or less empirical assumptions. Here we shall distinguish two broad groups of such theories.

The first semi-empirical group is typified in the theory of Rowe (1962, 1963, 1964 and 1972) and Horne (1965, 1969); see also Barden (1971), Proctor (1974) and Proctor and Barton (1974). This approach starts with the physical picture of granules sliding over each other against frictional resistance. It recognizes that the sliding tends to be localized at the boundaries of large groups of granules and it assumes that rolling at the contacts plays no significant role. The orientations of the contacts at which active sliding occurs are chosen so as to minimize the energy dissipation by friction. With the aid of this assumed principle, a "stress-dilatancy" relationship is derived which, in the case of a triaxial compression test, is of the form

$$\frac{\sigma_1}{\sigma_3} = \left(1 + \frac{1}{v}\frac{dv}{d\varepsilon_1}\right)\tan^2\left(\frac{\pi}{4} + \frac{\phi_\mu}{2}\right) \tag{7.3a}$$

$$\text{or } \frac{\sigma_1}{\sigma_3} = \left(1 + \frac{d\phi}{d\varepsilon_1}\right)\tan^2\left(\frac{\pi}{4} + \frac{\phi_\mu}{2}\right) \tag{7.3b}$$

where σ_1 and σ_3 are the axial and radial principal stresses, respectively (compression positive; in case of pore pressure, σ_1 and σ_3 are taken to be the conventional effective stresses), v is the specimen volume, ε_1 the axial strain (shortening positive), ϕ the porosity, and ϕ_μ the friction angle at the sliding contacts. Horne (1965) proceeds further and relates the dilatancy factor, $1 + d\phi/d\varepsilon_1$ for the triaxial compression test, to a fabric anisotropy measure which is viewed as changing in the course of the test and so accounting for the various stages in a test (initial hardening or softening and subsequent constant volume stages). However, direct microscopial observations have not been used in testing the theory. Oda (1972b, 1974) has attempted to combine microscopical fabric observations with mechanical and he has developed a theory along similar lines to Rowe and Horne, resulting in an expression analogous to (7.3a) but with a fabric factor in place of the dilatancy factor.

It is interesting to compare the form (7.3b) with the purely empirical Coulomb relationship for a cohesionless material

$$\frac{\sigma_1}{\sigma_3} = \tan^2\left(\frac{\pi}{4} + \frac{\phi_i}{2}\right) \tag{7.4}$$

where ϕ_i is the angle of internal friction (Paterson and Wong, 2005, p. 25). It is seen that $\phi_i > \phi_\mu$ during dilation and $\phi_i < \phi_\mu$ during compaction. In soil mechanics studies situations have often arisen where the observations do not appear to fit the form (7.3) when the real interparticle coefficient of friction tan ϕ_μ is used and so another angle ϕ_i is introduced in place of ϕ_μ (for example Rowe 1972). However, although an attempt is made to rationalize such a step in terms of a supposed physical model, the theory takes on a more or less purely empirical character at this point. Indeed, at no stage can the Rowe-Horne theory be regarded as one fully based on the microscopical observation of the flow mechanisms. Although there is reference to groups of grains sliding relative to each other, there is no recognition of the separation between the "chains" of main load-bearing contacts and the actual sites of sliding contacts, as revealed in the photoelastic and computer modelling experiments mentioned earlier, nor is any account taken of the possibly important role of the rotations of granules or groups of granules. Thus such an approach does not seem to be very promising from the point of view of further physical insight into mechanisms.

The second semi-empirical group is typified in the theories of Mandle (1947, 1966), de Josselin de Jong (1959, 1971, 1977), Spencer (1964, 1982), Mandl and Fernandez Luque (1970), Rudnicki and Rice (1975), Mehrabadi and Cowin (1978, 1980), Anand (1983) and Nemat-Nasser and coworkers, for example, Nemat-Nasser (1983, 1986); Christoffersen et al. (1981); Mehrabadi and Nemat-Nasser (1983); Lance and Nemat-Nasser (1986). Most of these theories have been described as "double slip" theories. They can be regarded as having a physical basis insofar as it is assumed that frictional sliding occurs on conjugate surfaces inclined at $\pm\left(\pi/4 - \phi_f/2\right)$ to the maximum principal compressive stress, but little attempt has been made to identify the sliding surfaces with microscopically observable features, and ϕ_f is treated as an empirical material parameter. In the more developed theories, the concept of a dilatation normal to the sliding surfaces is also introduced and described by a second material parameter v or ϕ_v while, in any case, the sliding on the surfaces is assumed to be controlled by a Coulomb criterion which specifies a limiting shear stress of $\tau_0 + \sigma \tan \phi_i$ (where σ is the normal stress across the sliding surface) incorporating two further material parameters, the angle of internal friction ϕ_i and the cohesion τ_0 (which is zero in the case of the pure particulate flows considered in this subsection). With up to three adjustable parameters apart from the cohesion parameter, such theories have considerable flexibility and can incorporate the observed non-axiality of principal strain rate and stress tensors and non-normality of strain rate direction and yield surface (Sect. 4.0.0). However, again, the development of these theories does little to aid physical insight into the deformation mechanisms and they can be regarded as essentially phenomenological.

A third group of theories that might also be viewed as being semi-empirical, although again distantly so, stem from attempts to combine concepts of real structure, such as granular structure, with those of continuum mechanics in treatments of so-called structured continua or Cosserat continua; for an introduction to this field, see Jaunzemis (1967, Chap. 11). In such theories, the interactions between granules that involve couples can be represented by couple stresses, and the granule dimension can be introduced as an absolute scaling parameter when specifying quantities such as the width of a shear band. As an example of such an application, see Mühlhaus and Vardoulakis (1987).

The role of pore fluid pressure also needs to be incorporated in granular flow theories. It is usually assumed that the influence of the pore fluid pressure is adequately dealt with by substituting conventional effective stresses $\sigma_{ij} - p\delta_{ij}$ for the actual macroscopic stresses σ_{ij} (where p is the pore fluid pressure and δ_{ij} the Kronecka delta), as for the case of brittle fracture (Paterson and Wong 2005, Chap. 7). This view largely derives from experimental studies in soil mechanics, especially those on flow in sand at low effective pressures. However, insofar as the resistance to flow arises from Amonton-type friction at elastically deforming contacts between granules, the conventional effective stress law could also be expected theoretically since the normal contact forces between granules would be proportional to the difference between the confining pressure and the pore fluid pressure.

7.2.3 Cataclastic Granular Flow

In the models of pure particulate flow (Sect. 7.2.2) it has been assumed that the granules or particles remain intact throughout the flow (except presumably for some wear on the surfaces, associated with the frictional sliding). Such a situation can only be assumed to apply at relatively small stresses, such as those commonly involved in soil mechanics or in the handling of particulate matter. In practice, grain crushing becomes significant when the mean normal stress exceeds a certain level which depends on mineralogy and grain size but for medium-grained sand is of the order of 1 MPa (Chaplin 1971; Hettler and Vardoulakis 1984; Vesic and Clough 1968).

We now consider the case of granular flow in which the intergranular forces are sufficiently large to lead to gross fracture of the granules. Such flow is here described as cataclastic granular flow, taking "cataclastic", from its etymological origins, to mean "pertaining to breaking down by fracturing". In the case of extensive fracturing, the term "comminutive granular flow" might alternatively be used. There are two distinct ways in which cataclastic effects can be of relevance to granular flow:

(1) In the first case, as in that of an initially more of less intact rock, the occurrence of cataclasis is a precondition to the possibility of granular flow, in that the cataclasis reduces the body to an assemblage of granules which can then move relative to each other in producing a granular flow. In geological rock mechanics, this combination of cataclasis and granular flow is commonly referred to as "cataclastic flow", without distinguishing between the two aspects. Often such a process is only one, possibly minor, component in a deformation that also involves crystal plasticity. In the latter case, if the fracturing is insufficiently pervasive to reduce the rock effectively to an assemblage of separate granules, the main contribution of the cataclastic component to the deformation will be dilatational rather than distortional.

(2) In the second case, that in which the body is initially already in a particulate state, the cataclasis may not only have a direct effect, in which the relative movement of groups of granules is facilitated by the fracturing or interlocking grains, but also an indirect or structural effect of changing the size distribution and packing density of the granules and hence influencing the potential ease with which further relative movement can occur. Thus, the fracturing of granules is important in relation to compatability or accommodation requirements and, consequently, to dilatancy.

The clearest example of the role of cataclastic effects is provided by the development and deformation of fault gouge in rocks, and its study serves to illustrate the principal aspects of cataclastic granular flow. The first of these aspects concerns the nature and the progressive comminution of the granular material that constitutes the gouge. From a study by Sammis et al. (1987) and earlier papers quoted by them, the following properties appear to apply to gouge:

(1) The fracturing of the granules consists largely of diametral splitting in response to compressive loading at contacts with granules of similar size and so is strongly influenced by the size distribution of neighbouring granules.
(2) The size distribution that develops is self-similar with respect to change in magnification, constituting a fractal distribution with a fractal dimension D of about 2.6. This observation suggests that the fragmentation process is scale invariant (Turcotte 1986).
(3) Assuming that each fracture event is accompanied by a local relative shearing displacement β (β = actual displacement divided by granule diameter), and using the averaging procedure of Molnar (1983) the mean macroscopic shear strain in the gouge is deduced to be about 6β in the case of $D = 2.6$. Applying this relationship gives quite large values of β, of the order of 10 or more, indicating that the amount of sliding and/or rolling that accompanies each fracture event must be quite large. Also the latter contribution possibly increases at larger strains relative to that of the local displacements associated with the fracturing itself (Biegel and Sammis 1988). These considereations give some idea of the local kinematics that may be involved in the mechanisms of cataclastic granular flow.

The second aspect of such flow on which one might seek guidance from studies on gouge is the dynamics. The dynamics of cataclastic granular flow can be expected to be dominated by some combination of friction and fracture effects. Since the friction itself will tend to reflect fracturing of asperities, the overall or macroscopic flow behaviour can thus probably be expected to follow a Coulomb-type law, analogous to that for macroscopic shear fracturing (Paterson and Wong 2005, Sect. 3.3) but with perhaps somewhat lower values for the coefficient of internal friction if rolling of granules is a significant factor. However, from the existing experimental literature, it is difficult to draw general conclusions and to develop a clear physically-based theory of cataclastic granular flow. The observations on gouge layers often refer to sliding at the gouge-slider interface (for example, Shimamoto and Logan 1981b). Even if the main body of the gouge is involved, as it tends to be at very large relative displacements (Blanpied et al. 1988), the deformation is often concentrated in Riedel shears (for example, Logan et al. 1981). Therefore, the stability of the deformation is likely to figure as a major aspect in any comprehensive theory. Complementary experimental observations relevant to cataclastic granular flow are also offered by studies on porous rocks, such as sandstones, with porosities of the order of 0.1 or higher (Borg et al. 1960; Hadizadeh and Rutter 1982, 1983; Handin and Hager 1957, 1958; Handin et al. 1963; Hirth and Tullis 1988). In such experiments, there tends to be a preliminary phase of deformation during which cataclasis associated with pore collapse contributes a component to the deformation. As noted in Sect. 7.1.2, large strains are needed for complete neighbour exchanges in granular flow and therefore steady-state conditions can generally only be expected at very large strains; transient behaviour, involving either strain hardening or strain softening, tends to be characteristic for the normal range of strains explored in axisymmetric tests. Such behaviour therefore needs to be an important part of theoretical studies, as well as the role of pore fluid pressure.

7.2.4 Cohesive Granular Flow

We now return to the consideration of granular flows in which the granules are envisaged as remaining intact during flow, but consider the situation in which the resistance to relative movement of granules includes a component other than the simple Amonton-type friction at contacts considered in Sect. 7.2.2. Two cases might be distinguished in respect of the source of the additional resistance.

In the first case there may be a cohesive interaction between the granules at their contacts. Such a type of interaction could arise from the presence of an intergranular film having finite shearing strength or viscosity, from electrical interactions between charged particles (especially important in clays), from long-range surface forces such as those studied by Israelischvili and coworkers (Israelachvili 1992), or from surface energy effects associated with variation in areas of interfacial contacts.

In the second case, there may be a finite shearing strength in the intergranular phase, which has to be overcome in accommodating the intergranular movements. A sandy sediment with a clay interstitial filling would be an example of such a case.

In the absence of a well-developed physical theory of the mechanism of pure particulate flow (Sect. 7.2.2), little can be added theoretically concerning the cohesive case. The simplest theoretical step would presumably be to replace the intergranular friction law $F_t = \mu F_n$ by one of a form such as

$$F_t = F_0 + \mu F_n \tag{7.5}$$

(Jaeger 1959), where F_t, F_n are the tangential and normal intergranular forces, respectively, and F_0 and μ ($= \tan \phi_\mu$) are material constants. It will be noted that the friction law (7.5) is formally equivalent to the Coulomb failure law

$$\tau = \tau_0 + \sigma \tan \phi_i \tag{7.6}$$

commonly used to describe the macroscopic behaviour, τ_0 and $\tan \phi_i$ being macroscopic material constants and τ, σ the shear stress and normal stress, respectively, acting on a notional plane inclined at $\pi/4 - \phi_i/2$ to the maximum compressive principal stress; however, in the absence of an adequate theory, the empirical constants τ_0 and $\tan \phi_i$ cannot be immediately related to the more physically meaningful constants F_0 and $\tan \phi_\mu$.

Pore fluid pressure would be expected to have an effect through its modification of the normal forces at intergranular contacts, in which case it should be possible to use conventional effective stresses again to express the effect, insofar as the pore fluid is chemically neutral. However, the presence of a pore fluid may also influence the values of the cohesion and, possibly, the friction parameters if the fluid is chemically active, as may well be the case if it is water.

If clays are present, it may be relevant, further, to distinguish between "free" pore water and adsorbed water (for example, Bombolakis et al. 1978).

7.2.5 *Effect of Temperature: Creep*

In the cases of pure particulate flow and of cataclastic granular flow the relative movement of the granules can be expected to be generally rather insensitive to temperature because of the relatively temperature-insensitive natures of the processes of friction and fracture. However, there are some temperature effects that can be measured and that can be of practical importance, and there is scope for additional temperature effects in the case of cohesive granular flow.

In general, there are two ways in which a deformation mechanism can be affected by a change in temperature. On the one hand, insofar as the process involves the local elastic distortion of the crystal structure, it will tend to depend on temperature in the way in which the elastic modulus depends on temperature, and no time-dependent or rate effects need be involved. On the other hand, insofar as the process involves the local disruption of the atomic bonding structure, it is likely to depend on thermal activation and so a time or rate dependence will be involved; in this case, the process will be characterized as a thermally-activated one and the possibility of creep or strain rate effects arises. The effects that we shall discuss in this section will be mainly of this second type.

In the case of models of pure particulate flow, the main physical parameter is the friction, and so temperature and time effects would be expected to have their origin in the temperature and sliding rate sensitivities of friction, which are generally relatively small. There are three main physical models for the origin of friction (Paterson and Wong 2005, Sect. 8.4.3):

(1) The adhesion theory of Bowden and Tabor (1950; 1964), which is based on the local plastic deformation of welded or adhering junctions.
(2) The dilatational work theory according to which the irreversible work done against the normal load through the dilatational component of the displacement (references in Paterson and Wong 2005, Sect. 8.4.3).
(3) The asperity fracture theory of Byerlee (1967) based on the local brittle fracture of interlocking asperities.

The asperity fracture model is probably the more widely applicable to rocks, at least at low to moderate temperatures, and in this case the temperature and time dependence would be similar to that for brittle fracture in rock and so be relatively unimportant (Paterson and Wong 2005, Sect. 3.4). However, there may be some high temperature or other situations in which the component minerals can readily deform plastically and in which the adhesion model of friction is therefore more relevant; temperature and time effects similar to those for plastic deformation might then be expected, although such a granular body may also be approaching the sintering threshold. In practice, the temperature and time aspects of pure particulate flow do not appear to have been much studied.

In modelling cataclastic granular flow, crack propagation within the granules plays an important role in addition to that of friction at their surfaces and may even become the rate controlling factor. The temperature and time effects would then be expected to be similar to those observed in crack propagation studies. While such effects are still relatively minor compared with similar effects in crystal plasticity and atom transfer flow, they are nevertheless readily observable and have been extensively studied in connection with static fatigue and subcritical crack growth (Anderson and Grew 1977; Atkinson 1984; Atkinson and Meredith 1987; Costin 1987; Hasselman and Venkataswaran 1983; Kranz 1979; Kranz 1980, 1983; Martin 1972; Ohnaka 1983; Scholz 1968, 1972; Swanson 1984). However, the application of such studies in the modelling of cataclastic granular flow has been less well developed.

The main application of the studies on the kinetics of crack-growth has been to so-called *brittle creep*, that is, the small-strain creep that is commonly observed as a precursor to macroscopic brittle fracture or to gross slippage on an already existing fault surface (Carter et al. 1981; Carter and Kirby 1978; Cruden 1970, 1974; Kranz 1980; Kranz and Scholz 1977; Lockner and Byerlee 1977a; Scholz 1968; Wu and Thomsen 1975; Yanagidani, et al., 1985). One description of deformation of this type has been as "elastic creep", on the grounds that it can be regarded as resulting from a decrease in the elastic modulus of the material because of the crack proliferation (for example, Hasselman and Venkataswaran 1983). However, this description, which seems to imply that the cracks close again on removal of the deviatoric stress, does not adequately cover the permanent strains (non-dilational as well as dilational) that may accompany the crack proliferation. There has been considerable progress in the understanding of the phenomenology and of the processes involved in brittle creep, such as its predominantly transient or primary character, with a relative lack of steady state behaviour but a strong tendency to a tertiary or failure stage, its relatively low temperature and strain-rate sensitivity, reflected in a low activation energy if suitably defined (generally less than 100 kJ mol^{-1}), and the importance of crack interaction; in addition to references already given above, see Price (1964), Robertson (1964), Lockner and Byerlee (1977b; 1980), Sasajima and Itô (1980), Ohnaka (1983), Segall (1984), Costin (1985)and Ortiz (1985). However, the theoretical basis for modelling brittle creep is still not well developed in a fully integrated way, incorporating both slow crack growth and evolving crack interaction aspects.

The other main aspect of cataclastic granular flow, complementary to the small-strain brittle creep, is the *large-strain* behaviour, typified in the deformation of fault gouge or of cataclasites or in phenomena such as hill creep. It seems likely that such cases will also commonly involve aspects of cohesive granular flow, especially when clays are present. Thus, whether the nature of the time and temperature dependences will be similar to those in small-strain brittle creep is questionable. Probably a complex mixture of basic flow mechanisms will be involved, the characterization of which has not yet been adequately achieved. Therefore, we shall not attempt to define models for the time and temperature dependence of granular flow in these situations. For a selection of studies on the time and temperature dependent aspects of flow in fault gouge and similar materials, see Summers and Byerlee (1977), (Bombolakis et al. 1978), Wu (1978), (Moody and Hundley-Goff 1980), (Wang et al. 1980), (Logan et al. 1981), Shimamoto and Logan (1981a, b), Teufel (1981), Chu and Wang (1982), (Morrow et al. 1982), (Moore et al. 1986), Raleigh and Marone (1986) and Rutter et al. (1986).

Closely related to such behaviour are the time and temperature dependent aspects of friction. These also often involve gouge to a greater or less extent and have been considerably studied in connection with stick–slip movement between sliding surfaces and its application to earthquakes. The variations in frictional resistance with displacement and with velocity have been of particular interest and may eventually have some application in modelling strain rate and temperature dependence in friction-controlled granular flow. For a selection of friction studies, see Dieterich (1972, 1978, 1979, 1981), Stesky et al. (1974), Stesky (1978), Johnson (1981), Dieterich and Conrad (1984), Wecks and Tullis (1985), Lockner et al. (1986), Tullis and Weeks (1986), Rudnicki (1988), Tullis (1988) and Rundle (1989).

7.3 Granular Flow Controlled by Thermally-Activated Processes

7.3.1 Introduction: superplasticity

Following the trend in materials science, the term "superplasticity" is taken here to be primarily a phenomenological one rather than a term referring to a particular deformation mechanism (Gilotti and Hull 1990; Paterson 1990). The term was originally used in metallurgy to refer to the development in tensile tests of very large elongations, often of the order of ten-fold (natural strains of 2 or more), and it has been used in the ceramics literature in the same sense (for example, Chen and Xue 1990, referring to natural strains in extension of 0.5–1 in many cases). The essential properties that promote large ductility in high temperature extension tests are a high strain rate sensitivity of the flow stress, which inhibits necking (Hart 1967), and the absence of processes such as microcracking and cavitation that lead to premature fracture. In practice, two other groups of properties are also observed to be associated with this behaviour:

(1) The flow stress depends strongly on the grain size and the superplastic regime is limited to grain sizes below a certain level, commonly of the order of 1–10 μm under experimental conditions (hence the terms "fine-grain super-plasticity" and "grain-size-sensitive flow")

(2) The grain shape tends to remain approximately equant rather than reflect the macroscopic strain, with no strong development of crystallographic preferred orientation; but there is a marked tendency for grain growth.

These properties have been established by many observations on metallic and ceramic materials (Baudelet and Suery 1985; Chen and Xue 1990; Edington and Melton 1976; Padmanabhan and Davies 1980; Paton and Hamilton 1982; Stowell 1983).

The term "superplasticity" has also been applied in situations where considerable ductility and/or a marked decrease in flow stress are exhibited under conditions close to those at which a phase transformation occurs or under which large internal stresses are generated. Such behaviour has been referred to as "transformational" or "environmental" superplasticity (Edington and Melton 1976; Sammis and Dein 1974; Sherby and Wadsworth 1990). We shall not consider transformational superplasticity here.

In rock deformation studies, where the uniaxial tensile test is little used and predominantly compressive stress regimes are normally involved, it is useful to extend the usage of " superplasticity" to embrace the general case of flow with high strain rate sensitivity, under the presumption that, where this condition applies, similar physical processes are involved and a potential for high ductility exists. Thus, in the following sections, superplasticity is taken to be *defined* by a high strain rate sensitivity of the flow stress, as measured by $\frac{d\ln\sigma}{d\ln\dot{\varepsilon}} = \frac{1}{n}$, where σ is the stress, $\dot{\varepsilon}$ is the strain rate and n may typically have a value between 1 and 2 (Paterson 1990); note that Gilotti and Hull (1990) prefer a definition in terms of "continuous" deformation to very high strains, which is less restrictive in terms of strain rate sensitivity and of implications for mechanisms. Superplasticity in this sense has been observed experimentally in calcite, anhydrite, olivine and feldspar aggregates by Schmid (1976), (Schmid et al. 1977), Schwenn and Goetze (1978), (Mueller et al. 1981), Vaughan and Coe (1981), Brodie and Rutter (1985), Chopra (1986), Ji and Mainprice (1986), Karato et al. (1986), Rutter and Brodie (1988), Stretton and Olgaard (1997), Walker et al. (1990). It has also been frequently proposed as being geologically important, at least in the sense that the flow stress is grain-size-sensitive and therefore low in fine grained rocks (for example, Boullier and Guéguen (1975), Guéguen and Boullier (1976), Twiss (1976), White (1976), Etheridge and Wilkie (1979), Evans et al. 1980), Schmid (1983), Behrmann (1985) and (Gilotti and Hull 1990).

7.3.2 Mechanisms of Temperature-Sensitive Granular Flow

The view is taken here that mechanistic models for superplastic flow can be based appropriately on the concept of granular flow (Sect. 7.1.1),. In particular, the approach in terms of granular flow is preferred to that of atom transfer models such as Nabarro-Herring or Coble creep (Chap. 5) for the following reasons:

(1) The granular flow model readily allows very large strain to be achieved without concomitant change in grain shape, as is observed, and it gives ready explanation of observed effects such as grain-neighbour switching and cavitation. In the case of atom transfer models, an extraordinary amount of grain boundary migration would have to be invoked and rationalized in order to explain retention of equant grain shape while increasing or decreasing the number of grains in a cross-section.

(2) The stress–strain rate relations are often non-Newtonian (Chokshi and Langdon 1985), in contrast to the Newtonian Nabarro-Herring or Coble models, an observation more readily accommodated in granular flow models.

(3) When the shear strength of grain interfaces is low, for reason of high temperature, partial melting, or the presence of a wetting phase, sliding on the interfaces can be expected to be the primary response to applied shear stress, although the accommodation processes brought into play by compatibility requirements are likely to be rate controlling.

As for atom transfer models (Chap. 5), the accommodation processes that are necessary for maintaining strain compatibility must, however, be seen as an inseparable aspect of the deformation process. The distinction emphasized here is that the relative grain movements play the primary role, in the sense of defining the overall deformation, and that the accommodation processes are brought in only to the extent necessary to maintain compatibility, whereas the reverse of this hierarchy of roles was envisaged in the atom transfer models. The accommodation processes can be either of an atom transfer type (any of the processes in Chap. 5) or of a crystal-plastic type (Chap. 6).

In the spirit of this approach, we shall attempt first to develop a general deformation model that is analogous to the temperature insensitive granular flow models of Sect. 7.2 but with thermally-activated processes playing the essential roles of accommodation and rate control. In this context, we shall then review a number of the more specific models for superplastic flow that have been proposed from time to time.

As a preliminary, certain aspects of the physics of superplastic deformation need closer examination. In particular, the concept of sliding on a migrating grain boundary and the constraints on accommodation processes present conceptual difficulties. However, there is a paucity of microstructural evidence reflecting these physical processes. This lack may be due, in part, to the inadequacy of the conventional observational methods, which are more appropriate to intracrystalline plasticity, and, in part, to there being actually fewer microstructural features remaining because of an increased tendency for them to be annealed out at the higher temperatures and smaller grain sizes generally involved. Dislocation densities tend to be low and substructures absent, while grain boundaries tend to be straight or only gently curved (for example, Ball and Hutchinson 1969; Kaibyshev et al. 1978; Padmanabhan and Davies 1980, pp. 91–95; Parayil et al. 1990). However, much more microstructural study in association with experimental work is clearly needed.

It is widely observed that pronounced grain growth occurs during deformation in the superplastic regime unless there are factors such as the presence of second phases, to inhibit it (Clark and Alden 1973; Suery and Baudelet 1973, 1980; Watts and Stowell 1971; Wilkinson and Cáceres 1984). It seems therefore that within a given phase there is some sort of coupling between the sliding on grain boundaries and their migration. Possibly, during the relative movement of the contacting grains, there are increased opportunities for the boundary atoms belonging to one grain to enter low energy sites belonging to the other grain, thereby promoting the migration of the boundary. On such a view, there might exist independently a driving force for the migration, deriving for example from interfacial energy decrease, and the grain boundary sliding would simply serve to accelerate the kinetics of the migration. However, it is possible that sometimes a driving force for migration may derive from the deformation, if there is an increase in the possibility of grain boundary sliding through bringing adjacent boundaries into coplanar configuration.

As far as granular flow itself is concerned, the only kinematic quantity of importance is the instantaneous relative movement of the main part of one grain relative to the main part of its neighbour, as reflected in the relative motion of the centres of mass existing at the instant. However, because of the grain boundary migration, what constitutes the grain has to be redefined from instant to instant. This circumstance should not present a conceptual problem if the deformation is viewed as a summation of successive increments, each taking place at one such instant and involving the grains as they exist at that instant.

The actual movement pattern of the grains in superplastic flow is not well known although observations have confirmed neighbour exchange (Ashby and Verrall 1973; Duclos 2004; Rai and Grant 1983). By analogy with pure particulate flow (Sect. 7.2.2), it could be expected that there would be both substantial short-range coordination of relative grain displacements and considerable irregularity. Associated with this pattern, the stress distribution could be expected to be highly irregular on the grain scale. Some groups of grains would be subjected to higher local stress than others and there would be pronounced stress concentrations or diminutions where relative grain movements tend to produce interference or create voids. These stress heterogeneities would provide the driving forces for the accommodation processes, whether diffusional or crystal-plastic.

If the accommodation is by diffusional transfer of material, the principles of this have been set out in Chap. 5. If it is by crystal plasticity, the processes are discussed in Chap. 6. The principal problem in either case is to elucidate and quantify the actual grain-scale driving forces and the detailed geometries of the processes.

In polyphase materials there are further constraints. Thus, grain growth may be limited for various reasons: for example, for geometric or topological reasons, depending on the proportions of the phases; or for surface energy reasons if a finely dispersed phase is distributed along the grain boundaries of another phase. Where accommodation by diffusive mass transfer is involved, the chemical identity of the phases must be retained by appropriate constraints on the fluxes of the individual components. Further, there may be differences in the diffusivities of these components along interphase boundaries and along grain boundaries in the same phase, the former being sometimes higher than the latter.

7.3.3 A Semi-empirical Theoretical Approach

In order to circumvent the difficulty of giving specific quantitative descriptions of the complex patterns of relative grain movements, we assume initially that the integrated effect of the accommodation processes is to prevent the dilatancy that would otherwise occur if the body behaved as an aggregate of unyielding grains of the same fixed shapes but zero inter-granular cohesion, and we assume that this dilatancy can be estimated approximately from the observations on densely-packed particulate masses (Sect. 7.2.2) (Paterson 1995a, 1995b, 2001). The dilatancy is expressed as

$$\frac{d\varepsilon_v}{d\varepsilon} = \tan\psi \tag{7.7}$$

where ε_v is the volumetric strain, ε the linear strain, and ψ the so-called dilatancy angle (Paterson and Wong 2005, p. 252). It is implicit that the actual deformation by granular flow occurs at constant volume under a sufficient confining pressure.

On the dynamical side, we assume that the intrinsic resistance to sliding on the grain boundaries is negligibly small and that the rate of sliding is therefore determined by the rate of accommodation at the sites of interference. Support for the assumption of negligible shear strength of grain boundaries at high temperature comes from observation on bicrystals (Ashby 1972) and from experience of practical failures such as filaments of light bulbs where they are crossed by grain boundaries (Koref 1926). The concept of such a low-strength regime corresponds to the early notion of the "equi-cohesive temperature", above which grain boundaries become weak relative to the grains (Jeffries and Archer 1924; Rosenhain and Ewen 1912) although, in rocks, the weakening may be achieved by the agency of fluids as well as by temperature alone.

Under the first assumption above, we take the quantity $d\varepsilon_v$ in (7.7) to be the relative volume of material on average to be displaced in a grain in the accommodation process. This displacement, which involves a change of shape of the grain, can alternatively be expressed as a linear strain $d\varepsilon_a$ given by

$$d\varepsilon_v = C_1 d\varepsilon_a \tag{7.8}$$

where C_1 is a geometric constant of order unity. Thus, from (7.7) and (7.8), and changing to strain rates, we have

$$\dot{\varepsilon} = \frac{C_1}{\tan\psi}\dot{\varepsilon}_a \tag{7.9}$$

The local accommodation strain rate $\dot{\varepsilon}_a$ will be determined by the local stresses brought into play and by the accommodation mechanism.

The local stress $(\sigma_1 - \sigma_3)_a$ driving the accommodation strain rate $\dot{\varepsilon}_a$ will derive from the macroscopic applied stress $\sigma_1 - \sigma_3$, subject to a degree of stress concentration, so that

$$(\sigma_1 - \sigma_3)_a = C_2(\sigma_1 - \sigma_3) \tag{7.10}$$

where C_2 is the stress concentration factor, $C_2 > 1$. The relationship between $(\sigma_1 - \sigma_3)_a$ and $\dot{\varepsilon}_a$ will then be determined by the nature of the accommodation mechanism. Of the various possible cases, we select the following for consideration:

1. *Accommodation by diffusion along grain boundaries, diffusion controlled.* This accommodation mechanism corresponds to Coble creep (Sect. 5.4), the local accommodation strain rate $\dot{\varepsilon}_a$ being governed by

$$\dot{\varepsilon}_a = C_{CO} \frac{V_m D_{gb} \delta}{RT} \frac{(\sigma_1 - \sigma_3)_a}{(d')^3} \tag{7.11}$$

in the notation of Sect. 5.4, where d' is now the average distance between source and sink, related to the grain size d by $d' = C_3 d$; the constant C_3 will be substantially less than unity because the accommodation adjustments do not involve the whole extent of the grain (cf. Ashby and Verrall 1973). From (7.9), (7.10) and (7.11) we then obtain the macroscopic flow law

$$\dot{\varepsilon} = \frac{C_1 C_2}{C_3^3 \tan \psi} C_{CO} \frac{V_m D_{gb} \delta}{RT} \frac{(\sigma_1 - \sigma_3)}{d^3} \tag{7.12}$$

This result is similar to that of (Raj and Chyung 1981); Wakai (1994) has developed an alternative theory based on the notion of migrating steps on the grain boundary from which higher stress dependence is possible.

2. *Accommodation by solution transfer along grain boundaries, solute diffusion controlled.* In the notation of § 5.5 the accommodation strain rate is now controlled by

$$\dot{\varepsilon}_a = C_{FT} \frac{V_m^2 c D_f \delta}{RT} \frac{(\sigma_1 - \sigma_3)_a}{(d')^3} \tag{7.13}$$

where $d' = C_3 d$, $C_3 < 1$ as for (7.11). From (7.9), (7.10) and (7.13) we obtain

$$\dot{\varepsilon} = \frac{C_1 C_2}{C_3^3 \tan \psi} C_{FT} \frac{V_m^2 c D_f \delta}{RT} \frac{\sigma_1 - \sigma_3}{d^3} \tag{7.14}$$

Rutter (1983) has proposed a similar relationship but that of Pharr and Ashby (1983) has a more complicated stress dependence.

3. *Accommodation by solution transfer along grain boundaries, reaction controlled.* In the notation of Sect. 5.6 the accommodation strain rate will be

$$\dot{\varepsilon}_a = C_R \frac{V_m bk}{RT} \frac{(\sigma_1 \sigma_3)_a}{d'} \tag{7.15}$$

where again $d' = C_3 d$, $C_3 < 1$ as for (8.11), leading, through (8.9) and (8.10), to

$$\dot{\varepsilon} = \frac{C_1 C_2}{C_3 \tan \psi} C_R \frac{V_m bk}{RT} \frac{\sigma_1 - \sigma_3}{d} \tag{7.16}$$

4. *Accommodation by dislocation glide.* The absence of subgrain structure in superplastically deforming materials suggests that any dislocation sources will tend to be within the grain boundaries (the subgrain size being, in effect, greater than the grainsize). Therefore, the density of dislocation sources and hence the mobile dislocation density ρ may reasonably be assumed to be proportional to the grain boundary area or the square of the grain size d. The mobile dislocation density may, in addition, depend on the rate at which dislocations can be extracted from the grain boundaries, as from a potential trough; that is, at low stresses (Sect. 3.2.4, Eq. 3.12d with $F = \tau bl$), ρ may also be postulated to be proportional to the local stress $(\sigma_1 - \sigma_3)_a$, then the Orowan relation 6.4b leads to

$$\dot{\varepsilon}_a \propto \frac{(\sigma_1 - \sigma_3)_a}{d^2}(\sigma_1 - \sigma_3)_a$$

$$or \; \dot{\varepsilon}_a = C_4 \frac{(\sigma_1 - \sigma_3)_a^2}{d^2} \exp\left(\frac{-Q}{RT}\right) \tag{7.17}$$

where Q is the sum of the activation energies for dislocation nucleation and dislocation motion and C_4 is a constant of dimensions $m^6 \, N^{-2}$. If the motion were limited by the rate of climb of the dislocations in the receiving grain boundary, dissipating pile-ups there, then the second part of the activation energy could be equal to that for grain boundary diffusion, making Q somewhat greater than for grain boundary diffusion. However, there are obviously other possibilities and no firm predictions about the level of the activation energy can be made at this stage. Substituting (7.17) and (7.10) in (7.9) leads to the macroscopic flow law

$$\dot{\varepsilon} = \frac{C_1 C_2}{\tan \psi} C_4 \exp\left(\frac{-Q}{RT}\right) \frac{(\sigma_1 - \sigma_3)^2}{d^2} \tag{7.18}$$

The exact value of the stress exponent in (7.18) is somewhat problematical because of the tenuous nature of the assumptions made above about the stress dependences of dislocation nucleation rate and velocity. However, the interesting point is that a stress exponent greater than unity can be rationalized, since observed values are often around 1.5–2.5. The stress exponent might also vary with strain rate or temperature if the nature of the factors limiting the dislocation velocity were to vary.

The above formulae (7.12)–(7.18) should be regarded as being of a generic nature, illustrating the general character of possible granular flow mechanisms rather than displaying all the properties of deformation in specific situations. However, some semi-quantitative conclusions can be arrived at through consideration of possible values of the constants C_1 to C_3 and $\tan\psi$, as follows:

C_1 The volumetric strain ε_v required for accommodation is most simply achieved by the displacement of a layer of relative thickness ε_a from one side of the grain to the other, in which case, from 7.8, $C_1 = 1$

C_2 This can be expected to be less than typical elastic stress concentration factors because of the non-elastic deformation involved: We make the subjective choice of $C_1 \approx 2$

C_3 In the cases 1, 2 and 3, it may suffice to postulate the diffusive transfer of material from one grain interface to an adjacent one; then d' in (8.11), (8.13) and (8.15) will be approximately equal to the diameter of a grain interface, that is, for a typical 12-faced grain, of the order of $\left(1/\sqrt{6}\right)d$, or $C_3 \approx 1/\sqrt{6}$ and $C_3^3 \approx \frac{1}{15}$

$\tan\psi$ The dilatancy that would occur in the granular flow of a low-porosity polycrystalline body in the absence of accommodation processes could be expected to be similar to that in dense sand, for which a typical value of $\tan\psi$ is around 0.7 (see Fig. 6.20c in Wood 1990). An alternative estimate is obtained by supposing that a transition from close packing of equal spheres (porosity ≈ 0.26) to loose random packing {porosity ≈ 0.38; Cargill 1984} requires a strain of about half that for complete neighbour exchange (Sect. 7.1.2), that is, about 0.2, leading to a value of $\tan\psi$ of $(0.38-0.26)/0.2 = 0.6$. We therefore choose $\tan\psi = 2/3$ as a typical value

Using the above estimates, we therefore obtain values for the first terms in (7.12), (7.14) and (7.16) of:

$$\frac{C_1 C_2}{C_3^3 \tan\psi} = 45 \text{ and } \frac{C_1 C_2}{C_3 \tan\psi} = 7$$

The strain rate predicted for the grain-boundary-diffusion-accommodated granular flow model (7.12) is thus around 45 times that for the classical Coble creep model (5–11), a large factor in favour of the granular flow model, as was first noted by Ashby and Verrall (1973). A granular flow view is similarly favoured in the case of diffusion-controlled solution transfer accommodation (7.14), and, to a lesser extent, in the case of the reaction-controlled accommodation (7.16).

There is no obviously simple way of estimating the quantity C_4 in (7.18) and so of assessing, at this generic level, the relative rates of granular flows with accommodation mechanisms involving atom transfer and dislocation glide, respectively. However, both types of model need to be borne in mind and their distinction will require extensive microstructural and rheological study, especially in respect to stress and grain size exponents. An example of such an analysis is that (Schmid et al. 1987) who establish a role for f-slip as an accommodation mechanism in the superplastic deformation of fine-grained calcite aggregates.

References

Allersma HGB (1982) Photo-elastic stress analysis and strains in simple shear. In: Vermeer PA, Luger HJ (eds) Deformations and failure of granular materials, IUTAM conference, Delft 1982, Rotterdam, A.A. Balkeema, pp 345–353

Anand L (1983) Plane deformations of ideal granular materials. J Mech Phys Solids 31:105–122

Anderson OL, Grew PC (1977) Stress corrosion theory of crack propagation with applications to geophysics. Rev Geophys Space Phys 15:77–104

Ashby MF (1972) Boundary defects and atomistic aspects of boundary sliding and diffusional creep. Surface Sci 31:498–542

Ashby MF, Verrall RA (1973) Diffusion-accommodated flow and superplasticity. Acta Metall 21:149–163

Atkinson BK (1984) Subcritical crack growth in geological materials. J Geophys Res 89:4077–4114

Atkinson BK, Meredith PG (1987) The theory of subcritical crack growth with applications to minerals and rocks. In: Atkinson BK (ed) Fracture mechanics of rock. Academic, London, pp 111–166

Ball A, Hutchinson MM (1969) Superplasticity in the aluminium-zinc eutectoid. Met Sci J 3:1–7

Barden L (1971) A quantitative treatment of the deformation behaviour of granular material in terms of basic particulate mechanics. In: Te-eni M (ed) Structure, solid mechanics and engineering design, London, Wiley, pp 599–612

Baudelet B, Suery M (eds) (1985) Superplasticity. Editions CNRS, Paris

Behrmann JH (1985) Crystal plasticity and superplasticity in quartzite: a natural example. Tectonophysics 115:101–129

Biegel RL, Sammis CG (1988) A ligand model for frictional behavior of a gouge layer. EOS Trans AGU 69:1463 (abs)

Blanpied ML, Tullis TE, Weeks JD (1988) Textural and mechanical evolution of granite gouge in high-displacement sliding experiments. EOS Trans AGU 69:1463

Bombolakis EG, Hepburn JC, Roy DC (1978) Fault creep and stress drops in saturated silt-clay gouge. J Geophys Res 83:818–829

Borg I, Friedman M, Handin J, Higgs DV (1960) Experimental deformation of St. Peter sand: a study of cataclastic flow. Geol Soc Am Bull 79:133–191

Boullier AM, Gueguen Y (1975) SP-mylonites: origin of some mylonites by superplastic flow. Contrib Mineral Petrol 50:93–104

Bowden FP, Tabor D (1950) The friction and lubrication of solids, Part I. Clarendon Press, Oxford, 337 pp

Bowden FP, Tabor D (1964) The friction and lubrication of solids. Part II. Clarendon Press, Oxford, 544 pp

Brewer R (1964) Fabric and mineral analysis of soils. Wiley, New York, pp 129–158

Brodie KH, Rutter EH (1985) On the relationship between deformation and metamorphism, with special reference to the behaviour of basic rocks. In: Thompson AB, Rubie DC (eds) Metamorphic reactions. Kinetics, textures and deformation. Springer, New York, pp 138–179

Byerlee JD (1967) Theory of friction based on brittle fracture. J Appl Phys 38:2928–2934

Cambou B (1993) From global to local variables in granular media, Powders and Grains 93. In: Thornton C (ed) 2nd international conference on micromechanics of granular media, A A Balkema, Birmingham, England

Cambou B (1982) Orientational distributions of contact forces as memory parameters in a granular material. In: IUTAM symposium: deformation and failure of granular materials, Delft, Rotterdam, A A Balkema, pp 3-12

Carter NL, Kirby SH (1978) Transient creep and semibrittle behavior of crystalline rocks. Pure Appl Geophys 116:807–839

Carter NL, Anderson DA, Hansen FD, Kranz RL (1981) Creep and creep rupture of granitic rocks. In: Carter NL, Friedman M, Logan JM, Stearns DW (eds) Mechanical behavior of crustal rocks. The Handin volume. American Geophysical Union, Washington, DC, pp 61–82

Chaplin TK (1971) Shearing dilatancy and crushing phenomena in granular materials. In: Structure, solid mechanics and engineering design. Proceedings of Southampton 1969 Civil Engineering Materials Conference, Wiley-Interscience, London, pp 951–962 (discussion 1163-1165)

Chen I-W, Xue LA (1990) Development of superplastic structural ceramics. J Am Ceram Soc 73:2585–2609

Chokshi AH, Langdon TG (1985) The role of interfaces in superplastic deformation. In: Baudelet B, Suery M (eds) Superplasticity. CNRS Editions, Paris, pp 2.1-2.15

Chopra PN (1986) The plasticity of some fine-grained aggregates of olivine at high pressure and temperature. In: Hobbs BE, Heard HC (eds) Mineral and rock deformation: laboratory studies. American Geophysical Union, Washington, pp 25–33

Christoffersen J, Mehrabadi MM, Nemat-Nasser S (1981) A micromechanical description of granular material behaviour. Trans ASME 81(48):339–344

Chu C-L, Wang C-Y (1982) Time-dependent volumetric consitutive relation for fault gouge and clay at high pressure. In: 23rd US symposium on rock mechanics, pp 270–278

Clark MA, Alden TH (1973) Deformation enhanced grain growth in a superplastic Sn-1 % Bi alloy. Acta Metall 21:1195–1206

Costin LS (1985) Damage mechanics in the post-failure regime. Mech Mat 4:149–160

Costin LS (1987) Time-dependent deformation and failure. In: Atkinson BK (ed) Fracture mechanics of rock. Academic, London, pp 167–215

Cruden DM (1970) A theory of brittle creep in rock under uniaxial compression. J Geophys Res 75:3431–3442

Cruden DM (1974) The static fatigue of brittle rock under uniaxial compression. Int J Rock Mech Min Sci 11:67–73

Cundall PA, Strack ODL (1979a) The development of constitutive laws for soil using the distinct element method. In: Numerical methods in geomechanics Aachen 1979. Proceedings of 3rd international conference on numerical methods in geomechanics, Rotterdam, A A Balkema, pp 289–298

Cundall PA, Strack ODL (1979b) A discrete numerical model for granular materials. Géotechnique 29:47–65

Cundall PA, Drescher A, Strack ODL (1982) Numerical experiments on granular assemblies: measurements and observations. In: Deformation and failure of granular material, IUTAM conference Delft 1982, Rotterdam, A A Balkema, pp 355–370

Cundall PA, Strack ODL (1983) Modeling of microscopic mechanisms in granular materials. In: Jenkins JT, Satake M (eds) Mechanics of granular materials—new models and constitutive relations. Elsevier, Amsterdam, pp 137–149

Cundall PA (1986) Distinct element methods of rock and soil structure. In: Cundall PA, Strack ODL (eds) Analytical and computational methods in engineering rock mechanics. Allen and Unwin, London, pp 129–163

Cundall PA (1988a) Computer simulations of dense sphere assemblies. In: Micromechanics of granular materials; Proceedings of US/Japan seminar, Sendai-Zao, Japan, Oct 26–30 1987 Elsevier, Amsterdam, pp 113–123

Cundall PA (1988b) Formulation of a three-dimensional distinct element model. Part 1. A scheme to detect and represent contacts in a system composed of many polyhedral blocks. Int J Rock Mech Min Sci 25:107–116

Cundall PA (1989) Numerical experiments on localization in frictional materials. Ingenieur-Archiv 59:148–159

Curie P (1894) Sur la symétrie dans les phénomènes physiques, symétrie d'un champ electrique et d'un champ magnetique. J de Phys 3:393–415

de Josselin de Jong G (1959) Statics and Kinematics of the Failable zone of a Granular Material (thesis), Delft, Uitgeverij Waltman

de Josselin de Jong G (1971) The double sliding, free rotating model for granular assemblies, Géotechnique 21:155–163

de Josselin de Jong, G, 1977, Mathematical elaboration of the double sliding, free rotating model, Archiv Mech, 29, 561-591

Deresiewicz H (1958) Mechanics of granular matter. In: Advances in applied m,echanics. Academic, London, pp 233–306

Dieterich JH (1972) Time-dependent friction in rocks. J Geophys Res 77:3690–3697

Dieterich JH (1978) Time-dependent friction and the mechanism of stick-slip. Pure Appl Geophys 116:790–806

Dieterich JH (1979) Modeling of rock friction. 1. Experimental results and constitutive equations. J Geophys Res 84:2161–2168

Dieterich JH (1981) Constitutive properties of faults with simulated gouge. In: Carter NL, Friedman M, Logan JM, Stearns DW (eds) Mechanical behavior of crustal rocks The Handin volume. American Geophysical Union, Washington, pp 103–120

Dieterich JH, Conrad G (1984) Effect of humidity on time- and velocity-dependent friction in rocks. J Geophys Res 89:4196–4202

Drescher A, de Josselin de Jong G (1972) Photoelastic verification of a mechanical model for the flow of a granular material. J Mech Phys Solids 20: 337–351

Drescher A (1976) An experimental investigation of flow rules for granular materials by means of optically sensitive glass particles. Géotechnique 26:591–601

Duclos R (2004) Direct observation of grain rearrangement during superplastic creep of a fine-grained zirconia. J European Ceramic Soc 24:3103–3110

Edington JW, Melton KN (1976) Superplasticity. Progress in materials science. Pergamon Press, Oxford, pp 61–170

Etheridge MA, Wilkie JG (1979) Grain size reduction, grain boundary sliding and the flow strength of mylonites. Tectonophysics 58:159–178

Evans B, Rowan M, Brace WF (1980) Grain-size sensitive deformation of a stretched conglomerate from Plymouth. Vermont, J Structural Geol 2:411–424

Feda J (1982) Mechanics of particulate materials: the principles. Elsevier, Amsterdam, 447 pp

Field WG (1963) Towards the statistical definition of a granular mass. In: Proceedings 4th Auststalia and New Zealand conference on soil mechanics, 143–148

Gilotti JA, Hull JM (1990) Phenomenological superplasticity in rocks. In: Knipe RJ, Rutter EH (eds) Deformation mechanisms, rheology and tectonics. The Geological Society, London, pp 229–240

Guéguen Y, Boullier AM (1976) Evidence of superplasticity in mantle peridotites. In: Stens, RGJ (ed) The physics and chemistry of minerals and rocks. NATO Institute, Newcastle upon Tyne, April 1974, Wiley, London, pp 19-33

Hadizadeh J, Rutter EH (1982) Experimental study of cataclastic deformation in quartzite. In: Issues in rock mechanics. 23rd Symposium on Rock mechanics. AIME, New York, pp 372-379

Hadizadeh J, Rutter EH (1983) The low temperature brittle-ductile transition in quartzite and the occurrence of cataclastic flow in nature. Geol Rundschau 72:493–509

Hamilton EL (1976) Variations of density and porosity with depth in deep-sea sediment. J Sediment Petrol 46:280–300

Handin J, Hager RV (1957) Experimental deformation of sedimentary rocks under confining pressure: tests at room temperature on dry samples. Am Assoc Petrol Geol Bull 41:1–50

Handin J, Hager RV (1958) Experimental deformation of sedimentary rocks under confining pressure: tests at high temperature. Am Soc Petrol Geol Bull 42:2892–2934

Handin J, Hager RV, Friedman M, Feather JN (1963) Experimental deformation of sedimentary rocks under confining pressure: pore pressure effects. Bull. Am. Assoc. Petrol. Geol 47:717–755

Hart EW (1967) The theory of the tensile test. Acta Metall 15:351–355

Hart RJ, Cundall PA, Lemos J (1988) Formulation of a three-dimensional distinct element model. Part II. Mechanical calculations for motion and interactions in a system composed of many polyhedral blocks. Int J Rock Mech Min Sci 25:117–125

Hasselman DPH, Venkataswaran A (1983) Role of cracks in the creep deformation of brittle polycrystalline ceramics. J Mater Sci 18:161–172

Hettler A, Vardoulakis I (1984) Behaviour of dry sand tested in a large triaxial apparatus. Géotechnique 34:183–198

Hirth G, Tullis J (1988) The transient nature of cataclastic flow in porous quartzite. EOS Trans AGU 68:1464 (abs)

Horne MR (1965) The behaviour of an assembly of rotund, rigid, cohesionless particles, I, II. Proc Roy Soc (London) A286, 62–78;79–97

Horne MR (1969) The behaviour of an assembly of rotund, rigid, cohesionless particles, III. Proc R Soc (London) A310:21–34

Israelachvili JN (1992) Intermolecular and Surface Forces. Academic, San Diego, 445 pp

Jaeger JC (1959) The frictional properties of joints in rock. Geofiz pur e appl 43:148–158

Jaunzemis W (1967) Continuum mechanics. Macmillan, New York, 604 pp

Jeffries Z, Archer RS (1924) The science of metals. McGraw-Hill, New York, 460 pp

Ji S, Mainprice D (1986) Transition from power law to Newtonian creep in experimentally deformed dry albite rock (abstract). Trans Am Geophys Union 67:1235

Johnson TL (1981) Time-dependent friction of granite: implications for precursory slip on faults. J Geophys Res 86:6017–6028

Kaibyshev OA, Kazachkov IV, Salikhov SY (1978) The influence of texture on superplasticity of the Zn-22 %Al alloy. Acta Metall 26:1877–1886

Karato S, Paterson MS, Fitz Gerald JD (1986) Rheology of synthetic olivine aggregates: effect of grain size and of water. J Geophys Res 91:8151–8176

Konishi J, Oda M, Nemat-Nasser S (1982) Inherent anisotropy and shear strength of assembly of oval cross-sectional rods. In: Conference on deformation and failure of granular materials, IUTAM conference, Delft 1982, Rotterdam, A A Balkema, pp 403–412

Koref F (1926) Ueber den Einfluss der Kristallstruktur auf die Formbeständigkeit von Wolfram-Leuchtkörpern, Zeitschrift für technische Physik 7:544–547

Kranz RL, Scholz CH (1977) Critical dilatant volume of rocks at the onset of creep. J Geophys Res 82:4893–4898

Kranz RL (1979) Crack growth and development during creep in barre granite. Int J Rock Mech Min Sci 16:23–36

Kranz RL (1980) The effects of confining pressure and stress difference on static fatigue of granite. J Geophys Res 85:1854–1866

Kranz RL (1983) Microcracks in rocks: a review. Tectonophysics 100:449–480

Lance GL, Nemat-Nasser S (1986) Slip-induced plastic flow of geomaterials and crystals. Mech Materials 5:1–11

Lockner D, Byerlee JD (1977a) Acoustic emission and creep in rock at high confining pressure and differential stress. Bull Seismol Soc Am 67:247–258

Lockner D, Byerlee JD (1977b) Acoustic emission and fault formation in rocks. In First conference on acoustic emission/microseismic activity in geologic structures and materials, Penn. State Univ., June 1975, Clausthal, Trans Tech Publ, pp 99–107

Lockner DA, Byerlee JD (1980) Development of fracture planes during creep in granite. In: Proceedings of 2nd conference on AE/microseismic activity in geological structures, and materials, Clausthal, Trans Tech Pub, pp 11–25

Lockner DA, Summers R, Byerlee JD (1986) Effects of temperature and sliding rate on frictional strength of granite. Pure Appl Geophys 124:445–469

Logan JM, Higgs NG, Friedman M (1981) Laboratory studies on natural gouge from the U. S. Geological survey Dry Lake Valley No 1 well, San Andreas fault zone. In: Carter NL et al (eds) Mechanical behavior of crustal rocks. The Handin volume. American Geophysical Union, Washington, DC, pp 121–134

Mandel J (1947) Sur les lignes de glissement at le calcul des déplacements dans la déformation plastique. Compt Rend Acad Sci (Paris) 225:1272–1273

Mandel J (1966) Conditions de stabilité et postulat de Drucker. In: Rheology and soil mechanics, IUTAM symposium, Grenoble, April 1-8, 1964, Springer, Berlin, pp 58–68

Mandl G, Fernández Luque R (1970) Fully developed plastic shear flow of granular materials. Géotechnique 20:277–307

Martin RJ (1972) Time-dependent crack growth in quartz and its application to the creep of rocks. J Geophys Res 77:1406–1419

Mehrabadi MM, Cowin SC (1978) Initial planar deformation of granular materials. J Mech Phys Solids 26:269–284

Mehrabadi MM, Cowin SC (1980) Pre-failure and post-failure soil lasticity models. J Engng Mech Div, Proc Am Soc Civ. S E, 106 (EM5):991–1003

Mehrabadi MM, Nemat-Nasser S (1983) Stress, dilatancy and fabric in granular materials. Mech Materials 2:155–161

Mogami T (1965) A statistical approach to the mechanics of granular materials. Soils Found 5:26–36

Mogami T, Imai G (1967) On the failure of the granular material. Soils Found 7:1–19

Mogami T (1969) Mechanics of granular material as a particulated mass. In: Proceedings of 7th international conference on soil mechanics and foundation engineering, vol 1. Mexico, Sociedad. Mexicana de Mecánica de Suelos Suelos, pp 281–285

Mogami T, Imai G (1969) Influence of grain-to-grain friction an shear phenomena of granular material. Soils Found 9:1–15

Molnar P (1983) Average regional strain due to slip on numerous faults of different orientations. J Geoph Res 88:6430–6432

Moody JB, Hundley-Goff EM (1980) Microscopic characteristics of orthoquartzite from sliding friction. II Gouge, Tectonophysics 62:301–319

Moore DE, Summers R, Byerlee JD (1986) The effects of sliding velocity on the frictional and physical properties of heated fault gouge. Pure Appl Geophys 124:31–52

Morrow CA, Shi LQ, Byerlee JD (1982) Strain hardening and strength of clay-rich fault gouges. J Geophys Res 87:6771–6780

Mueller WH, Schmid SM, Briegel U (1981) Deformation experiments on anhydrite rocks of different grain-sizes: rheology and microfabric. Tectonophysics 78:527–543

Myer LR, Kemeny JM, Zheng Z, Suarez R, Ewy RT, Cook NGW (1992) Extensile cracking in porous rock under differential compressive stress. Appl Mech Rev 45:263–280

Mühlhaus H-B, Vardoulakis I (1987) The thickness of shear bands in granular materials. Géotechnique 37:271–283

Nemat-Nasser S (1982) Fabric and its influence on mechanical behavior of granular materials. In: Deformation and failure of granular material, IUTAM conference Delft 1982, Rotterdam, A A Balkema, pp 37–42

Nemat-Nasser S (1983) On the finite plastic flow of crystalline solidis and geomaterials. Appl Mech 50:1114–1126

Nemat-Nasser S (1986) Generalization of the Mandel-Spencer double-slip model. In: Gittus J, Zarka J, Nemat-Nasser S (eds) Large deformations of solids; physical basis and modelling. Elsevier, London, pp 269–282

Oda M (1972a) Initial fabrics and their relations to mechanical properties of granular material. Soils Found 12:17–36

Oda M (1972b) Deformation mechanism of sand in triaxial compression tests. Soils Found 12:45–63

Oda M (1972c) The mechanism of fabric changes during compressional deformation of sand. Soils Found 12:1–18

Oda M (1974) A mechanical amd statistical model of granular material. Soils Found 14:13–27

Oda M, Konishi J (1974a) Microscopic deformation mechanism of granular material in simple shear. Soils Found 14:25–38

Oda M, Konishi J (1974b) Rotation of principal axes in granular material during simple shear. Soils Found 14:39

Oda M (1982) Fabric tensor for discontinuous geologic materials. Soils Found 22:96–108

Ohnaka M (1983) Acoustic emission during creep of brittle rock. Int J Rock Mech Min Sci 20:121–134

Ortiz M (1985) A constitutive theory for the inelastic behaviour of concrete. Mech Mat 4:67

Padmanabhan KA, Davies GJ (1980) Superplasticity. Springer, Berlin, 312 pp

Parayil TR, Dunlop GL, Howell PR (1990) The role of dislocations in the superplastic deformation of a duplex stainless steel. In: Superplasticity in metals, ceramics, and intermetallics. Materials Research Society, Pittsburgh, pp 93–98

Parkin AK (1965) On the strength of packed spheres. J Australian Math Soc 5:443–452

Paterson MS, Weiss LE (1961) Symmetry concepts in the structural analysis of deformed rocks. Geol Soc Am Bull 72:841–882

Paterson MS, (1990) Superplasticity in geological materials. In: Superplasticity in metals, ceramics, and intermetallics. Materials Research Society, Pittsburgh, pp 303-312

Paterson, M S, 1995a, A granular flow approach to fine-grain superplasticity. In: Plastic deformation of ceramics. Plenum Press, New York, pp 279–283

Paterson MS (1995b) A theory of granular flow accommodated by material transfer via an intergranular fluid. Tectonophysics 245:135–151

Paterson MS (2001) A granular flow theory for the deformation of partially molten rock. Tectonophysics 335:51–61

Paterson MS, Wong T-f (2005) Experimental rock deformation—the brittle field, 2nd edn. Springer, Berlin, 347 pp

Paton NE, Hamilton C H (eds) (1982) Superplastic forming of structural alloys. In: The minerals, metals and materials society, Symposium held in San Diego, California, June 21-24 1982, Warrendale, 420 pp

Pharr GM, Ashby MF (1983) On creep enhanced by a liquid phase. Acta Metall 31:129–138

Price NJ (1964) A study of the time-strain behaviour of coal-measure rocks. Int J Rock Mech Min Sci 1:277–303

Proctor DC (1974) An upper bound for phi in the stress-dilatancy equation. Géotechnique 24:269–288

Proctor DC, Barton RR (1974) Measurements of the angle of interparticle friction. Géotechnique 24:581–604

Rai G, Grant NJ (1983) Obervations of grain boundary sliding during superplastic deformation. Met Trans A 14:1451–1458

Raj R, Chyung CK (1981) Solution-precipitation creep in glass ceramics. Acta Metall 29:159–166

Raleigh B, Marone C (1986) Dilatancy of quartz gouge in pure shear. In: Hobbs BE, Heard HE (eds) Mineral and rock deformation: laboratory studies. The Paterson volume. American Geophysical Union, Washington, pp 1-10

Rennie BC (1959) On the strength of sand. J Australian Math Soc 1:71–79

Robertson EC (1964) Viscoelasticity of rocks. In: Judd WR (ed) State of stress in the earth's crust. Elsevier, New York, pp 181–224

Rosenhain W, Ewen D (1912) Intercrystalline cohesion in metals. J Inst Metals 8:149–173

Rowe PW (1962) The stress-dilatancy relation for static equilibrium of an assembly of particles in contact. Proc Roy Soc Lond A269:500–527

Rowe PW (1963) Stress-dilatancy, earth pressure, and slopes. J Soil Mech Fdns Div Amer Soc Civ Engrs 89:37–61

Rowe PW (1964) Reply to discussion on stress dilatancy, earth pressure and slopes. J Soil Mech Fdns Div Amer Soc Civ Engrs 90:145–180

Rowe PW (1972) Theoretical meaning and observed values of deformation parameters for soil. In: Roscoe memorial symposium on stress-strain behaviour of soils. Henley-on-Thames, G T Foulis and Co, pp 143–194, 277–281

Rudnicki JW, Rice JR (1975) Conditions for the localization of deformation in pressure-sensitive dilatant materials. J Mech Phys Solids 23:371–394

Rudnicki JW (1988) Physical models of earthquake instability and precursory processes. Pure Appl Geophys 126:531–554

Rundle JB (1989) A physical model for earthquakes. 3. Thermodynamical approach and its relation to nonclassical theories of nucleation. J Geophys Res 94:2839–2855

Rutter EH (1983) Pressure solution in nature, theory and experiment. J Geol Soc London 140:725–740

Rutter EH, Maddock RH, Hall SH, White SH (1986) Comparative microstructures of naturally and experimentally produced clay-bearing fault gouges. Pure Appl Geophys 124:3–30

Rutter EH, Brodie KH (1988) Experimental "syntectonic" dehydration of serpentinite under controlled pore water pressure. J Geophys Res 93:4907–4932

Sammis CG, Dein JL (1974) On the possibility of transformational superplasticity on the earth's mantle. J Geophys Res 79:2961–2965

Sammis CG, King G, Biegel RL (1987) The kinematics of gouge deformation. Pure Appl Geophys 125:777–812

Sasajima S, Ito H (1980) Long-term creep experiment of rock with small deviator of stress under high confining pressure and temperature. Tectonophysics 68:183–198

Satake M (1982) Fabric tensor in granular materials. In: IUTAM conference on deformation and failure of granular materials, Delft 1982, Rotterdam, A A Balkema, pp 63–68

Schmid SM (1976) Rheological evidence for changes in the deformation mechanism of Solenhofen limestone towards low stresses. Tectonophysics 31:T21–T28

Schmid SM, Boland JN, Paterson MS (1977) Superplastic flow in fine grained limestone. Tectonophysics 43:257–291

Schmid SM (1983) Microfabric studies as indicators of deformation mechanisms and flow laws operative in mountain building. In: Hsu KJ (ed) Mountain building processes. Academic, London, pp 95–110

Schmid SM, Panozzo R, Bauer S (1987) Simple shear experiments on calcite rocks: rheology and microfabric. J Structural Geol 9:747–778

Scholz CH (1968) Mechanism of creep in brittle rock. J Geophys Res 73:3295–3302

Scholz CH (1972) Static fatigue of quartz. J Geophys Res 77:2104–2114

Schwenn MB, Goetze C (1978) Creep of olivine during hot pressing. Tectonophysics 48:41–60

Segall P (1984) Rate-dependent extensional deformation resulting from crack growth in rock. J Geophys Res 89:4185–4195

Sherby OD, Wadsworth J (1990) Observations on historical and contemporary developments in superplasticity. In: Superplasticity in metals, ceramics, and intermetallics. Materials Research Society, Pttsburgh, pp 3–14

Shimamoto T, Logan JM (1981a) Effects of simulated fault gouge on the sliding behavior of Tennessee sandstone. Tectonophysics 75:243–255

Shimamoto T, Logan JM (1981b) Effects of simulated fault gouge on the sliding behavior of Tennessee sandstone: nonclay gouges. J Geophys Res 86:2902–2914

Spencer AJM (1964) A theory of the kinematics of ideal soils under plane strain conditions. J Mech Phys Solids 12:337–351

Spencer AJM (1982) Deformation of ideally granular materials. In: Hopkins HG, Sewell MJ (eds) Mechanics of solids: Rodney Hill 60th anniversary volume. Pergamon Press, Oxford, pp 607-652

Stesky RM, Brace WF, Riley DK, Robin P-YF (1974) Friction in faulted rock at high temperature and pressure. Tectonophysics 23:177–203

Stesky RM (1978) Mechanisms of high temperature frictional sliding in Westerly granite. Can J Earth Sci 15:361–375

Stowell MJ (1983) Design of superplastic alloys. In: Proceedings on 4th risø international symposium on metallurgy and materials science, Roskilde, Denmark, Risø National Lab, pp 119–129

Stretton I, Olgaard DL (1997) A transition in deformation mechanism through dynamic recrystallization — evidence from high strain, high temperature torsion experiments. EOS Trans AGU 78:F723

Suery M, Baudelet B (1973) Flow stress and microstructure in superplastic 60/40 brass. J Mater Sci 8:363–369

Suery M, Baudelet B (1980) Hydrodynamical behaviour of a two-phase superplastic alloy: alpha/beta brass. Phil Mag 41:41–64

Summers R, Byerlee J (1977) A note on the effect of fault gouge composition on the stability of frictional sliding. Int J Rock Mech Min Sci 14:155–160

Swanson PL (1984) Subcritical crack growth and other time- and environmental-dependent behaviour in crustal rocks. J Geophys Res 89:4137–4152

Teufel LW (1981) Pore volume changes during frictional sliding of simulated faults. In: Carter NL, Friedman M, Logan JM, Stearns DW (eds) Mechanical behavior of crustal rocks. The Handin volume, American Geophysical Union, Washington, pp 135–145

Thornton C (1979) The conditions of failure of a face-centred cubic array of uniform rigid spheres. Géotechnique 29:441–459

Thornton C, Barnes DJ (1982) On the mechanics of granular material. In: IUTAM Conference on deformation and failure of granular materials 1982 , Rotterdam, A A Balkema, pp 69–77

Thornton C, Barnes DJ (1986a) Computer simulated deformation of compact granular assemblies. Acta Mech 64:45–61

Thornton C, Barnes DJ (1986b) Evolution of stress and structure in particulate material. In: Tooth AS, Spence J (eds) Applied solid mechanics, vol 1. Elsevier Applied Science, London, pp 191–204

Thurston CW, Deresiewicz H (1959) Analysis of a compression test of a model of a granular material. Trans ASME 81(26):251–258

Trollope DH (1968) The mechanics of discontinua or clastic mechanics in rock problems. In: Stagg KG, Zienkiewicz OC (eds) Rock mechanics in engineering practice. Wiley, London, pp 275–320

Tullis TE, Weeks JD (1986) Constitutive behavior and stability of frictional sliding of granite. Pure and Appl Geophys 124:383–414

Tullis TE (1988) Rock friction constitutive behaviour from laboratory experiments and its implications for an earhtquake prediciton field monitoring program. Pure Appl Geophys 126:555–588

Turcotte DL (1986) Fractals and fragmentation. J Geophys Res 91:1921–1926

Twiss RD (1976) Structural superplastic creep and linear viscosity in the earth's mantle. Earth Planet Sci Lett 33:86–100

Vaišnys JR, Pilbeam CC (1975) Mechanical properties of granular media. Annual Rev Earth Planet Sci 3:343–360

Vaughan PJ, Coe RS (1981) Creep mechanism in Mg_2GeO_4: effects of a phase transition. J Geophys Res 86:389–404

Vesic AS, Clough GW (1968) Behaviour of granular materials under high stress, J. Soil Mech. Found. Div. Proc. Am. Soc. Civil Eng 94:661–688

Wakai F (1994) Step model of solution-precipitation creep. Acta Metall Mater 42:1163–1172

Walker AN, Rutter EH, Brodie KH (1990) Experimental study of grain-size sensitive flow of synthetic, hot-pressed calcite rocks. In: Knipe RJ, Rutter EH (eds) Deformation mechanisms, rheology and tectonics. The Geological Society, London, pp 259–284

Wang C-Y, Mao N-H, Wu FT (1980) Mechanical properties of clays at high pressure. J Geophys Res 85:1462–1468

Watts BM, Stowell MJ (1971) The variation in flow stress and microstructure during superplastic deformation of the Al-Cu eutectic. J Mater Sci 6:228–237

Weeks JD, Tullis TE (1985) Frictional sliding of dolomite: a variation in constitutive behavior. J Geophys Res 90:7821–7826

White SH (1976) The effects of strain on the microstructures, fabrics and deformation mechanisms in quartzites. Phil Trans Roy Soc London A238:69–86

Wilkinson DS, Cáceres CH (1984) On the mechanism of strain-enhanced grain growth during superplastic deformation. Acta Metall 32:1335–1345

Wood DM (1990) Soil Behaviour and Critical State Soil Mechanics. Cambridge University Press, Cambridge, 462 pp

Wu FT, Thomsen L (1975) Microfracturing and deformation of Westerly granite under creep condition. Int J Rock Mech Min Sci 12:167–173

Wu FT (1978) Mineralogy and physical nature of clay gouge. Pure Appl Geophys 116:681–689

Yanagidani T, Ehara S, Nishizawa O, Kusunose K, Terada M (1985) Localization of dilatancy in Ohshima granite under constant uniaxial stress. J Geophys Res 90:6840–6858

Index

A

Activation area, 81, 136, 137, 152, 154, 164
Activation volume, 81, 193–197
Athermal creep, 163–165
Athermal dislocation interaction
  models, 159–163
Athermal dislocation models, 149–165
Athermal field, 74, 75, 78
Atomic theory of diffusion, 54
Atomic transfer flow, 87, 91, 92, 95

B

Bailey-Orowan equation, 170
Brittle creep, 226
Burgers vector, 11, 96, 102, 103, 104, 113–117,
  119, 120, 122–129, 134, 135, 137–139,
  154, 158, 160, 176, 181, 183

C

Cataclastic granular flow, 222, 223,
  225, 226
Climb velocity, 139, 140, 171
Coble creep, 99–101, 228, 232
Cohesive granular flow, 224–226
Compatibility, 8, 88, 96–101, 102, 104, 110,
  177–179, 186, 209, 212, 214, 229
Compatibility problem, 177
Conductivity, 5, 13, 46, 50, 56, 59, 60,
  62, 64, 127
Covalent bonding, 2, 3
Creep, 75, 97, 99, 100, 102, 163, 225

Cross-slip, 116, 124, 134, 138, 139, 146, 158,
  160, 162, 169, 172, 192, 196
Crystal defects, 5, 44, 53
CSL lattice, 9–11

D

Diffusion, 48, 51, 54, 58, 97
Diffusion coefficient, 47–52, 54–58, 65, 92,
  95–103, 131, 139, 140, 167, 171, 172
Diffusion control, 95, 97, 102, 171, 232
Diffusion creep, 91, 95–99, 104, 176
Dihedral angle, 14, 15
Dilatancy, 88, 130, 181, 190, 192, 209, 213,
  214, 220, 231, 234
Dislocation, 102, 113, 114, 121, 123–126, 131,
  132, 136, 137, 139, 145, 147, 157, 166
Dislocation assemblages, 141, 143, 150, 192
Dislocation climb creep, 102, 103
Dislocation density, 25, 42, 81, 102–104, 115,
  117, 118, 137, 144–148, 154, 155, 158,
  160, 161, 164, 166, 168, 171, 174, 176,
  183, 191, 193–195, 233
Dislocation interactions, 128, 136, 158, 161,
  164, 167, 169
Dislocation interaction with
  dispersed phases, 153
Dislocation multiplication, 103, 158, 161, 168
Dislocation reaction, 125
Dislocation velocity, 115, 134–137, 152, 154,
  165, 166, 174, 234
DSC lattice, 10, 11
Dynamics of mechanical twinning, 176

M. S. Paterson, *Materials Science for Structural Geology*,
Springer Geochemistry/Mineralogy, DOI: 10.1007/978-94-007-5545-1,
© Springer Science+Business Media Dordrecht 2013